Acquisition Editor:	Kristin Klinger
Senior Managing Editor:	Jennifer Neidig
Managing Editor:	Sara Reed
Development Editor:	Kristin Roth
Copy Editor:	Amanda Appicello
Typesetter:	Jennifer Neidig
Cover Design:	Lisa Tosheff
Printed at:	Yurchak Printing Inc.

Published in the United States of America by
        IGI Publishing (an imprint of IGI Global)
        701 E. Chocolate Avenue
        Hershey PA 17033
        Tel: 717-533-8845
        Fax: 717-533-8661
        E-mail: cust@igi-pub.com
        Web site: http://www.igi-global.com

and in the United Kingdom by
        IGI Publishing (an imprint of IGI Global)
        3 Henrietta Street
        Covent Garden
        London WC2E 8LU
        Tel: 44 20 7240 0856
        Fax: 44 20 7379 0609
        Web site: http://www.eurospanonline.com

Library of Congress Cataloging-in-Publication Data

Basden, Andrew, 1948-
 Philosophical frameworks for understanding information systems / Andrew Basden, author.
   p. cm.
 Summary: "There are five main areas in which humans relate to information and communications technology: the nature of computers and information, the creation of information technologies, the development of artifacts for human use, the usage of information systems, and IT as our environment. This book strives to develop philosophical frameworks for these areas"--Provided by publisher.
 Includes bibliographical references and index.
 ISBN 978-1-59904-036-3 (hardcover) -- ISBN 978-1-59904-038-7 (ebook)
 1. Technology--Philosophy. 2. Technology--Social aspects. 3. System theory. I. Title.
 T14.B3625 2007
 303.48'33--dc22
                2007024491

British Cataloguing in Publication Data
A Cataloguing in Publication record for this book is available from the British Library.

Philosophical Frameworks for Understanding Information Systems

Andrew Basden, University of Salford, UK

IGI PUBLISHING

Hershey • New York

Philosophical Frameworks for Understanding Information Systems

Table of Contents

Preface

The aim of this book is to introduce a different way of looking at IT and IS (information technology, information systems), and to suggest some tools to help us do this. The tools derive from philosophy but are orientated to everyday experience of IT and IS.

Those tools will be applied to five areas of research and practice in IT/IS. Allow me to explain, by way of five personal, autobiographical vignettes, what has motivated my involvement in IT/IS and why philosophy is important. This will give a feeling for how this book approaches the seemingly rather heavy topic of 'Philosophical Frameworks for Understanding Information Systems'.

Vignette 1. The Diversity of the World

Working with computer applications in computer-aided design, the health sector, the chemical industry and the surveying profession from 1970 to 1987 it seemed to me that there were four irreducibly different 'aspects of knowledge' that needed to be encapsulated in computer programs or knowledge bases:

1. Items and relationships, such as patients and problems that each patient might have

2. Quantitative and qualitative values, such as the strength of dose of a drug, the date when a problem started, the name of the problem

3. Spatiality, such as proximity to infection

4. Change: events and processes, such as accidents and healing

Each of these aspects of knowledge seemed to deserve a different fundamental approach to representing them, otherwise programming errors would increase. But I doubted whether I had the full set of such aspects, so I sought a fuller set after returning to academic life as a lecturer in 1987.

I did not find one. Most of my colleagues in computer science and artificial intelligence assumed that standard computer languages like C or PROLOG were sufficient: so what's my problem? At the other extreme were those who, like a professor who knew some philosophy and whom I approached with the question of what 'the' correct set of aspects is, replied "There aren't any, they are socially constructed."

Perhaps because of mild Asperger's Syndrome, I could not accept this. Even though I agreed that that human beings do socially construct their categories, I also believed that there is a reality that transcends us and cannot be socially constructed, and that aspects of knowledge were part of that. Privately, I was still curious, and continued to ponder these aspects of knowledge—what aspects there might be, how to distinguish them, how to implement knowledge representation formalisms in computer terms based on each, and the question that underlies these: what aspects as such are. Eventually I expressed these ideas in Basden (1993).

I had always taken an everyday, lifeworld attitude and was a little wary of theorizing, whether of a rationalist, positivist, social constructivist or any other kind. Though I did not realise it at the time, my difficulties related to philosophy: the radical difference between the everyday and the theoretical attitudes of thought and the need to integrate ontology with epistemology. Not until I found a philosophy that treated the everyday attitude with due respect and did not force me into either social constructivism or positivism or naïve realism, did I find an answer that satisfied me.

Vignette 2. Usefulness

Working in the UK chemical industry in the early 1980s, I had a manager who had been given a top-of-the-range PC with the latest software. He found it easy to use, but I remember him standing in his office and asking, "But what the heck do I use it for?" Of course nobody would ask that today. But it cemented into my mind the difference between usability (and other technical qualities) on one hand and usefulness on the other—a difference I had felt for the decade before but not expressed clearly. Basden (1983) began to express these issues.

Over the next dozen years, I experienced a succession of new technologies from within—knowledge based systems (KBS), multimedia and virtual reality—all of which failed to develop their expected potential. This was not because of any lack of technical excellence but because of too little attention given to usefulness. Usefulness seemed to me a crucial issue in the human factors and knowledge-based systems (KBS) communities. Yet it did not enter their academic debates and most researchers responded with "So what!" Our paper on it (Castell, Basden, Erdos, Barrows, & Brandon, 1992), though it received an award, elicited no general response.

Suddenly, in reaction to such books as Thomas Landauer's (1996) *The Trouble With Computers* usefulness came onto the agenda. But the debate leapt straight from "So what!" to "I have the solution to ensure usefulness." The intervening stage of trying to understand what usefulness is was bypassed.

What is usefulness? Why is it that some information systems seem to bring both beneficial and detrimental impact, often to different stakeholders? Which stakeholders are important? How do we address unanticipated impacts (whether beneficial or detrimental)? What about longer-term impacts? How can we design for usefulness, predict usefulness, evaluate usefulness? In short, how do we differentiate success from failure?

To address such issues demands a way of understanding that recognises dynamic diversity and also has a strong basis for differentiating right from wrong (benefit from detriment). Objectivist approaches like cost-benefit analysis are too narrow while subjectivist approaches cannot cope with unanticipated impacts, especially of the long-term variety, nor can they differentiate benefit from detriment. Just like knowledge representation, usefulness and the success or failure of IS demands an everyday rather than theoretical approach.

During that period I became an avid player of computer games, both personal games like *Moria*, *ZAngband* and *The Settlers*, and MUDs (multi-user dungeons) on the Internet. Real playability is like usefulness, depending more on content, story and humour than on technical issues like user interface or graphics. So I wanted to find an understanding of usefulness that was not restricted to work life and organisational use but can address the types of issues encountered in all kinds of computer use.

I discovered that the philosophy I referred to above provides a basis for understanding dynamic diversity, has a strong basis for differentiating right from wrong (normativity) and allowed me to go beyond work applications. It transcends objectivism and subjectivism. It was in regard to use of IS that I first found it immensely practical in application, so much so that I have perhaps been captivated by it ever since.

Vignette 3. Knowledge Elicitation and IS Development

I have always enjoyed programming computers. It is creative. What I produce can be elegant and beautiful, as well as doing some good. Developing KBSs, which involves knowledge elicitation, expanded programming to include the challenge of getting to grips with knowledge in all its diversity and coherence.

Knowledge elicitation involves interviewing experts in a field to obtain some of their knowledge with which to construct the KBS. But I found that, rather then taking the conventional approach at the time of eliciting heuristics (rules of thumb that are actually used in practical expertise), better results are obtained by separating the general 'understanding' and 'laws' underlying the heuristics from the contextual and personal factors.

Moreover, the expert is not just a source of knowledge but a human being, whose expertise is part of who s/he is as a person, so knowledge elicitation is not merely knowledge transfer but involves an intimate relationship of mutual trust and respect, in which the experts feel free to open up.

These two approaches to knowledge elicitation was, in retrospect, one reason for what was an embarrassingly high success rate in developing KBSs. But could it be transferred to others? I wanted to understand why seeking 'understanding' and developing a relationship of trust led to success. Again, philosophy was called for. The strategy of seeking 'understanding' presupposes there is in fact something to understand that we have a hope of finding. Subjectivism would say there is nothing, ultimately, 'out there' to understand. Objectivism might allow 'out there' but immediately tries to reduce understanding of it to logic, especially that of the natural sciences. We (Attarwala & Basden, 1985) rejected both, but at the time we had no philosophical basis for doing so. Now, it appears, the philosophy that helped me understand aspects of knowledge and usefulness can also help here.

From the mid 1980s I had to raise my sight above knowledge elicitation to the wider project that is IS development (ISD). While software engineering methods were becoming increasingly structured in the 1970s and 1980s—analyse, specify, design, implement, test: all aiming for automated, formal proofs of a program's correctness—my own practical development went in the opposite direction because the domains for which I developed KBSs, databases, etc.—medical patients, stress corrosion cracking in stainless steel, agricultural planning, business strategy, budget-setting in the construction industry—were ill-structured. No prior specification could be drawn up since the very activity of implementation reveals new knowledge and stimulates users to change their minds about what they want. So we developed our own 'client centred methodology' (Basden, Watson, & Brandon, 1995).

Of course, I was not alone in finding these characteristics of the process of development. But where our approach differed lay firstly in its key notion of responsibility, not just to the 'customer' but to all stakeholders, to the community and society, to the future, to the domain of application, to the nature of reality itself, and even (some would add) to God. Secondly, it lay in its sensitivity to everyday life of the application, the stakeholders and the entire ISD team. Though until the mid 1990s I did not attempt a philosophical understanding, it now transpires that the philosophy I found so helpful in the first two areas above is helpful in this area too.

Vignette 4. Nature of Computers

When I read Allen Newell's paper *The Knowledge Level* soon after it was published in 1982, I immediately thought, "Yes!" It put into concepts and words what I had intuitively held to be so: that there is a fundamental difference between symbols of our computer programs, databases or KBSs and the knowledge they 'hold' as their content. From a study of citations of this paper over the next 20 years, it seems that many others found the same intuitive agreement. *The Knowledge Level* freed a whole generation of us who worked in the AI and HCI (artificial intelligence, human computer interaction) fields to talk about computer systems at 'the knowledge level', in terms of their content, independently of how it is represented in symbols and their manipulations.

But Newell had in fact taken us further. He argued that the symbol and knowledge levels were just two in a sequence of levels at which computer systems may be described—the physics, the electronics, the digital signals and bits, the symbols and the knowledge. (Later, several of us added a sixth level, variously called tacit or social.) Thus Newell proposed in effect a multi-level understanding of the nature of computers. I have used it as an educational tool to separate out distinct types of issues in HCI, KBS, multimedia, virtual reality; my students find it very useful.

Newell tried to ground the knowledge-symbol difference in philosophy but not very successfully (see Chapter V), and he merely assumed the other levels. However he made a curious philosophical claim that few have noticed:

Computer system levels really exist, as much as anything exists. They are not just a point of view. Thus, to claim that the knowledge level exists is to make a scientific claim, which can range from dead wrong to slightly askew, in the manner of all scientific claims. (Newell, 1982, p. 99)

This is a strong ontological claim. But on what basis could he be right? And if someone else suggests a new level (as for example I did with a 'tacit level') on what basis do we judge the candidate new level? Yet again, the very philosophy I found helpful in the three other areas is one that is able to provide a philosophical underpinning for Newell's levels. More: it can throw new light on the AI question of whether a computer can 'think'.

Vignette 5. The Information Society

What damage will IT wreak on humanity and the wider environment? Or what good? That is, how do we understand what has been called the information society? It is too early to tell. Not many predicted climate change from cars and planes until recently. If we cannot predict 'scientifically' what the real problems of IT will be, maybe we can at least take note of the ethical dimension? For example, is it not true that selfishness and self-interest mainly bring problems (refer to business writers since the year 2000)? So, maybe humanity's development of IT could be guided by eschewing self-interest, rather than by seeking 'scientific' attempts to plan it or by assuming that we can continue to please ourselves?

On a personal note, I have worked in the green movement for many years. For me green activism followed, and was a result of, my conversion to Christ and a filling with the Holy Spirit around 1970. Over the years I have become convinced that what has needlessly driven us towards destruction of the earth is not just big business nor a conspiracy, but our idolatrous world-view and self-centred attitudes, which pervade every aspect of the way we live and do business. It even pervades the way we carry out research. The root of our problem is 'religious', even while at the same time there are economic and political factors at work. Those in the feminist and anti-globalisation movements know this very well (though they usually use terms other than 'religious'). I use the word 'religious' rather widely, almost as a synonym for 'ideological'; see Chapter II.

Likewise, when we take a societal, global perspective on IT, a 'religious' aspect is inescapable, both for those involved in the practice of the area and for the content of our theories. To understand the issues in this area, I wanted an approach that acknowledges religious commitments and presuppositions, not only in humanity as studied but also in we who study it. Lo! I found that the philosophy that helped me so much in the four other areas acknowledges and opens up religious issues.

'The Whole Story'

Each vignette indicates a distinct area of research and practice in IT/IS in which I have been involved. In each, a particular issue emerged:

- **Vignette 1:** The diversity of meaningful reality that we want to represent or model in computers
- **Vignette 2:** Everyday diverse normativity and repercussions of computer use
- **Vignette 3:** Responsibility to both those we work with and to a diverse reality that transcends us and yet which we can understand
- **Vignette 4:** The multi-level nature of computers
- **Vignette 5:** Religious root of IT, of those who practise and research it, and of society

Each indicated the need for philosophy—and that one philosophy in particular has been useful to the author.

Because I have been intimately involved in all these areas, I have tended to see them as closely related to each other, to form what we might call 'the whole story that is information technology'. Almost from the start, when programming for my PhD in the 1970s, I would try to reach beyond my immediate area and feel for the others. In the everyday 'whole story', the areas interweave and while we may conceptually distinguish them, we need a way of understanding each area that acknowledges the others. Yet in each, a different community of practice has developed, a different research agenda and way of thinking and different research communities, which seldom speak to or understand each other. What each finds meaningful the others find meaningless. As I argue in Chapter I, this is why philosophy is needed: philosophy not only helps us understand the issues in each area, but it is the discipline that allows us to acknowledge a variety of spheres of meaning in relation to each other.

Unfortunately, I did not find any philosophy that was on offer either useful or attractive. Ancient Greek thinking opposed form and matter. Mediaeval thinking opposed the sacred and secular. Modern thinking opposes control and freedom, being and norms, thing and thought, subject and object. Or it tries to think them together in ways that ultimately are arbitrary.

Throughout the vignettes, I have mentioned a philosophy that has helped me in each area. It is the philosophy of the late Herman Dooyeweerd, a mid-20th century Dutch thinker. It manages to integrate form with matter, sacred with secular, control with freedom, being with norms, thought with thing, and subject with object in a way that does not denature any of them and yet is not arbitrary. At the same time it deliberately takes an everyday attitude even while it also acknowledges and welcomes the results of scientific work. The reason it can integrate things we have long assumed to be incompatible is because Dooyeweerd questioned the most fundamental presuppositions that have lain at the roots of Western thinking over the last 2,500 years.

Purpose of This Work

In this work I try to show that Dooyeweerd is at least interesting enough, to those who work or research in any of these areas of IT/IS, to be considered alongside other approaches. I do not seek to show Dooyeweerd is superior to any of these, let alone replace them. Rather I simply recommend him for further study. To do this I make a proposal (or set of proposals) for the adoption, development, testing and refinement of his approach by IS researchers and practitioners. It is only once this has been achieved that it is right to even begin to evaluate Dooyeweerd against other thinkers, because before then his thought would not have been properly understood.

As mentioned at the start, the central purpose of this work is to propose a new way of looking at IT/IS, which is philosophically sound and yet practical. That is, it develops five philosophical frameworks for understanding IS, one for each area. But it might be of value in other ways too:

- It is unusual in advocating a lifeworld approach in all areas of research and practice, reinterpreting the issues in each in the light of everyday experience

- It aims at integrable frameworks for understanding the various areas with a view to being able to richly understand the 'whole story that is information systems' in a coherent way

- It addresses both technical and non-technical areas at the micro and macro levels

- It demonstrates how philosophy in general may be used to construct (lifeworld-oriented) frameworks for understanding

- It is the only general introduction to Dooyeweerd's philosophy available in the field of IT/IS

For the last reason, that part which explains Dooyeweerd (Chapters II, III) is designed to be used as a reference work relevant to IS which gives pointers into his thought for further study.

What the work specifically offers includes:

- It indicates how philosophy in general can be employed in working out frameworks for understanding in several areas of research and practice in IS.

- Though the framework it develops for each area is new, it discusses how each links to a selection of extant frameworks and significant issues in its area.

- It throws fresh light on some issues in each area, which, even if the reader might not wish to adopt either the framework offered or this particular philosophy, could be useful in stimulating new ideas or strategic directions in research or practice. It also provides a number of practical devices for each area.

- It systematically explains and critically discusses Dooyeweerd's philosophy, placing it in the context of other philosophies.

- It reviews current research in each area that has made use of this philosophy, and sets the research direction for its further application in IS.

- It makes a number of suggestions for critiquing and refining the philosophy itself.

Despite an emphasis on philosophy, this work (especially Chapters IV-VIII) should be readable by those who have only limited understanding of conventional philosophy. Indeed, knowledge of conventional philosophy might not help, because of Dooyeweerd's very different approach. Reference is made to other philosophers, not because the reader is assumed to understand them, but mainly in order to situate Dooyeweerd among them for those readers who do know their work.

The Chapters

Chapter I is preparatory to formulating philosophical frameworks for understanding information systems. It clarifies what is meant by 'information systems', setting out the five areas referred to in the vignettes. It discusses what is meant by 'understanding' and outlines some characteristics of the everyday 'lifeworld'. It explains what is to be expected in 'frameworks', and finally discusses what is meant by 'philosophical' and what the role of philosophy should be in such a work. In doing this, it gives initial reasons why Dooyeweerd is worth exploring.

Chapter II sets out Dooyeweerd's general approach to philosophy and explains in what ways it is so different from most Western thinking. In particular it makes clear that Dooyeweerd offered both a critical and a positive philosophy, first deconstructing three millennia of Western thought and then having enough courage to construct something to be itself critiqued.

Chapter III explains those portions of Dooyeweerd's positive proposal that will be used in formulating frameworks for understanding. It is necessary to introduce this philosophy systematically because the reader needs to be able to grasp it to the extent of being able to apply it in their own fields in ways not discussed here, to understand how it relates to other streams of thinking, and even to properly test and refine it. It explains Dooyeweerd's 'general theory of modal spheres', his approach to being and things, to knowledge, and to human life. It reviews some extant critique of Dooyeweerd.

Chapters IV to VIII then explore how Dooyeweerd's ideas can be used to formulate frameworks for understanding each of the areas indicated by the vignettes (though in a different order). Each chapter begins with a brief discussion of what 'everyday experience' or 'lifeworld' means in the area, which surfaces several main issues that need to be addressed. Application of Dooyeweerd's thought is then explored in relation to these issues, including offering practical devices that might assist practitioners in the area. Also each chapter contains discussion of some extant ways of seeing the area. The aim of each chapter is not only to formulate a framework for understanding, but also to demonstrate a general approach that anyone might be able to follow, using whatever is their favourite philosophy.

Chapter IV begins with human use of IT artefacts and systems, as introduced in Vignette 2, because we want to place the human being, rather than the technology, at the centre. Three usage relationships are discussed: human computer interaction, engagement with represented content and human living with computers (the first two do not feature in the Vignette), each of which is seen as multi-aspectual human functioning but each exhibiting a different set of issues. (The Dooyeweerdian notion of multi-aspectual functioning is explained in Chapter III.) Both the structure (i.e. nature) of each and the normativity that is operative in each is explored. The framework thus developed provides a way of addressing the tricky problems mentioned in Vignette 2. How this framework can engage with extant frameworks for understanding is discussed, including Walsham's 'Making a World of Difference' and Winograd and Flores' 'Language Action Perspective'.

Chapter V then discusses the nature of computers (Vignette 4) and also information, because the reflective user will at some time ponder this, and because we need some basis for understanding what computers can and cannot do. The latter centres on the artificial intelligence question of whether computers can think or not. Employing Dooyeweerd's notion of the multi-aspectual meaningful whole (explained in Chapter III), it sees computers as multi-level systems, not dissimilar to Newell's theory. Indeed, the framework developed here shows how Dooyeweerd can provide a sound philosophical underpinning for Newell's theory, including his strong ontological claim. Fresh light is shone into Searle's Chinese Room thought-experiment. Dooyeweerd's treatment of performance art is used to understand the nature of computer programs.

Chapter VI formulates a framework for understanding IS development (Vignette 3) as multi-aspectual human functioning in which the post-social aspects are prominent. In fact, four distinct but interwoven multi-aspectual functionings are explored, two reflecting the issues encountered in Vignette 3. The history of perspectives on methodologies in ISD is explained from a Dooyeweerdian perspective, and it is shown how Dooyeweerd can enrich Checkland's well-known soft systems methodology. It is in this area that most application of Dooyeweerd to IS has occurred so far, and the work of several thinkers is discussed.

Chapter VII discusses the technological resources that IS developers make use of—the programming or knowledge representation (KR) languages, code libraries and inter-program protocols. It begins with Brachman's call for 'KR to the people' (KR languages so natural that anyone could, in principle, use them to develop their own IS). This call is answered by reference to Dooyeweerd's approach to the diversity of the world, with which a detailed proposal is presented for a multi-aspectual development toolkit. My own premature proposal for 'aspects of knowledge' (Vignette 1) is seen as a small subset of this. The framework formulated here acknowledges a transcending reality of immense diversity, and discusses how to future-proof the toolkit to cope with unforeseen requirements. It also, perhaps more usefully, shows how this proposal can be used as a yardstick against which to measure extant proposals, from the relational data model, object orientation, Wand and Weber's proposal, and the employment of Alexander's design patterns in software.

Chapter VIII discusses how Dooyeweerd's philosophy can address the 'macro' level of IS: the global, societal issues. I have called this our technological ecology, in which human living is 'inside' IT (Vignette 5). First, Schuurman's use of Dooyeweerd to define a 'liberating vision for technology' is discussed, which establishes the conditions under which technological development is a blessing rather than curse for humanity. Then the circular relationship between us and IT is examined: though created by us IT nevertheless changes not only the way we live but how we see ourselves. Critiques by feminism that IT has become inscribed with masculinity and by others that it has become inscribed with Western values are outlined and shown to be commensurable with, but a subset of, what Dooyeweerd could offer. The root of problems in this area of technological ecology is traced to various types of religious dysfunction, for which mere economic, social or technological solutions will be ineffective. Throughout this area is the issue of the ultimate Destiny of IT and humankind.

Chapter IX reflects on the proposals and on our use of Dooyeweerd's philosophy. It summarises the five frameworks for understanding, and discusses how they can cohere to help us understand the 'whole story' that is IS. The benefits and limitations of having used Dooyeweerd are discussed, and various suggestions made for critiquing and refining Dooyeweerd's philosophy itself are collected together. The degree to which the whole exercise has been able to meet the requirements set out in Chapter I is discussed, including the issue of whether one overarching framework should be sought. Finally, the very process of our exploration is discussed, including how the approach adopted in this volume could be used by those who wish to employ a different set of aspects, different areas of research and practice, or even an alternative philosophy to that of Dooyeweerd. The book ends with a brief suggestion for the future.

References

Attarwala, F. T., & Basden, A. (1985). A methodology for constructing expert systems. *R&D Management, 15*(2), 141-149.

Basden, A. (1983). On the application of expert systems. *International Journal of Man-Machine Studies, 19*, 461-477.

Basden, A. (1993). Appropriateness. In M. A. Bramer & A. L. Macintosh (Eds.), *Research and development in expert systems X* (pp. 315-328). Cranfield, UK: BHR Group.

Basden, A., Watson, I. D., & Brandon, P. S. (1995). *Client centred: An approach to developing knowledge based systems.* Chilton, UK: Council for the Central Laboratory of the Research Councils.

Castell, A. C., Basden, A., Erdos, G., Barrows, P., & Brandon, P. S. (1992). Knowledge based systems in use: A case study. In *British Computer Society Specialist Group for Knowledge Based Systems, Proceedings from Expert Systems 92 (Applications Stream)*. Swindon, UK: British Computer Society.

Landauer, T. K. (1996). *The trouble with computers: Usefulness, usability and productivity.* Cambridge, MA: Bradford Books, MIT Press.

Newell, A. (1982). The knowledge level. *Artificial Intelligence, 18*, 87-127.

Note

Trademarks

Amiga is a registered trademark of Amiga Inc.

Apple and Macintosh are registered trademarks of Apple Computer, Inc.

eBay is a registered trademark of eBay, Inc.

IBM is a registered trademark of International Business Machines Corporation.

Java is registered trademark of Sun Microsystems, Inc.

Lemmings is a registered trademark of Psygnosis Limited.

Microsoft is a registered trademark of Microsoft Corporation.

SNCF is a registered trademark of Société nationale chemins de fer France.

TEX is a trademark of the American Mathematical Society.

All other trademarks are the property of their respective owners.

Copyrights

Acknowledgment

First and foremost I would like to thank Martin and Margaret Ansdell-Smith, without whom this book would never have been written. They always encouraged me through difficult periods in writing, helping me to get the book into shape several times by suggesting new ways of looking at the work, new messages to give and even some new ideas. They also undertook much of the painstaking work of ensuring many details were correct and commenting on its readability.

Next I would like to thank Henk Geertsema, *emeritus* holder of the Dooyeweerd Chair at the VU-Amsterdam in the Netherlands, and Roy Clouser, *emeritus* professor of philosophy at Trenton State College, New Jersey, both internationally acknowledged experts in Dooyeweerdian philosophy. They read and commented on my exposition and use of Dooyeweerd from the perspective of Dooyeweerdian scholarship, correcting some of my misunderstandings, suggesting new ways of expressing Dooyeweerd's ideas and giving much encouragement.

I would like to thank my friend and colleague, Heinz Klein, professor *emeritus* SUNY Binghamton, USA, and internationally recognised as one of the major thinkers in information systems research over the past 20 years, whose knowledge of the phenomenological, hermeneutic, linguistic and critical turns in philosophy as applied to IS is second to none. Despite many pressures, including a very heavy work schedule, he gave me continual inspiration and, with his customary incisive way of thinking challenged my beliefs and stimulated me to clarify a number of issues, especially about the lifeworld.

Special thanks are due to my colleagues at the University of Salford, Dr. Helen Richardson and Professor Alison Adam, who both encouraged me to keep going when I felt like giving up. Helen read a draft of the manuscript and saw more clearly than I did what the value of the work might be. Professor Michael Myers, University of Auckland, New Zealand, might be surprised to be included in my acknowledgments, but on one occasion with one comment he helped me see more clearly what the book was to be about.

I would like to thank my friend Lyndon Fallows for useful advice about the style of writing to use in the book.

I would like to thank Dr. Michael Winfield, recently of the University of Central England, for permission to adapt a diagram from his thesis and for pushing me, years ago, to publish philosophical works.

Thanks are due to the University of Salford for allowing me 6 months' study leave in 2003, which enabled me to begin the project of writing this book, and to the Herman Dooyeweerd Foundation, which funded travel to North America to visit a number of people who knew Dooyeweerd's ideas. Thanks are also due to the publishers of this book for permission, despite their house style, to add numbers to the headings within chapters in order to facilitate cross-referencing. In particular, Kristin Roth, managing development editor, has been very effective in helping me prepare the book for publication. I would like especially to thank three anonymous reviewers whose challenging questions and most helpful suggestions helped me shape the argument of the book.

Finally, I would like to thank my wife Ruth for her patience, support and encouragement over several years when the writing of this book dominated much of our lives.

Andrew Basden
June 2007

Chapter I

Introduction

In 2005, the total spending on IT (information technology) was around $1,000 billion (Network News, 2006). Does the world obtain $1,000 billion worth of benefits from this spending?

According to studies made over the past 20 years, the failure rate of information systems (IS) using IT has consistently been over 50%.[1] So why do we—humanity—pour $1,000 bn every year into it? Would that money not be better spent on health, overcoming poverty, or reducing climate change and environmental damage?

Such considerations challenge us in seeking to understand what is going on and why the benefit from IT or ICT (information and communication technology) seems so much less than the amount we are so willing to spend.

- Perhaps it is because our technologies are not yet good enough?
- Perhaps it is because we do not use IT in the right way or for the right things?

- Perhaps it is because information systems (IS) development is inefficient?
- Perhaps we too readily accept the way IT controls our lives, and the assumption that IT is the solution to everything?
- Perhaps our very assumptions about the nature of computers and what they can do for us are mistaken?

The purpose of this book is to introduce a different way of looking at IT and IS, and to suggest some tools to help us do this and to take action. The tools derive from philosophy but are oriented to everyday experience of IT/IS. And those tools will be applied to five areas of research and practice in IT/IS to which these questions relate. They are also the areas related to the five vignettes in the Preface (Vignettes 1, 2, 3, 5, 4, respectively).

But wait! Why do we need a different way of looking at IT/IS? An alternative question may be raised:

- Perhaps the very posing of the question above is itself to be questioned. After all, what might we mean by $1,000 bn worth of benefit? Do we not obtain benefit that cannot be measured in dollars? On the other hand, does not IT/IS do immense damage that likewise cannot be measured in dollars?

Walsham (2001, p. 251) says his whole professional interest may be encapsulated in the question, "Are we making a better world with IT?" But then, on the next page, he asks, "What is meant by a better world?"

The different way of seeing IT/IS introduced in this book aims to provide the basis for answering all these questions. It will work at both levels, addressing the five areas mentioned earlier and Walsham's first question, and also the question of what it is meaningful to consider as "better" or "problem". As such, this work must necessarily reinterpret what we might call "the whole story that is information systems".

The usual way to argue that a different way of thinking is needed is first to show there are problems which the old ways of thinking cannot solve and then show that the new way can solve them. Under such a strategy the first half of this book would be devoted to examining each of the extant ways of understanding IT/IS to argue conclusively that they cannot solve the problem. But such a strategy has two weaknesses.

One is demonstrated by the alternative question itself. The very posing of a problem depends on, and comes from, a particular way of looking at things. The types of reasons one must marshal to argue that we need a different way of thinking, are not objective facts, pre-given, but are themselves generated by what is on offer as ways of thinking. To even discuss whether we obtain $1,000 bn worth of benefit already means we are thinking financially rather than, for example, technically or in terms of justice. This is not to say there might after all be no financial problem, but it underlines that our very way of looking at things dictates how we see them

ε ⎸ over a problem only shows that
s ⎸ a new way of thinking is neces-
s ⎸ and diverse area, there are many
v ⎸ way of looking at it can ever be
tı

T ⎸ a new way of looking at IT/IS
pı ⎸ ake room for the new way. That
is ⎸ to take an adversarial stance. It
m ⎸ inking, but it can also, instead,
en ⎸ ways of looking at IS and IT
ra ⎸ his work is not so much a *new*
wε ⎸ IT/IS as a *different* way (con-
no ⎸
So ⎸ tead, it presents a proposal to
be ⎸ d perhaps eventually rejected
— ⎸ existing ways must be replaced. The book does not

have a substantial section devoted to demolition but begins construction almost from the start. During the process of construction—that is, of developing and exploring the different way of understanding IT/IS—indications will be given of where it might have advantages over existing ways of thinking. Some of these indications become meaningful because of the different approach itself, though most are issues that have already been recognised and commented upon by researchers and practitioners in the various fields of IT or IS. Some readers might then use it to formulate new frameworks to replace extant ones, but others might use it to support, critique, and enrich them.

Now let us return to the phenomenon of IT/IS with which we began. The title of the book is *Philosophical Frameworks for Understanding Information Systems*. Most of the rest of this chapter is devoted to clarifying what is meant by "philosophical", "frameworks", "understanding", and "information systems". They will be examined in reverse order.

The issues in this chapter deserve a whole book to themselves, because they have not been adequately discussed in this way. However, as the vignettes in the Preface make clear, the author's main contribution is the content of such frameworks rather than the notion of frameworks as such, and so it is the content that is discussed in Chapters II to VIII. This chapter could thus be seen not only as explaining these concepts in preparation for Chapters II to VIII, but also as a summary of what should become a much fuller future discussion.

1.1 Information Systems:
What are We Trying to Understand?

"The field of computing appears to be in a state of transition," said Glass, Ramesh and Vessey (2004). "Although historically it has evolved as several stovepipes of knowledge—predominantly ... CS, SE and IS [computer science, software engineering and information systems]—there is now some impetus for amalgamation." In an analysis of research and thinking in these three fields, they noted that an increasing number of "integrated schools" of computing are being formed.

The "stovepipes of knowledge" evolve. It began to be clear during the 1980s that, while traditional software engineering techniques might be suited to the research laboratories and to the generation of well-defined software for well-structured applications, a radically different approach was needed to cope with the ill-structured applications in business and such approaches as soft systems methodology (Checkland, 1981) have become popular.

Yet these are not the only "stovepipes". Others have emerged, distinct from their three, such as the information society (Lyon, 1988) and what Landauer (1996) called "the trouble with computers": the benefits or detrimental impact of information technology used as part of human living and working. Those working in artificial intelligence (AI) often differentiate themselves from computer science because their central interest is not algorithms and data structures as such but the nature of computers and especially in what way they resemble human beings.

The five questions posed at the start and the five vignettes in the Preface indicate five "stovepipes of knowledge", each about a different type of relationship humanity has with the technology, respectively as its shaper, its user, its developer, its conceiver, and its denizens.

1.1.1 Areas of Research and Practice in IT/IS

Each such "stovepipe of knowledge", each type of relationship, constitutes a distinct area of research and practice, in which a different set of issues is meaningful, and a different set of assumptions is made about what is important, what is deemed a "problem" that needs solving, and what is good practice and research.

To take just two examples, those working in computer science, for example developing better graphics-handling facilities, are concerned with algorithms, while those trying to gain benefit from employing computers in business, such as for insurance brokers using an electronic placement system (Walsham, 2001), are concerned with the benefits or detrimental impact the system has on themselves as insurance brokers.

Neither area is of much interest to the other. Those who practise in each area see things in different ways and are driven by different motivations. Research methods in each area are different—those of computer science, logic, and psychology in one, those of sociology, business, and economics in the other. Different types of theory are generated and different conceptual frameworks assumed.

Should they be of interest to each other? It seems unreasonable to demand that insurance brokers should concern themselves with graphics algorithms (even assuming the IS they are using displays its information graphically). It seems unrealistic to expect those who design graphics algorithms to think of all possible beneficial and detrimental impacts arising from use of all possible programs that might conceivably use their algorithms for insurance brokers.

And yet there remains a thin thread of responsibility here in each direction, which implies that the two areas of research and practice that concern themselves with use of IS and algorithm design are not to be seen as entirely independent of each other. Insurance brokers might see their data as graphs but if the graphics algorithms are wrong those graphs might fail or, worse, mislead. In the other direction, insurance brokers must tell the algorithm designers what they need to do with graphics, or the appropriate algorithms will not be developed.

Over the last few decades, a differentiation of areas from each other has occurred. There has been a working-out of the rich diversity and coherence of each area. Researchers and practitioners in one area share largely the same set of concepts, issues, and concerns—and these differ markedly from those in other areas. In each area, debate arises about what may be deemed appropriate research criteria and methods for the area, and what issues or problems are within the remit of the area, within its horizon of meaning. As this occurs, each area becomes conscious of itself and discussion ensues on whether each constitutes a discipline (e.g., Shaw & Garlan, 1996).

The delineation of areas cannot, therefore, be an exercise in seeking complete independence between them. Nor is it one of differentiating theory from practice, because in each area there is both theory and practice. For example, while it might be tempting to see computer science as "theory" being employed in business "practice", there is practice as well as theory in computer science and there is theory as well as practice in business use of IT.

1.1.2 Problems and Issues in IT/IS

Each area exhibits different types of problem and different everyday issues, all of which contribute to the overall problem in "the whole story that is IS". Some examples of problems in each area include:

- **Human use of computers (HUC):** An electronic placement system designed to help insurance brokers was under-used because it undermined the principle of Utmost Good Faith. The Socrate rail ticketing system inconvenienced French rail passengers in a major way. A Swedish stock control system re-engineered business processes successfully—but led to a fall in profits rather than a rise. Moving outside business, a computer game might have stunning graphics—but awful gameplay. (The first three are discussed in Chapter IV.)

- **Nature of computers (NoC):** The Japanese "Fifth Generation" project of the early 1980s sought to harness the "intelligence" of computers to solving major problems of society, but its hopes remain unfulfilled because computers are not intelligent in the way people are. The artificial intelligence question of whether computers can truly understand remains unresolved, and has degenerated into dogmatic positions.

- **IS development (ISD):** "The delivery process itself can be expensive and error prone. ... Engineering standards have been slow to develop and even slower to be adopted. Project management performance has been abysmally inconsistent. ... All too often organizations have sunk millions of dollars into 'runaway' projects that deliver less functionality than promised, significantly later and for considerably greater expenditure than planned. In the worst cases, some projects end up being cancelled." (Thorp, 1998, p. 11)

- **Information technology resources (ITR):** Graphics algorithms might not allow a full range of colours or textures to be drawn or allow a graph to be modified once drawn. Computer languages (like PROLOG), while excellent for expressing and manipulating distinct pieces of data logically, can make it difficult to do justice to continuous spatial phenomena such as overlapping or linguistic phenomena like nuanced similarities of meaning—which are important, respectively, in geographic information systems and knowledge management.

- **Information technology as ecology (ITE):** Skilled craft-workers become button-pushers (Noble, 1979). The wide availability of personal computers led to "the calculator widow" (Ceruzzi, 1996). IT is too "male" (Adam, 1998). "On the Internet nobody knows you're a fraud" (Crawford, 1996, p. 595). "Do Western-origin IS and methodologies embed features that inhibit cross-cultural working?" (Walsham, 2001, p. 229). Noble (1997) wrote of the religion of technology.

The reason for this order is that it reflects how everyday experience of IT/IS might present itself to us. Most people's experience of computers is of using them. When using them, users might reflect on what computers are and even what information is. The used IS must be developed. IS developers need technological resources. Finally, all of us live in a technological environment, which we also shape. Thus this order is neither what Strijbos and Basden (2006) call "bottom up", namely from technology to society, nor "top down", but from the everyday outwards.

It is these five areas for which philosophical frameworks for understanding will be formulated in Chapters IV to VIII. Other areas doubtless exist, and it is hoped that the approaches exhibited in this work will show readers how to formulate frameworks for understanding them.

Though such differentiation and delineation of areas continues, as Glass et al. (2004) have indicated there is also a move towards integration. It is clear, for example, in the work of the late Enid Mumford (Avison et al., 2006). The discussion in this book recognises there is a "whole story that is IS". The challenge to integration is that what is meaningful in one area is often completely meaningless in others, as the earlier example of insurance brokers and graphics algorithms demonstrates.

What both researchers and practitioners need is to have some way of understanding, or looking at, the area(s) in which they work, which respects other areas. But what is "understanding"?

1.2 Understanding

The range of issues that are meaningful in an area is vast. For example, issues meaningful to the use of computers include: human computer interaction, user interfaces, ergonomics, user-computer dialogue, proximal use of tools, use by individuals to accomplish their tasks, use by organisations to reap benefit, playing computer games (such as MUDs) in which the "user" engages with virtual characters and items, immersion and presence in virtual reality, social relationships around the computer, impacts of using IS and how it comes about, unexpected and indirect impacts, diversity of impacts such as communicational, social, economic, legal, or ethical, multiple stakeholders, benefit versus detrimental impact, success versus failure, use of IT to change social structures, and to challenge assumptions, and much more.

In each area a similarly rich collection of issues has become meaningful. What does it mean to understand them? Researchers and practitioners need a basis for understanding not only the issues they know are relevant to them, but also the relationships between them, and also how to expand their understanding as new issues arrive. How may we—how can we—obtain a way of understanding all potentially relevant issues of an area of research and practice in a coherent way?

1.2.1 Problems with Theoretical Frameworks

A time-honoured way is to construct theories that explains some of the issues, and/or methodologies that guide them. For example, ergonomics, Davis' (1986) technology acceptance model (TAM) and Latour's (1987) actor-network theory (ANT) each raise and address important issues. But such ways of understanding usage focus on a narrow range of issues. This is not a deficiency in them because most do not claim to do more than that, but it is a danger. The danger is that that very focus can lead researchers and practitioners to assume that nothing else is meaningful, and so other issues become downplayed, suppressed, and ignored. For example, ergonomics downplays the higher levels of HCI, and completely ignores issues of benefit in use. TAM is much wider, purporting to cover both ease of use and usefulness, but it tends to do so in a managerial way. ANT claims to escape this, and especially to focus on both micro and macro levels, but can downplay the normative difference between benefit and detriment. The same tendency to divert attention away from some important issues attends theoretical frameworks in any area.

Information security is not an area that is covered in this book; that is left to another occasion, when a framework for understanding security may be developed by following the methods sketched here. But security experts Bruce Schneier and Niels Ferguson understand the problem of narrow approaches. As they say in the preface of their book *Practical Cryptography* (Ferguson & Schneier, 2003):

Arguing about whether a key should be 112 bits or 128 bits long is rather like pounding a huge stake into the ground and hoping the attacker runs right into it. You can argue whether the stake should a mile or a mile-and-a-half high, but the attacker is simply going to walk around the stake. Security is a broad stockade: it's the things around the cryptography that make the cryptography effective.

In development of IS to support business tasks the knowledge that must be elicited is multi-faceted (Jacob & Ebrahimpur, 2001, p. 78):

The majority of the managers have a long history in the company and successful leadership of projects is dependent on long tenure since the preference is for managers with a generalist competence profile. In the words of one interviewee: One needs to have as broad a knowledge base as possible. It is the outer parameters that one must have knowledge about.

Though loosely defined, "outer parameters" indicates facets that are often overlooked. In interdisciplinary applications of IS "outer parameters" abound.

The "everyday" of both practice and research is rich and full of surprises many of which arise from "outer parameters". If frameworks are to help us understand any area of information systems, then they must be able to cope with the diversity of the everyday and be open to surprises as they arrive. Theories and theoretical models can provide insight into specific issues, but they should never be allowed to divert attention away from other important issues. For this reason, for example, cost-benefit analysis is grossly insufficient as way of understanding IS use.

1.2.2 Lifeworld-Oriented Frameworks for Understanding

The type of understanding sought in this work is that of what philosophers have called the lifeworld, or "pre-theoretical" or "naïve" experience (which are all used here almost synonymously with "everyday" without any negative connotation). While much of our understanding is theoretical and involves explicit conceptual structures, some understanding is intuitive, cultural, embodied in aspiration and attitude, much of which cannot be made fully explicit. This book seeks to respect the importance of all such things in formulating frameworks for understanding, and to formulate frameworks that will be sensitive to such things in the everyday experience of practice and research in each area. As far as this author is aware, such an approach to understanding is not common in the five areas delineated earlier.

There are two ways in which the lifeworld can be used as empirical input to inform our attempts to understand an area: we may be guided by the content of everyday experience, or its structure. The content is the actual beliefs and assumptions people hold, and it varies with times and cultures: the lifeworld content of medieval Europe was very different from that of today. But the structure of the lifeworld is its nature and characteristics regardless of culture and content. Sometimes it might not be clear whether something is structure or content, but in most cases, in this book it is the structure that must guide us rather than the content. Therefore, it is useful to make explicit some the characteristics of such a lifeworld, everyday approach to understanding each area of research and practice.

Husserl (1970) first used the term "life-world" to differentiate it from the "worlds" of sciences like physics, psychology, sociology, and to highlight its importance. For centuries, the "everyday" way of knowing had been thought inferior to the theoretical or scientific way, and everyday life had often been deemed unworthy of attention. But Husserl argued that the sciences cannot operate without a stock of shared life-experiences which give meaning to the concepts used in science.

Following Husserl's lead, many thinkers have reflected upon the lifeworld and highlighted other characteristics. Some of the more important characteristics are as follows, and some implications for lifeworld-oriented frameworks for understanding (LOFFU) are mentioned. They are very much summarised here, but are expanded later.

Husserl's view implies a LOFFU should be open to such stocks of shared experiences within the field it tries to understand and not be limited to those seen through the "lens" of a particular theory. As will be argued in Chapter IV, for example, Foucault's notion of power used as a lens blinds us to important aspects.

"The life-world is above all the province of practice, of action" (Schutz & Luckmann, 1973, p. 18). This implies a LOFFU should concern itself with practice as well as research.

Husserl called our attention "to the things themselves." The lifeworld is a "pre-given reality with which we must cope" (Schutz & Luckmann, 1989, p. 1). This givenness extends to types as well as individuals (Schutz & Luckmann, 1973, p. 229). This implies a LOFFU should enable the researcher and practitioner to "listen" to the world of their area as it presents itself to them, without trying to force it into a priori conceptual structures they bring to it. Subjectivist ways of looking at IT/IS can have difficulty here; see Chapters V, VII.

Lifeworld is engagement. Heidegger (1962), though he did not use the term "lifeworld", stressed that in everyday life we engage with things rather than distance ourselves from them as "rational" thinkers or actors. This implies a LOFFU should not assume that either researchers or practitioners in its area are "rational actors" who always deliberately plan and rationally deduce, taking a stance "above" the area, but that they are closely engaged with it.

The lifeworld attitude differs fundamentally from the theoretical attitude. Several thinkers have stressed this (Schutz called it a "natural attitude"). A theoretical attitude observes or acts at a distance, tries to formulate an explicit theory or method, into which it tries to squeeze all that is experienced, and evaluates or justifies things by reference to that theory or method. An everyday, natural, lifeworld attitude can understand things more intuitively, can evaluate or justify by reference to common sense (including norms), and does not try to reduce the diversity experienced to a theory or method. A case study in Chapter IV highlights the problems of a theoretical attitude. But theory is not forbidden.

The lifeworld resists being made explicit. Husserl, Heidegger, and others emphasised its background character. The lifeworld is that curious thing that dissolves as we try to take it up piece by piece (Habermas, cited in Honneth, Knodler-Bunte, & Windmann, 1981, p. 16). This implies a LOFFU should not assume that researchers and especially practitioners in the area always do, or even can, express what is relevant or important explicitly. LOFFUs should always be suspicious of explicit statements. Positivist frameworks can fail here. So, in Chapters IV, VI, and VII the author's own intuitions are referred to.

The lifeworld exhibits meaning and normativity. Habermas (1987) stressed the normativity and meaning that we experience in the lifeworld. By contrast "systems" (such as the economic system of money or the political system of voting) work by mechanical rules that are supposedly devoid of meaning and normativity. This implies a LOFFU should have place within it for meaning and normativity, as well as structure and process. Objectivist and subjectivist approaches both ignore normativity (Chapters IV, VI, VIII) and positivism ignores meaning (Chapters V, VII).

But this does not exhaust the characteristics of the lifeworld. Rogers (1998) is nervous that, in focusing only on what is intersubjectively believed about the world, most discourse about the lifeworld ignores the world as such. Also, as will be seen later, especially in Chapter VII, everyday experience presents a wide diversity that coheres. The relationship between everyday and critical attitudes has also not been given sufficient attention. Finally, it has been assumed that lifeworld meaning is generated by intersubjective activity, but this needs to be questioned because it rests on certain presuppositions. These issues will be addressed toward the end of Chapter III, when the notion of "everyday" will be expanded.

1.3 Frameworks

A framework for understanding (FFU) an area is a way of seeing the area. But that involves many things, including the actual (social) activity of practice and research within the area, the implicit understanding that functions within this, making some understanding explicit, interpreting it conceptually, discussing the appropriateness of conceptual frameworks, and proposing better conceptual frameworks.

An FFU guides both research and practice in the area and itself emerges out of and may be refined by such research and practice over the years. Some have arisen from practice, others from research. Some frameworks for understanding an area are explicitly stated while others might be tacit. Some explicit frameworks arising from research appeal to philosophy.

1.3.1 Frameworks in Each Area

In each of the five areas mentioned earlier a number of frameworks for understanding the area have arisen, of which the following are referred to in Chapters IV to VIII.

Human use of computers has been seen as:

*   Dialogue with a distal agent
*   Proximal use of a tool
*   Psychological behaviour

- A matter of cost versus benefits
- Power, conflict, and emancipation

Nature of computers and information has been understood as:

- Electronic hardware
- Programmed bit manipulation machine; information is the bit
- Stored symbols manipulated; information is symbols
- Agent; information is knowledge
- All these at different levels
- Like human beings (artificial intelligence)
- A new reality of bodiless mind (cyberspace)

Information systems development (ISD) has been seen as:

- Programming,
- Structured development (e.g., waterfall model)
- Iterative development
- Knowledge elicitation and knowledge representation
- "Orchestrating changes of appreciation through a cyclic learning process"

Shaping of IT resources (algorithms, computer languages, etc.) used by IS developers has been seen in the following terms:

- Developers can perform wonders with: first-order predicate logic (logic programming), or relations (relational data model), objects and classes (object-orientation), and so on.
- They need a complete KR ontology.
- We should aim for "KR to the people".
- Alexander's design patterns are the way to go.

The information society or technological ecology has been addressed in terms of:

- Technological determinism
- Social shaping of technology
- Structuration theory
- "Revealment" (Heidegger)
- "Silicon idol" or a "religion"
- A "liberating vision for technology"

Some of these ways of seeing things in their area have been made explicit, thought out, and given philosophical basis, while others have not been linked with philosophy, not been fully thought out, or not even been made explicit. But the latter are, nevertheless, frameworks for understanding because those who research and practise in an area share a set of beliefs, concepts, norms, and so forth, that are meaningful, and these could be linked to philosophy.

Frameworks within an area often disagree with each other over some methodological or philosophical point, and indeed many have arisen by means of a paradigmatic reaction against earlier ones. But frameworks from different areas are more fundamentally incommensurable with each other because they operate in completely different horizons of meaning.

1.3.2 Issues that Constitute Frameworks for Understanding

The framework for understanding an area adopted by us influences or determines the conceptual apparatus we employ when working in the area—how we classify things, what theories we devise and kinds of methodologies and rules we formulate to guide our research or practice, what we see as important in the area, what types of questions we find ourselves asking, what we see as problematic and what we allow as possible solutions. (This is why, earlier, it was argued that it is not appropriate to try, in this book, to prove conclusively that a different way of looking at IT and IS is necessary.) An FFU cannot be proven either correct or incorrect by theoretical means because it is held as a pre-theoretical commitment as a set of beliefs and assumptions about the area that is adopted or adoptable by those working in the area.

Scientific paradigms, as discussed by Kuhn (1970), may be part of an FFU. But those who practise in an area need, in addition, an awareness of the normativity of the area, which defines what is deemed to be "problem" and what drives aspirations of those working in the area. Behind any area flows a history, at its root lies a motivation, and at its centre stands an assumption of the position human beings take in relation to the area and other presuppositions about the nature of reality. A good FFU will have some place for all such things, reaching to the horizon of meaning of the area. Such things are addressed by philosophy.

To discuss and justify research and/or practice in an area fully, and how the area relates to others, requires the FFU to be explicit. To make an FFU explicit involves surfacing a number of things assumed by those working in the area in both research and practice.

1.3.3 Characteristics of Frameworks

The kind of framework needed for the purposes of this work may be characterised as follows.

- **A framework:** Not necessarily formal, nor fully expressed in written form, nor even fully explicit. But its main themes or principles are made explicit in a reasonably systematic way. It involves not just concepts related together but also attitudes, commitments, beliefs, aspirations, and cultures.

- **Open:** An FFU should be open to extension, but in a way that is true to its nature, rather than by merely bolting new pieces on.

- **Coherent rather than logical:** The elements and characteristics of an FFU should fit together comfortably even though it might be impossible to present an unassailable logic to justify it. For example, the framework developed in Chapter IV for computer use tries to do justice to both the structure and the normativity of various human-computer relationships. Logic alone would be content with doing justice to structure without normativity. But the very notion of "do justice to" only makes sense if we do justice to all that is relevant, to normativity as well as structure.

- **Guiding:** The frameworks should provide a basis for guiding both research and practice. It need not include a methodology as such but should enable methodologies to be developed by researchers and practitioners and critically evaluated. For example, soft systems thinking is a framework while soft systems methodology (Checkland, 1981) is a methodology therein. Practical devices will be discussed.

- **Whole story:** The frameworks should be able to link to other frameworks, not by accident but self-consciously. It should be able to link to alternative frameworks for the same area in order that they should enrich and support rather than replace existing ones. It should link to frameworks from other areas if we are to have any chance of being able to address the "whole story that is IS".

- **Everyday understanding:** As mentioned previously, the type of understanding that frameworks should support, and encourage researchers and practitioners to seek and develop, is that of lifeworld of the area. That is, it should be able to (be expandable to) accommodate all the diversity of the everyday life of the area. This can include theoretical, scientific knowledge, but not as its primary form.

Philosophical thinking is necessary to ensure such characteristics.

1.3.4 A Single Unitary Framework?

As can be seen, there is immense diversity in the ways we understand IS. In each area frameworks offer different capabilities, and some compete for supremacy over others. But the differences are even greater between areas. For example actor-network theory (Latour, 1987) concerns itself with the interaction between people and IS and is usually completely indifferent to whether the IS is programmed using logic programming or object-orientation. And yet there is a link, in that logic programming tends to assume all tasks for which we use computers can be expressed as propositions and predicates in logic, which assumption can constrain our tasks.

There is a move among some (e.g., Shaw & Garlan, 1996) to define "the" discipline of information systems as a single, unitary area of research and practice, and therefore to seek a single, over-arching framework by which to understand it. But could or should we seek a single over-arching framework for understanding the "whole story"? Jones (2000) calls this "mission impossible". Lyytinen (2003) argues that it is neither possible nor desirable: such a quest is "hopeless"; there is no agreement about what constitutes good theory nor about the appropriate criteria for progress in a field. He points to the incommensurability between positivist and "postpositivist" approaches.

This book does not presuppose that we can find one "ultimate foundation" for understanding IT/IS. Rather, it seeks a way of formulating frameworks for understanding each area that cohere with each other. It seeks to accept that things which might be meaningless to one might be meaningful to others, and accord them respect. It is in this way, not by seeking a single framework, that it approaches "the whole story that is IS".

But immediately the challenge arises—and Lyytinen does not seem to have met it—of how to account for both the incommensurabilities between extant frameworks and the links between areas, and to find a way of converting incommensurability into mutual respect. Such things, yet again, are addressed by philosophy.

1.4 Philosophy

"One critical, but largely unrecognized aspect in these debates," remarked Lyytinen (2003), "is the necessity to draw upon philosophical studies as they relate to the nature of scientific knowledge and its foundations. Therefore it is interesting to explore the content of the underlying philosophical argument in these debates and what role they assume to the philosophy as a field of inquiry." It is more than "interesting"; it is useful, and even necessary.

1.4.1 Philosophy and Frameworks for Understanding

Midgley (2000), in *Systemic Intervention*, gives several reasons why philosophy is necessary, even—and especially—when considering practice. Among the reasons he offers are:

- Philosophical assumptions can be used to justify (and critique) practice.
- Philosophy can be used to help define alternatives, and explain why they should be considered.
- Philosophical arguments can assist debates about methodology.
- Philosophy helps us select and defend guidelines for practice.
- Fluency in philosophical debate could help critique and justify intuitive notions and ethical stances in scientific methodology.
- Philosophy helps us see practice in a different light.
- Philosophy reveals why different approaches are incompatible and cannot simply be "compared" (he gives the example of utilitarian and rights-based approaches to managing the National Parks in the USA).

While Midgley specially wished to counter anti-philosophical stances in relation to practice, his reasons why philosophy is important can be broadened if we replace the word "practice" with "research and practice in IS" in most of the above.

Philosophy has a number of main branches, and these are what is necessary to address the issues that constitute FFUs mentioned earlier:

- **Ontology:** What is the nature of the world with which those who work in the area engage?

- **Epistemology:** How do we come to understand and know our area? What constitutes good research? How do we form conceptual structures or a good theory?

- **Philosophical ethics:** What is good and evil in the area: normative or anti-normative, problematic or to be aspired to? What is the root motivation?

- **Methodology:** What methods should we use in research? In practice how do we overcome problems or achieve our aspirations?

- **Anthropology:** What is the role of human beings in the area?

- **Critical philosophy:** What presuppositions lie at the root of any FFU, and what transcendental conditions are necessary for things to be possible?

Therefore, to consider, discuss, formulate an FFU for any area requires some reference to philosophy. There are many different philosophies, and many views on what philosophy is (even this taxonomy of branches of philosophy might not meet with universal agreement). So we must clarify what is meant by philosophy.

1.4.2 Philosophy: A Sketch

Philosophy has a long pedigree, throughout which there have been many ideas on what philosophy is and what its task and role are. To the pre-Socratic Greeks, philosophy linked religious reverence with thinking about the principles of the natural world. Then it became moral and political theorizing, and under Plato and Aristotle, became vast metaphysical constructions. In the medieval period, Western philosophy concerned itself with the relationship between faith and reason, and was the "handmaiden of theology". After the Renaissance philosophy again took up the task of exploring the foundations of physical science, and then turned to examination of the human mind.

Kant believed that philosophy's task was to determine what reason can and cannot do. Hegel, on the other hand, believed that the philosopher's vocation was to approach the absolute through consciousness. Comte restricted his "positive" philosophy to employ only accepted scientific methods, and logical positivism focuses on the activity rather than the content of philosophy. Bergson, in contrast, believed that the deepest and most important knowledge came from what he called philosophical intuition. For Whitehead, the duty of philosophy was primarily to "frame a coherent, logical, necessary system of general ideas in terms of which every element of our experience can be interpreted" (Levi, 1975, p.272). To the early Wittgenstein, "The object of philosophy is the logical clarification of thoughts." Marxist views of philosophy centre on practical issues and stress how class interests affect our world views. To Dewey, philosophy embraced ethics and methodology as well, and he saw its role as helping us understand the unity and interrelatedness of all that is.

It is clear that philosophy is many-sided and can serve many purposes. "Part of what makes it difficult to find a consensus among philosophers about the definition of their discipline," suggests Levi (1975, p. 248) "is precisely that they have frequently come to it from different fields, with different interests and concerns, and that they therefore have different areas of experience upon which they find it especially necessary or meaningful to reflect." He gives several examples, drawn across history of philosophy:

Thomas Aquinas (a Dominican friar of the 13ᵗʰ century), George Berkeley (a bishop of the Irish Church in the 18ᵗʰ century), and Søren Kierkegaard (a Danish divinity student in the 19ᵗʰ century) all saw philosophy as a means to assert the truths of religion and to dispel the Materialistic or Rationalistic errors that, in their opinion, had led to its decline. Pythagoras in ancient south Italy, Rene Descartes in the late Renaissance, and Bertrand Russell in the 20ᵗʰ century have been primarily mathematicians whose views of the universe and of human knowledge have been vastly influenced by the concept of number and by the method of deductive thinking. Some philosophers, such as Plato or the British philosophers Thomas Hobbes and John Stuart Mill, have been obsessed by problems of political arrangement and social living, so that whatever else they have done in philosophy has been stimulated by a desire to understand and, ultimately, to change the social and political behaviour of men. And still others—such as the Milesians (the first philosophers of Greece), Francis Bacon, an Elizabethan philosopher, and, in the 20ᵗʰ century, Alfred North Whitehead, a process metaphysician—have begun with an interest in the physical composition of the natural world, so that their philosophies resemble more closely the generalizations of physical science than those of religion or sociology. (Levi, 1975, p. 248)

The implication of this diverse picture is that our response to any statement of what philosophy is must take account of the characteristics of the human being who philosophises.

Following his quick survey of the diversity of views, Levi (1975, p. 248) said, "But if any single opposition is taken as central throughout the history of Western philosophy at every level and in every field, it is probably that between the critical and the speculative impulses." The speculative impulse is often driven by a desire to construct positive philosophical proposals for understanding the nature of reality. It builds a proposal, which could then be subjected to the critical impulse. The critical impulse seeks to question assumptions in views held about issues (such as about the nature of computers or what is necessary in IS development). [We may criticise Levi's own speculative use of the word "speculative" (rather than, for example, "constructive") as perhaps betraying a presupposition that there are no firm grounds for such construction, a presupposition that will be questioned later.]

1.4.3 Application of Philosophy in Information Systems

Philosophy has been applied within the field of information systems for 50 years. From the 1950s, artificial intelligence referred to philosophy to address the question of whether computers can think, by asking philosophical questions about what thinking and computers are. But the nature of computers is only one of the five areas explored here, and the questions raised in the others go far beyond this very limited AI question.

The author is aware of four other attempts to apply philosophy widely to IS to move toward frameworks for understanding.

Philosophical Aspects of Information Systems (Winder, Probert, & Beeson, 1997) is one work that contains a number of useful papers in which different thinkers try to apply philosophy more widely in IS. It is a collection of chapters by various authors but, though some are very stimulating, one cannot obtain an overall picture.

More cohesive is Mingers and Willcocks (2004) *Social Theory and Philosophy for Information Systems*. Each chapter exposes a particular philosophical position in relation to IS, and is written by an acknowledged expert in that thinking:

- Social theory and philosophy in general
- Functionalism and neo-functionalism
- Phenomenology of Husserl and Heidegger
- Hermeneutics
- Adorno's critical theory
- Habermas' critical theory
- Foucault's notion of power/knowledge
- Structuration theory
- Social shaping of technology
- Critical realism
- Complexity and IS

As the authors admit, however, many other philosophers were omitted, including Bourdieu, Luhmann, Merleau-Ponty, and Maturana. Early (pre-Kantian) thought is almost entirely absent. Overall, this book is an interesting attempt to bring together several philosophical approaches to IS in a way that is helpful to readers. Yet even this does not completely fulfil all our requirements in all areas. It does not show how these could work together to help us understand "the whole story" and the work as a whole seems to reflect the tendency to downplay ontology.

One of the major philosophical streams omitted by Mingers and Willcocks is process philosophy. Midgley (2000) attempts to show how process philosophy can help us understand "systemic intervention". Though not specifically aimed at IS, the development and application of IS (the areas of ISD and HUC) could be seen in these terms. It is more systematic than the others and makes an attempt at an FFU, but it is not for IS.

Winograd and Flores (1986) undertook a similar exercise, attempting to formulate a coherent, philosophical FFU based on Heidegger. Their argument is, in the main, coherent and persuasive, and their work in now classic and widely referenced. But there have been criticisms of it.

None of these works entirely fulfils the requirements set out earlier: a framework for understanding that is open, coherent, guiding practice and able to address the "whole story" that is IS. Winograd and Flores work perhaps comes closest and so is examined later (Chapter IV). Critical examination of the others must wait for another occasion (and the author hopes to contribute to this in due course).

But before this occurs the author wants to place another philosophy, of a radically different type, on the table, so that all these types of philosophy (and perhaps yet others) may be considered together, and, using this philosophy as an exemplar, to explore the process by which LOFFUs may be formulated for each area. In the rest of this chapter, a reasonably systematic discussion will be made of what roles philosophy should play in relation to IS, the suitability of philosophy to IS, the philosophical issues that need to be addressed, prior to introducing the different philosophy.

1.4.4 Roles of Philosophy in Information Systems

In an article entitled "Some philosophical and logical aspects of information systems", Robb (1997) suggests (p. 8), "To me, the key philosophical question is 'What argument or evidence would justify me in asserting this conclusion to be true?'" This could apply to any area. But to create FFUs or find philosophical grounding for them requires much more from philosophy than this. We do not start with this or that "conclusion" to be tested by argument or evidence because not only have we still to determine what conclusions we wish to so test, but we have not even worked out a strategy by which we could do so. Is it appropriate to even seek testable conclusions?

The Cambridge Dictionary of Philosophy (Audi, 1999) goes further, suggesting it is the role of science to find out what is true, but the role of philosophy to find out what truth is. According to this view, Robb's view earlier would be a scientific rather than fully philosophical one. Philosophy stands back from the sciences to gain a wider, cross-disciplinary standpoint that enables us to understand all such scientific views of truth in relation to each other. This can help form FFUs in areas in which several sciences interact; for example psychology, linguistics, sociology, and economics are all relevant in understanding different aspects of computer use. While that nicely articulates the difference between science and philosophy, FFUs are concerned not only with truth but also with how we should live in each area (normativity) and with presuppositions that are not "true" so much as "held", by a deep, often tacit, commitment.

Lyytinen (2003) characterises such views as the earlier mentioned as "a positivist, or Popperian concept of what should be the role of philosophy as a reference discipline," which assumes that "Philosophy should provide the ultimate language and a logic for the justification of IS knowledge." He continues:

This view, as we all know, has been largely contested in the philosophy in the post modern era. Another, "postpositivist", view of the role of philosophy in the IS field would be to view it as a critical voice in a conversation about information systems that seeks to undermine any quest for a ultimate foundation of the IS discipline. Philosophy deconstructs any language and logic that has been developed for such purposes.

The "critical voice" is vital to question assumptions and expose presuppositions in each area of IS. But Lyytinen does not give much specific guidance.

Hodges' (1995) does, suggesting five specific ways in which philosophy can inform information systems research:

- As a reference discipline *per se* in the sense that it would contribute a subject and a methodology, such as dialectic method (to critically analyse the assumptions of a theory), ethics, or general systems theory;

- By helping to lay conceptual foundations on which to build theory, perhaps alongside scientific work;

- Epistemologically informed studies of issues in research and practice, since epistemology is a branch of philosophy;

- Employing the approaches and insights of hermeneutics and phenomenology in order to develop an understanding of information technology in a broader context;

- Employing the approaches and insights of existentialism to move away from a presumed intellectualism and give due regard to experience and being.

The first three are indeed partly how philosophy is used here, but these still do not fully meet our needs. First, we are not primarily concerned to use philosophy as a reference discipline in itself, since what we are seeking here is not so much a philosophy *of* information systems, but a philosophy *for* information systems. We are seeking to find philosophy that can be useful in helping us understand information systems, but we are not restricted to a purely philosophical type of understanding. Rather, we want philosophical support for everyday understanding.

Secondly, as mentioned previously, the foundations we want are not merely conceptual, but must also be normative and presuppositional.

Hodges' third role, Wernick, Shearer, and Loomes (2001) suggest, should be broadened beyond epistemology, especially when considering software, its development, evolution, and use. Ontology is important, especially to Poli (2001), in order to philosophically inform our understanding of types of technology.

Wernick et al. (2001) have also criticised the last two, suggesting that they are better categorised as examples of the application of specific philosophical directions, while the first three map out areas of information systems research and practice which can be supported by reference to philosophy. But Courtney and Porra (2000) justify the last two as yielding unconventional and imaginative ideas, in contrast to the first three, which they see as reflecting a "somewhat conventional, cognitive processing view of IS research." Moreover, it is phenomenology, and thought derived from it, that drew attention to the lifeworld. The philosophy used in this work is neither phenomenological nor existential, though it has been influenced by both. Rather, it critically questions the very deepest roots of Western thought and it is this critique that generates the "different way of looking at IT/IS" with which this work opened.

1.4.5 The Suitability of Philosophy

It is these roles of philosophy that qualify it to address the characteristics of frameworks outlined earlier. As a means of laying conceptual foundations it enables the formulation of frameworks in the first place. As a guide to reasoning it ensures their quality. Transcending different sciences, philosophy enables the whole story to be addressed. Its critical voice makes frameworks open and provides normative direction. Phenomenology and existentialism, as seen earlier, made sure the lifeworld is no longer taken for granted. But the desire for a coherence that is distinct from logical consistency, might pose a problem, especially in the light of Lyytinen's desire to "undermine any quest for an ultimate foundation of the IS discipline."

But, just as the discussion of IT/IS should not be constrained by Robb's positivist view, neither should it be constrained *a priori* by Lyytinen's "postpositivist" anti-foundationism. As illustrated in the vignettes in the Preface, foundations of an appropriate kind are vital for an everyday, lifeworld approach. Deconstruction will indeed be employed, but Lyytinen's language of "seeks to undermine any quest" speaks of a pre-theoretical dogma against "ultimate foundations", and the need to adhere to this dogma must be questioned. Lyytinen's dogma seems to have two roots. One is a fear of unimaginative conceptual frameworks (often of a positivist-realist kind) that are inappropriately forced on us. The other is dislike of the battles that have raged between paradigms vying for supremacy over each other.

The approach in this work is to employ a single philosophy, but in a way that should not arouse Lyytinen's fears. It is one that does see a difference between coherence and logical consistency. From it, a diversity of conceptual frameworks and paradigms can emerge and, as mentioned at the start, it does not seek supremacy over others. It will still be the kind of critical voice that Lyytinen desires. It will also ensure proper reasoning, help us stand back from the sciences, act as reference discipline where necessary, lay conceptual foundations and stimulate unconventional ideas. It will do these with a respect for the everyday lifeworld. Unlike some recent turns in philosophy, it can cover all the main branches of philosophy listed previously: ontology, epistemology, philosophical ethics, methodology, philosophical anthropology, and critical philosophy, and so it is likely to be able to respect what is meaningful in all five areas of research and practice in IS.

1.4.6 Philosophical Issues in IT/IS

Different philosophies have been most appealed to in different areas of IS. Though this might to some extent be an accident of history, there might be a more substantial reason: in each area different sets of issues are important, especially in its lifeworld. Table 1-1 shows a few issues in each area with the philosophical issues they raise.

It should be clear from this why (with Lyytinen) this work does not seek a single, unitary framework for all areas. Because most extant philosophies, especially during the last 500 years, tend to focus on a subset of these issues, each area has appealed to a different type of philosophy in order to understand the issues it faces. Expanding on Table 1-1, the difficulty with many philosophies becomes clear:

• Human use of computers involves humans and IT, and so requires philosophy that acknowledges the possibility of a genuine point of contact between technology and human beings. Being mostly of the lifeworld, with the human being in a social context, use requires a philosophy that affords dignity to everyday life and to what it means to be fully and socially human. So materialist and rationalist philosophies are unlikely to be helpful. To deal with impact of use, especially unexpected impact, including on non-human stakeholders like animals or the environment, requires a philosophy that can transcend and yet acknowledge the perspectives of human stakeholders. To differentiate benefit

Table 1-1. Philosophical issues raised by IT/IS

Area	IS Issue	Issue in Philosophy
Human Use of Computers	User-computer relationship Human activity with computer Impact in use Benefit versus detriment Variety of impacts	Self and world Subject-object relation Meaning and repercussions Good and evil Diversity and unity
Nature of Computers	Computer as experienced Nature of computer, info The AI question Hardware, bits, symbols	Self-world relation Ontology Anthropology, ontology Meaning and reductionism
Information Systems Development	Teamwork Lifecycle methods Guidelines Human creativity Requirements analysis Knowledge elicitation Conflict	Social theory Process Normativity Freedom and intention Possibility, responsibility Knowledge, epistemology Perspectives
Information Technology Resources	KR languages Types and classes Inappropriateness	Philosophical linguistics Universals and individuals Diversity of reality
Information Technology as Ecology	Validity of ICT as human endeavour ICT as our environment Gender issues Modern dominance of ICT	Philosophical ethics and destiny Structural relations Diversity of meaning, norms Perspectives, progress

from detriment, especially when both occur, demands an intrinsic normativity within the philosophy, and, again, one that transcends, rather than being derived from, the value systems of those involved. Therefore subjectivist philosophies are unlikely to be sufficient.

• The nature of computers requires a philosophy that takes the nature of things seriously. So ontology is important. It is difficult to see how philosophies of the nominalist tradition can provide a basis for understanding the nature of computers (as opposed to our beliefs about the nature of computers). Moreover, it is not helpful when a philosophy presupposes either that computers and humans are basically the same or that they are so radically different that there can be no comparison between them. It is difficult to see how either material-ist, romanticist, or Scholastic philosophies can, ultimately, do more than fall back on dogma when confronted with these issues. We need a philosophy that enables us to discuss the ontic status of both humans and computers, and of the relationship between them.

- ISD requires a philosophy that can give a genuine account of human knowledge of, interpretation of, and perspectives on, situations in which IS is to be applied. This suggests subjectivist philosophies or those of the linguistic turn—until the requirement to address possibility and responsibility is acknowledged. Then the need for a non-relative normativity becomes clear. This is why critical social theory has become increasingly popular, with its over-riding norm of emancipation. But what is emancipation? Philosophy should be able to speak to the management aspects of ISD, and not just relegate them to the realm of practice.

- The shaping of IT resources requires not only ontology, but an ontology of diversity. Otherwise, ultimately, we have no basis for discussing the variety we encounter when we take a pre-theoretical attitude to information technology, and are driven back to reductionism. But the ontology we need must also be able to speak of the coherence we experience within this diversity. That suggests we need a pluralistic ontology in which the ontic categories are irreducible to each other and yet there are ontic relationships between them. Reviewing the history of philosophy, we can generally observe that when ontology has presupposed that there is some "substance" or "process" in terms of which everything should be explained, it has usually tended toward reductionism, either of a monist or dualist type. Historicism and subjectivism seem to allow for diversity but these find it difficult to avoid fragmentation.

- To understand the information society, as a technological ecology, requires a philosophy that enables us to analyse the circular dependency between environment and technology. It must see both inscription and societal structures as meaningful and mutually irreducible. Philosophy that presupposes environment (including society) can be reduced to interactions between individuals is unlikely to be useful. It calls for a philosophy that is comfortable not only with diversity but especially with the notion of long-term destiny of the technological project of humankind. Such philosophies are rare outside Scholasticism or Marxism.

Most streams of philosophy cannot address all these—whether Plato, Aristotle, Scholasticism, early rationalism or empiricism, Kantian, Hegelian philosophy, process philosophy, phenomenology, existentialism, linguistic turn, critical turn or postmodernism. In the author's experience, some of which is outlined in the Preface, the radical critical-positive philosophy pioneered by the Dutch thinker, the late Herman Dooyeweerd (1894-1975), is one that is able to provide these kinds of insights in all areas.

1.4.7 Dooyeweerd's Philosophy

Herman Dooyeweerd came from the realm of politics and jural science and was a Christian believer of the Reformed tradition. But he assiduously held that philosophy must not be reduced to either politics or theology, nor to social theory, nor to linguistics, nor to logic, nor to mathematics, nor to any other such special scientific arena. In order to avoid such reduction even in his own thinking, he undertook a transcendental critique of theoretical thought in order to establish the necessary "transcendental" conditions that make philosophy as such possible. His conclusions included:

- Everyday life must be given due respect, as a given, and never treated as a theory.
- All theoretical thought, including philosophy, is rooted in pre-theoretical ("religious") stances, and so can never be neutral—including his own thought.
- Philosophy is an integrative discipline that enables us to think about the relationships between not only the sciences but also the distinct aspects of our experience, and thus the diversity and coherence of the lifeworld.
- Philosophy is thus open to issues that are found in all disciplines, such as things and processes, laws and norms, method, knowledge and intuition, the place of the human being, diversity, unity and origins—that is, ontology, normativity, methodology, epistemology, anthropology, and so forth.

It is this view of philosophy that commends itself to those who wish to formulate LOFFUs for the diverse areas of the coherence that is information systems.

1.5 Our Approach

Such a claim may be difficult to believe and some readers will already have reacted against it, or some, perhaps, against the word "Christian". I, the author, ask you, the reader, not to react, but to hear the case I put in this work. Dooyeweerd's philosophy should be critically examined in order to properly explore its potential. A surface examination will not do.

As mentioned at the start, the standard practice in academic writing is to begin by arguing there is a problem that needs solving, then to make a proposal that solves the problem. This may be appropriate for a proposed theory or methodology, but it is not appropriate for proposing a framework for understanding because it is the framework itself that, as a way of seeing things, informs us what is problematic. When we state problems at the start of an argument, we have already adopted a framework for understanding the area in which we are working, even if tacitly. This is especially the case when looking at a whole field. To argue conclusively there is a problem that requires a different philosophy already presupposes that philosophy, and hence cannot prove either the validity of, or need for, it.

Therefore it is inappropriate to attempt any prior argument for Dooyeweerd. So the reader is introduced directly to Dooyeweerd's philosophy in Chapters II and III, to the extent that will be needed to explore each area of IT/IS. Chapter II is an overview that outlines some ways in which Dooyeweerd is different, and offers a dozen reasons why his philosophy should be of interest to us. Chapter III introduces portions of Dooyeweerd's philosophy that will be needed to formulate LOFFUs. Chapter III may be treated as a resource for this exploration, but it is designed also to give a little deeper understanding of the "why" as well as the "what" of Dooyeweerd so that the reader can take it further if they choose. A brief critique of Dooyeweerd is included in Chapter III.

Then, Chapters IV to VIII explore in some depth how Dooyeweerd may be used in each of five areas of research and practice to formulate frameworks for under-standing:

- Chapter IV: Human use of computers
- Chapter V: The nature of computers and information
- Chapter VI: IS development
- Chapter VII: Information technology resources
- Chapter VIII: Information technology as ecology

These chapters will use a mixture of philosophical, theoretical, and practical dis-course, drawing on a range of published material. Each chapter will also discuss how Dooyeweerd's philosophy can engage with some extant frameworks and major issues in each area. Note that research methodology in each area is not discussed, only research content, strategic directions for research, and some practical devices for use in practice.

The final chapter will reflect on the exploration. It summarises the frameworks developed, reflects on the frameworks, reflects on how Dooyeweerd has been used, then reflects on the process of this whole exploration. It makes suggestions for how readers might benefit from this exploration even if they do not wish to use Dooye-weerd. But it ends with a plea to take Dooyeweerd seriously.

From time to time, other philosophies and philosophers will be referred to, especially ones that have been mentioned in various IT/IS discourse. The reader does not need to understand those in order to benefit from the argument here and may, if they wish, simply ignore such references. The purpose of such references is twofold. One is to situate Dooyeweerd among other ways of thinking, for those who are aware of them, showing the ways his thought is either similar to or differs from theirs. The other is to suggest directions for further study by those who know philosophy and wish to explore Dooyeweerd more deeply in relation to others. A more penetrating analysis of the relationship between Dooyeweerd's thought and that of others must be left to philosophers and is beyond the scope of this work.

The next two chapters, therefore, seek to provide a comprehensible understanding of Dooyeweerd's thought sufficiently comprehensive to allow the formulation of frameworks for understanding and their subsequent critique and refinement.

To assist understanding and further study, parts and subparts of chapters are numbered and many cross references have been included, both from the framework chapters into those explaining Dooyeweerd, so that the relevant portions of Dooyeweerdian thinking can be located quickly and understood in the broader context of his thought, and also from Dooyeweerd out to the framework chapters, so that when trying to understand Dooyeweerd examples of how each point applies can be found quickly. Copious references to Dooyeweerd's own writings have been provided in order to facilitate deeper study.

References

Adam, A. (1998). *Artificial knowing: Gender and the thinking machine.* London: Routledge.

Audi, R. (Ed.). (1999). *The Cambridge dictionary of philosophy* (2nd ed.). Cambridge, UK: Cambridge University Press.

Avison, D., Bjørn-Andersen, N., Coakes, E., Davis, G. B., Earl, M. J., Elbanna, A., et al. (2006). Enid Mumford: A tribute. *Information Systems Journal, 16*, 343-382.

Butterfield, J., & Pendegraft, N. (1996). Cultural analysis in IS planning & management. *Journal of Systems Management, 47*, 14-17.

Ceruzzi, P. (1996). From scientific instrument to everyday appliance: The emergence of personal computers, 1970-1977. *History and Technology, 13*(1), 1-31.

Checkland, P. (1981). *Systems thinking, systems practice.* New York: Wiley.

Cotterill, T., & Law, N. (1993). EIS: A practical approach. *Proceedings of the 1993 Hewlett-Packard Computer Users' European Conference HPECU/HPCUA,* Harrow, UK. HP Computer Users Association.

Courtney, J. F., & Porra, J. (2000, August). Mini-track on the philosophical foundations of information systems. Paper presented at the *Americas Conference on Information Systems, AMCIS 2000,* Long Beach, CA.

Crawford, W. (1996). I heard it through the Internet. In R. Kling (Ed.), *Computerization and controversy: Value conflicts and social choices* (pp. 594-596). San Diego, CA: Academic Press.

Davis, F. D. (1986). *A technology acceptance model for empirically testing new end-user information systems: Theory and results.* Unpublished doctoral dissertation. MIT Sloan School of Management, Cambridge, MA.

Ferguson, N., & Schneier, B. (2003). *Practical cryptography.* Indianapolis, IN: Wiley.

Gladden, G. R. (1982). Stop the life-cycle, I want to get off. *ACM SIGSOFT Software Engineering Notes, 7*(2), 35-39.

Glass, R. L., Ramesh V., & Vessey, I. (2004, June). An analysis of research in computing disciplines. *Communications of the ACM, 47*(6), 89-94.

Habermas, J. (1987). *The theory of communicative action: Vol. 2. The critique of functionalist reason* (T. McCarthy, Trans.). Cambridge, UK: Polity Press.

Heidegger, M. (1962). *Being and time* (J. Macquarrie & E. Robinson, Trans.). Oxford, UK: Blackwell. (Original work published 1927)

Hodges, W. S. (1995). Five roles that philosophy can play in MIS research. In *Proceedings of the 1995 Meeting of the Americas Conference on Information Systems, AMCIS 1995,* Pittsburgh, PA. Atlanta, GA: Association for Information Systems.

Honneth, A., Knodler-Bunte, E., & Windmann, A. (1981). The dialectics of rationalization: An interview with Jürgen Habermas. *Telos, 49,* 3-31.

Husserl, E. (1970). *The crisis of European sciences and transcendental phenomenology* (D. Carr, Trans.). Evanston, IL: Northwestern University Press.

Jacob, M., & Ebrahimpur, G. (2001). Experience vs. expertise: The role of implicit understanding of knowledge in determining the nature of knowledge transfer in two companies. *Journal of Intellectual Capital, 2*(1), 74-88.

Jones, M. (2000). Mission impossible: Pluralism and 'multi-paradigm' IS research. *Information Systems Review, 1*(1), 217-232.

Kuhn, T. S. (1970). *The structure of scientific revolutions.* Chicago: University of Chicago Press.

Landauer, T. K. (1996). *The trouble with computers: Usefulness, usability and productivity.* Cambridge, MA: MIT Press/Bradford Books.

Latour, B. (1987). *Science in action.* Cambridge, MA: Harvard University Press.

Levi, A. W. (1975). Philosophy, history of Western. In *Encyclopaedia Britannica* (Vol. 14, pp. 247-275). Chicago: Encyclopaedia Britannica.

Lyon, D. (1988). *The information society: Issues and illusions.* Cambridge, UK: Polity Press.

Lyytinen, K. J. (2003). Information systems and philosophy: the hopeless search for ultimate foundations. In J. I. DeGross (Ed.), *Proceedings of the Americas Conference on Information Systems: AMCIS 2003.* Atlanta, GA: Association for Information Systems.

Lyytinen, K., & Hirschheim, R. (1987). Information systems failures: A survey and classification of the empirical literature. *Oxford Surveys in Information Technology, 4,* 257-309.

Midgley, G. (2000). *Systemic intervention: philosophy, methodology and practice.* New York: Kluwer/Plenum.

Mingers, J., & Willcocks, L. P. (Eds.). (2004). *Social theory and philosophy for information systems.* Chichester, UK: Wiley.

Network News. (2006, March 27). IT spending to increase 6.3% in 2006, IDC says. *Network World, 23*(12), 8.

Noble, D. F. (1979). Social choice in machine design: The case of automatically controlled machine tools. In A. Zimbalist (Ed.), *Case studies on the labor process* (pp. 18-50). New York: Monthly Review Press.

Noble, D. F. (1997). *The religion of technology: The divinity of man and the spirit of invention.* New York: Alfred A. Knopf.

Poli, R. (2001). *ALWIS: Ontology for knowledge engineers.* Unpublished doctoral thesis, University of Utrecht, Netherlands.

Robb, F. F. (1997). Some philosophical and logical aspects of information systems. In R. L. Winder, S. K. Probert, & I. A. Beeson (Eds.), *Philosophical aspects of information systems* (pp. 7-21). London: Taylor and Francis.

Rogers, W. K. (1998). *Human life and world: On the insufficiency of the phenomenological concept of the life-world.* Paper presented at the Twentieth World Congress of Philosophy. Retrieved April 2, 2007, from *http://www.bu.edu/wcp/Papers/TKno/TKnoRoge.htm*

Schutz, A., & Luckmann, T. (1973). *Structures of the life-world, volume I.* Evanston, IL: Northwestern University Press.

Schutz, A., & Luckmann, T. (1989). *Structures of the life-World, volume II.* Evanston, IL: Northwestern University Press.

Shaw, M., & Garlan, D. (1996). *Software architectures: Perspectives on an emerging discipline.* Upper Saddle River, NJ: Prentice Hall.

Strijbos, S., & Basden, A. (2006). In search of an integrative vision of technology. In S. Strijbos & A. Basden (Eds.), *In search of an integrative vision for technology: Interdisciplinary studies in information systems* (pp. 1-16). New York: Springer.

Thorp, J. (1998). *The information paradox: Realizing the business benefits of information technology.* Toronto, Ontario, Canada: McGraw-Hill.

Walsham, G. (2001). *Making a world of difference: IT in a global context.* Chichester, UK: Wiley.

Wernick, P., Shearer, D. W., & Loomes, M. J. (2001). Categorising philosophically-based research into software development, evolution, and use. In *Proceedings of the Seventh Americas Conference on Information Systems, AMCIS 2001*, Boston (pp. 1999-2005). Atlanta, GA: Association for Information Systems.

Whyte, G., & Bytheway, A. (1996). Factors affecting information systems' success. *International Journal of Service Industry Management, 7*(1), 74-93.

Winder, R. L., Probert, S. K., & Beeson, I. A. (Eds.). (1997). *Philosophical aspects of information systems.* London: Taylor and Francis.

Winograd, T., & Flores, F. (1986). *Understanding computers and cognition: A new foundation for design.* Reading, MA: Addison-Wesley.

Endnote

[1] At least 50% (Lyytinen & Hirschheim, 1987), 60% (Cotterill & Law, 1993), 50%, (Whyte & Bytheway, 1996), 60% (Butterfield & Pendegraft, 1996), 75% (Gladden, 1982).

Section I
Dooyeweerd's Philosophy

Chapter II

Overview of Dooyeweerd's Philosophy

The aim of this chapter is to introduce the thought of Herman Dooyeweerd, giving a basic overview of the main themes and ways of thinking that are relevant to understanding information systems. More detailed explanation and discussion of specific points of his philosophy will be offered where they are first needed in later chapters, where these basic themes will be used to formulate frameworks for understanding research and practice in several areas of IS. So the reader should not feel required to absorb reams of philosophy before addressing IS issues.

Not all of Dooyeweerd's thought is explained, only that needed for understanding information systems. But what is provided should be sufficient not only to explain the frameworks developed but also to enable the reader to take this work further. Some links are made to other philosophic thinkers referred to by the IS communities.

For a more complete rendering of Dooyeweerd, see summaries by Kalsbeek (1975), Choi (2000) or Clouser (2005), or see *A New Critique of Theoretical Thought* (Dooyeweerd, 1984), which is Dooyeweerd's own four-volume *magnum opus*. This was first published in 1953-5, and it extended an earlier similar work published in Dutch (1935) by responding to criticisms thereof. We will also draw upon Dooyeweerd (1979), (1986), and (1999).

2.1 Dooyeweerd's Approach to Philosophy

Dooyeweerd was unusual as a philosopher. It has been said[1] that Dooyeweerd aimed, not so much to construct a philosophical theory about reality as to "open up the structure of reality so that once we 'get' it, we can run with it. He clears away all the things that keep us from seeing it." In the author's experience this is exactly what happens, even among those with no knowledge of philosophy. Dooyeweerd's way of thinking is immensely practical and what at first sight appears complex seems rather to express the complexity that is everyday reality.

Dooyeweerd's interest was in the whole breadth and depth of reality: in what there is and occurs, in what might become, in how we know, in what we believe and what we have presupposed, in what is right and wrong, in meaning, in humanity, divinity and mundanity, and in everyday experience—all that the branches of philosophy mentioned earlier. For example, IS in use can succeed and fail in many ways for many reasons; Dooyeweerd "clears away" that which stops us taking seriously the diversity of ways in which success and failure occur and the very real and yet subtle difference between success and failure.

Yet this clearing away was not nihilism, nor even an extreme form of subjectivism. Dooyeweerd believed that 2,500 years of theoretical, philosophical, and scientific thinking have obfuscated rather than revealed the structure of reality—from the Greek philosophers, through the medieval periods, through the Reformation, Renaissance and Enlightenment, into the modern period, up to the middle of the twentieth century. Yet he also believed that this obfuscation was mainly inadvertent rather than deliberate, and he respected many thinkers as offering genuine insight. He believed that this whole 2,500-year-old river of thinking itself needed to be properly understood as part of reality.

The root of the problem, he argued, was deep presuppositions at the root of Western thinking, such as the dividing of form from matter (especially in Greek thought), sacred from secular (medieval) and control from freedom (modern). Neither philosophical nor scientific thought is ever neutral nor absolutely true; all is seen through the "lens" of our presuppositions. Dooyeweerd prefigured many "critical" thinkers in IS and management today in admitting they see reality through a particular "lens", such as Adam et al. (2006). But, being a true philosopher, he always applied his thinking to himself. So he openly declared his own presuppositions, which differed from those of most Western thought, and actively engaged in self-critique and invited the critique of others.

But he did not assume, as most after Kant have done, that all lenses must be deemed equally valid. He undertook a critical examination not only of the various "lenses" on offer but also of what constitutes any lens, the lens that is the theoretical attitude of thought itself, going deeper than either Kant or Husserl did. As a result, there is reason to believe that the "lens" he used is clearer and less distorting of the structure of reality than others are. This is the first reason for interest in Dooyeweerd.

So Dooyeweerd was a critical philosopher *par excellence*. But he was not only critical. After deconstructing Western theoretical thinking, he was courageous enough to make a positive proposal that might itself be subjected to critique. This positive proposal arises from his different presuppositions; it is the way reality might appear to anyone who looks at it through the type of lens he used. Support for the first reason for interest in Dooyeweerd comes when we find that all the philosophical issues relevant to IS tabulated in Chapter I are visible through his "lens" and may be usefully examined by it.

This chapter first examines Dooyeweerd's critical approach and recounts what Dooyeweerd consequently believed to be the root of philosophical thinking, and then outlines some main characteristics of Dooyeweerd's proposal, in preparation for examining it in detail in Chapter III.

2.2 Dooyeweerd's Critical Approach

In his critical approach, which occupies most of Volume I of Dooyeweerd (1984), he undertook a deep critique of theoretical thought, especially as it is found in philosophy. If we want to use tools, we must first be sure they are suited to our task; Dooyeweerd was not convinced that theoretical thought as it has been understood and practised for the past 2,500 years is suited to understanding the complexity of the world as it presents itself to us in our everyday experience.

This is the second reason for interest in Dooyeweerd: he prepared tools suited to thinking about everyday experience. A note about terminology: Dooyeweerd used the words "everyday", "naïve", and "pre-theoretical" almost synonymously as adjectives for thought, experience, and attitude, and all three may often be equivalent to "lifeworld", a word Dooyeweerd never used. In most places we will use "everyday", but the others will be used occasionally instead, as synonyms, where style seems to require it.

2.2.1 Dooyeweerd's Immanent Critique

Dooyeweerd's main critical task was to determine what the conditions are that make a theoretical attitude in thought possible, distinct from an everyday attitude. From this, its capabilities and limits may be understood so that we do not expect too much of it. He reviewed how it has developed since the early Greek thinkers, showing how the theoretical attitude has always been given undue preference over the everyday attitude. Theoretical thought has always tended to narrow and distort our understanding of everyday experience (an example is cost-benefit analysis, yet gender-based evaluations and even interpretive approaches like actor-network theory can be just as narrow). He argued that the main mistake was to assume theoretical thought is autonomous—that is, it can be used as a neutral foundation for, or judge of, everything else and the route to "true" knowledge.

Dooyeweerd did not reject theoretical thought as such but examined sympathetically ("immanently") the millennia of theoretical thinking, seeking to understand and expose its presuppositions and demonstrate that these presuppositions led to deep incoherency or other problems. Clouser (2005) clarifies three types of incoherency to which Dooyeweerd paid attention:

- **Self-referential incoherency:** When a theory makes a claim that denies itself as a theory (for example, Freud's claim that every belief arises from unconscious emotional needs makes that claim a product of Freud's own needs, rendering it useless as a theory);

- **Self-assumptive incoherency:** When a theory is incompatible with beliefs we must make for it to be true (for example, extreme materialism denies there is anything other than physicality, hence no logic nor language, hence no possibility of any theory);

- **Self-performative incoherency:** "A theory must be compatible with any state that would have to be true of a thinker, or any activity the thinker would have to perform, in order to have formulated the theory's claims" (Clouser, 2005, p. 85).

Clouser explains (p. 83), "Two of these incoherencies have been noticed by philosophers in the past, but are not yet taken seriously enough in my opinion. The third is relatively new, having been first defined and deployed by Herman Dooyeweerd about fifty years ago." It emphasises the centrality of the human person who is doing theoretical thinking. Dooyeweerd's thought prefigures much since then, including for example Polanyi's (1962) "personal knowledge", Habermas' early (1972) insistence on "human interests".

Dooyeweerd undertook extensive immanent critiques of thinkers from the early Greek period, through the medieval period, through Reformation, Renaissance and Enlightenment, then Kant and phenomenology in particular, through to the middle of the twentieth century, examining such types of incoherency. It was the third type especially that led him to the conclusion that philosophy, even so-called critical philosophy, was not critical enough (1999, p. 6):

Neither Kant, the founder of the so-called critical transcendental philosophy, nor Edmund Husserl, the founder of modern phenomenology, who called his phenomenological philosophy "the most radical critique of knowledge", have made the theoretical attitude of thought into a critical problem. Both of them started from the autonomy of theoretical thinking as an axiom which needs no further justification.

(Specifically, while Kant posed the critical question of how we are able to think, he took for granted that we can take a theoretical attitude, and while Husserl went deeper and showed that all theorizing depends on a framework of meaning that is found in everyday life he still assumed that everyday life generates that meaning. Dooyeweerd went deeper than both, and explored what enables us to take a theoretical attitude and whence the meaning ultimately arises.)

If Dooyeweerd is right this implies a deep incoherency at the root of such philosophy. Since much philosophical thinking in IS stems ultimately from this (via Heidegger, Gadamer, Habermas, Foucault, and others) it may be that the foundations of philosophical thinking in IS are shaky. This is a third reason for interest in Dooyeweerd.

2.2.2 Dooyeweerd's Transcendental Critiques

In response, Dooyeweerd did not assume the autonomy of theoretical thinking. He did not even presuppose that it is possible to take a theoretical attitude, but first asked what is necessary in order to do so, by the method of transcendental critique.

While Kant, Husserl, Habermas, Bhaskar, and others have used transcendental critique, it may be that Dooyeweerd has new insights to offer, both because he worked this notion out in considerable detail and also because he came from a different direction (see the following) and therefore might uncover issues overlooked by others.

Developing not just one but two transcendental critiques of theoretical thought, which he called the first and second ways of critique and are explained by Choi (2000), he showed not only that theoretical thought can never be neutral, but that the non-neutrality is religious in nature (rather than, for example, arising from mere subjective opinion). By "religious" he did not mean relating to any particular religion or creed, explaining (1984, I, p. 57):

To the question, what is understood here by religion? I reply: the innate impulse of human selfhood to direct itself toward the true *or toward a* pretended *absolute Origin of all temporal diversity of meaning, which it finds focused concentrically in itself.*

The problems in philosophy, Dooyeweerd argued, stem from failure to acknowledge its "religious" root: theoretical thought is driven by presuppositions that are of a religious nature.

In the first way of critique, he showed that our religious stance concerning an absolute Origin influences our pre-theoretical choice of Archimedean Point of reference from which to reflect on the diversity that we experience. But this was criticised as depending on a particular view of the role of philosophy. In his second way of critique, which made no such assumption, he argued that the human self chooses pre-theoretically how to synthesise our theoretical thinking with what is thought about, and the nature of this choice is religious, especially governed by ground-motives (see the following). As Geertsema suggests (2000, p. 85):

The main aim of his philosophy has always been to show how the religious starting point controls philosophical and scientific thought, both in relation to humanistic and Scholastic thinking and in relation to his own reformational conviction.

As Geertsema points out later (2000, p. 99), "He is not satisfied with an argument that shows that in fact philosophy always has been influenced by religious convictions." Rather, "He wants to show that it cannot be otherwise, because it is part of the nature of philosophy or theoretical thought. For that reason he called his critical analysis a transcendental critique."

The extent to which he succeeded in the eyes of philosophers is still being debated, Geertsema says, but many of his insights are particularly useful in information systems, starting with the notion of ground-motives. This is a fourth reason for interest in Dooyeweerd.

2.3 The Religious Root of Philosophical Thought

In his study of Western thinking, Dooyeweerd found that four religious "ground-motives" ("*grondmotieven*") have been the "spiritual driving force that acts as the absolutely central mainspring of human society" (Dooyeweerd, 1979, p. 9).

2.3.1 Ground-Motives

A ground-motive is the "moving power or spirit at the very roots of man, who so captured works it out with fear and trembling, and curiosity" (Dooyeweerd, 1984, I, p. 58). It generates in us supra-theoretical presuppositions that we make about the nature of reality, including theoretical thinking itself, and affects how we go about obtaining knowledge and solving problems. Ground-motives encapsulate and determine all the branches of philosophy that a reflective society develops, including ontology, epistemology, anthropology, methodology, and philosophical ethics.

It [a ground-motive] thus not only places an indelible stamp on the culture, science, and social structure of a given period but determines profoundly one's whole world view. If one cannot point to this kind of leading cultural power in society, a power that lends a clear direction to historical development, then a real crisis looms at the foundations of culture. Such a crisis is always accompanied by spiritual uprootedness. (Dooyeweerd, 1979,p. 9)

He discussed four ground-motives that have driven Western thought over the past 2,500 years:

- The Greek ground-motive of form-matter (FMGM)
- The Judeo-Christian ground-motive of creation, fall, and redemption (CFR)
- The medieval ground-motive of nature-grace (NGGM), which arose from a synthesis of them and itself gave rise to
- The humanist ground-motive of nature-freedom (NFGM), within which arose the Science Ideal and the Personality Ideal as dialectically opposing poles

Figure 2-1. Development of Western thought

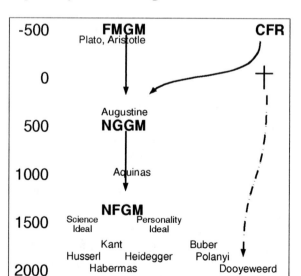

These have interacted historically, as shown in Figure 2-1.

Dooyeweerd did allow for others, himself mentioning that of the Zoroastrian religion (Dooyeweerd, 1979, p. 112), and Choi (2000) discussing those of Korean thought, but he discussed these four in some depth. Since they will be referred to throughout this work, they are briefly explained.

2.3.1.1 The Form-Matter Ground-Motive (FMGM)

The Form-Matter ground-motive presupposes that all being, occurrence (activity, behaviour), knowing, and good and evil may be explained in terms of form and matter. Dooyeweerd traced its origin in the coming together of the early Greek nature religions which deified a formless, cyclical stream of life and blind fate, *Ananke*, with the culture religion of form, measure, and harmony.

The Form-Matter ground-motive itself emerged as Greek thinking became more organised at the time Plato and Aristotle so that, for example, the being of a thing like a computer might be explained as matter (silicon, copper, etc.) in the form of a computer.

Form-Matter became a dualism, which elevated form (eternal, spiritual, reliable, pure) over matter (temporal, material, decaying, changing, impure). Form was seen as Good and matter, Evil (or, for those who enjoyed carousing, the other way round!), and the remedy was to rid one's life of as much matter as possible. Philosophers, as experts in Form should rule the State. We still feel its influence today in the mind-body dualism, in the assumed superiority of working with the mind over working with the hands, and also in the assumption of the autonomy of theoretical thought.

But it has inherent problems for understanding IS. Not only does it tend, ultimately, to reduce everyday experience to a theory, but everyday thinking as such was dismissed as being untutored. More specifically, it makes it difficult to resolve the artificial intelligence (AI) question of how computers compare with humans (see Chapter V). Dooyeweerd argued (1984, II, p. 417ff.) that under FMGM universality and individuality cannot be reconciled—which is important for creating general technological resources as discussed in Chapter VII.

2.3.1.2 The Creation-Fall-Redemption Ground-Motive (CFR)

The Creation-Fall-Redemption ground-motive was held by culture that had been informed by the Jewish and Christian Scriptures, and emerged from the Biblical idea that the Cosmos is separate from, created by, and depends on, a Divine being, God. In this view, the Divine is personal and good, and so all reality is intrinsically good, and may be enjoyed. In particular, both matter and form are Good, both dynamic and static, both body and soul, both hand-work and mind-work. Reliability is no longer founded on Form, but on a covenant-keeping God. Evil is located, not in one or other half of reality, but in the heart of humankind; not in the structure of the cosmos but in the response that we make within it. The remedy for evil is that the Divine pro-actively steps in to save the Creation (such as the various rescues experienced by the people of Israel, but supremely by Jesus Christ). Christian versions add that the Divine indwells human beings to make them a "new creation".

The form-matter opposition is annulled, as is the mind-body dualism. Universality and individuality are reconciled, displaying "fullness and splendour" (Dooyeweerd, 1984, II, p. 418).

The influence of the CFR ground-motive is found more in everyday life than in academic life. Its lack of influence in academic circles is partly because of the long influence, since 500 AD, of the NGGM (see the following), which held that faith and reason are in separate compartments, and so faith has little part to play in genuine academic discourse except to pronounce dogma. Therefore attempts to work out a CFR-based theoretical approach have been sparse.

CFR is the ground-motive from which Dooyeweerd worked, Polanyi (1962) also explicitly worked from it, and Friesen (2003) suggests that von Baader started from the same motive, but neither worked it out in as much detail as Dooyeweerd did.

2.3.1.3 The Nature-Grace Ground-Motive (NGGM)

The Nature-Grace ground-motive (NGGM) was a synthesis of the Form-Matter and Creation-Fall-Redemption ground-motives, which emerged around 500 AD when European Christian thinkers recognised the glories of ancient Greece. Grace refers to the realm of the sacred and nature to the secular. To some extent one might see nature and grace as "Christened" versions of matter and form, but reason, which used to be in form, became part of the nature realm; the type of knowing that we find in the grace realm is faith. Theology became the "queen of the sciences". Originally a duality, it became a dualism in which nature was seen to be a hindrance to grace and even inimical to it. This led to that oppression and injustice that characterised the pre-Reformation period in Europe.

As a philosopher, Dooyeweerd saw such problems as an inevitable result of the dualistic divorcing of secular from sacred, elevating religion into a separate sphere which has no relevance to ordinary life. Ordinary life became seen as of lower value, a necessary evil, and, as with FMGM, not worthy of respect or study. Another problem was that it separated natural reality from humankind and made God the ultimate cause and end of it, independent of humankind (Dooyeweerd, 1984, II, pp.52-3). This opened the way to the humanistic notion that humankind could do with Nature whatever it wished.

As Nature became dualisticaly divorced from Grace, the Church became ever more powerful and religious oppression became rife, until reactions occurred in the Renaissance and Reformation. The Reformation believed that the root of the problem was the synthesis that gave rise to NGGM in the first place, and sought to return to *Sola Scriptura*, while the Renaissance saw the root of the problem more in religion itself, and from this emerged the Nature-Freedom ground-motive.

2.3.1.4 The Nature-Freedom Ground-Motive (NFGM)

The Nature-Freedom ground-motive (NFGM) emerged from the Nature-Grace motive, around 500 years ago, with God being replaced by the free human ego (Freedom pole). "Nature" was redefined to refer to the non-free and non-human, to that which is determined, to mechanism, control, causality, rationality. What Dooyeweerd called the Ideal of Science became opposed to the Ideal of Personality.

Dooyeweerd (1979) traced its effects on Western thought and culture in pages 148-206. What follows is a highly-simplified picture, which even so shows many intertwined strands, all of which are driven by an antinomy between Nature and Freedom that cannot be resolved from within this ground-motive. An early expression of Freedom was the desire for emancipation from church hegemony and feudalism, leading eventually to such things as the French Revolution, which then became absolutist (Nature pole). Another was a desire for freedom from natural disasters and diseases, to which end the natural sciences were co-opted. Science was immensely successful and, as Dooyeweerd (1979, p.150) put it:

Modern man saw "nature" as an expansive arena for the exploration of his free personality, as a field of infinite possibilities in which the sovereignty of human personality must be revealed by a complete mastery *of the phenomena of nature.*

But (p. 153):

When it became apparent that science determined *all of reality as a flawless chain of cause and effect, it was clear that nothing in reality offered a place for human freedom. ... Nature and freedom, science ideal and personality ideal—they became enemies. ... Humanism had no choice but to assign religious priority or primacy to one or the other.*

This has led to many fractures, including material body versus thinking soul and autonomous thinking subject versus non-autonomous object (Descartes), being versus morality (Hume), science versus faith, public versus private, and thought versus thing (Kant), and so on, including variants in various fields like law, politics and sociology examined by Dooyeweerd. But everyday experience does not recognise these splits, and the human spirit eventually reacts against them. Hobbes reacted against Descartes into materialism, Marx against the being-morals split into absolutism, Husserl against the thought-thing split into phenomenology, Heidegger against the subject-object split into existentialism, and so on.

But then further fractures would appear, because the NFGM cannot allow Nature and Freedom to integrate. According to Dooyeweerd (1984, I, pp. 64-65), Hegel tried "think together" the two poles but ultimately failed because "this antinomy cannot be resolved." (For a possible Dooyeweerdian reinterpretation of Hegel, see Basden, 1999.) The influence of NFGM can also be seen in dialectical movements from rationalism to irrationalism and from positivism to constructivism.

The problems of NFGM, at least from a Dooyeweerdian perspective, will become apparent throughout this work. It is NFGM that holds sway today and most philosophic thought appealed to in IS is a product of NFGM. That is why it cannot address "the whole story that is information systems".

2.3.2 Effect of Ground-Motives on Understanding Information Systems

A ground-motive affects the thinking at a deep level, determining not so much the outcome of debates and research as the very ground on which these take place. For example, as will be discussed in Chapter V, the AI question of similarity between computer and human is different under different ground-motives:

- **FMGM:** As physical versus mental
- **NGGM:** As sacred human spirit versus secular machine
- **NFGM:** As determined versus non-determined behaviour
- **CFR:** As fulfilling God's cosmic purposes, which are spoiled by humanity

These ground-motives are not unique to Dooyeweerd, though part of Dooyeweerd's contribution was to see them as a spiritual driving force and to show how they relate to each other. Vollenhoven (1950) spoke of three periods of Western thinking: pre-synthesis, synthesis, and post-synthesis, in which the synthesis is between Greek and Hebrew thought. Heidegger, in relation to the synthesis that led to NGGM, spoke of "matter and form borrowed from an alien philosophy" (1971, p. 29). Habermas seems to have understood something similar when he speaks (2002, p. 157) of synthesis between "Athens" and "Jerusalem", and has remarked (1992, p. 12):

I do not believe that we, as Europeans, can seriously understand concepts like morality and ethical life, persons and individuality, or freedom and emancipation, without appropriating the substance of the Judeo-Christian understanding of history in terms of salvation. And these concepts are, perhaps, nearer to our hearts than the conceptual resources of Platonic thought, centering on order and revolving around the cathartic intuition of ideas.

Nearer our hearts, perhaps, but not nearer our minds. Dooyeweerd wanted to change this, and understand the impact of the religious root of humanity—whether Judeo-Christian or any other—on theoretical thought itself. He argued that even the so-called Christian medieval philosophies, for example, of Aquinas, were not based on creation, fall and redemption, but rather presupposed a dialectic between nature and grace, or sacred and secular.

But it is important to avoid assuming any of the ground-motive presuppositions are a "truth"; we always see through a "lens". That control and freedom are fundamentally incompatible is a presupposition, not a truth. As Chesterton once aptly pointed out (1908, p. 35), "The ordinary man ... has always believed that there was such a thing as fate, but such a thing as free will also." That is the lifeworld view, which the NFGM arrogantly denies and leads us to attempt to explain away, usually in terms of reductionism or mysticism. Dooyeweerd rejected such an approach and offered philosophical grounds to support Chesterton's lifeworld observation by arguing that the NFGM is merely a pre-theoretical presupposition, and Dooyeweerd proposed a deeper understanding in which both control and freedom can co-exist and all these fractures can be healed. This is a fifth reason for interest in Dooyeweerd.

In considering the nature of computers we need to bring form and matter together, but the FMGM prevents us doing so. In considering the information society we need to bring sacred and secular together, but the NGGM prevents us doing so. In considering IS development and use we need to bring control and freedom together, but the NFGM prevents us doing so. But the CFR is the only non-dualistic ground-motive of the four, and thus might not prevent such things. CFR allows healing of the fractures and thus might be useful as a basis for formulating frameworks for understanding information systems in all these areas.

2.3.3 Immanence-Standpoint

The three dualistic ground-motives—FMGM, NGGM, NFGM—put theoretical thinking into dialectical swings while they are in force, and it never comes to rest. Through his critical survey of theoretical thought Dooyeweerd found that underneath these three ground-motives is what he called the immanence-standpoint or immanence-philosophy. This presupposes that the basic, self-dependent principle that explains and generates all else may be sought within temporal reality itself. Clouser (2005) explains this more clearly than Dooyeweerd does.

The strands of philosophy that we, in the 21st century, believe to be radically different—such as positivism and subjectivism—are all in the same camp, to Dooyeweerd, variations of immanence-philosophy: the only difference between them lying in what each takes to be self-dependent. For example (1984, I):

The age-old development of immanence-philosophy displays the most divergent nuances. It varies from metaphysical rationalism to modern logical positivism and the irrationalist philosophy of life. It is disclosed also in the form of modern existentialism. The latter has broken with the Cartesian (rationalistic) "cogito" as Archimedean point and has replaced it by existential thought, conceived of in an immanent subjectivistic historical sense. (p.13)

Immanence-philosophy in all its nuances stands or falls with the dogma of the autonomy of theoretical thought. ... Not only traditional metaphysics, but also Kantian epistemology, modern phenomenology and phenomenological ontology in the style of Nicolai Hartmann continued in this respect to be involved in a theoretical dogmatism. (p.35)

Dooyeweerd also found it in such thinkers as Heidegger who, though opposing much that preceded him, nevertheless "moves in the paths of immanence philosophy; his Archimedean point is in 'existential thought', thus making the 'transcendental ego' sovereign" (1984, IV, p. 88).

Dooyeweerd's conclusion, expressed in Dooyeweerd (1984, III, p. 169) was:

Our general transcendental critique of theoretical thought has brought to light that the philosophical immanence-standpoint can only result in absolutizations of specific modal aspects of human experience.

Absolutization will be very evident in almost all areas of research and practice in IS. Especially relevant to the information society, is Dooyeweerd's continuation:

Similarly we may establish that on this standpoint every total view of human society is bound to absolutizations both of specific modal aspects and of specific types of individual totality.

The importance here is that most philosophers referred to in IS research, in all areas, are of the immanence-standpoint.

Dooyeweerd argued that the immanence-standpoint inevitably results in a number of problems, some of which we note here because they affect our ability to formulate frameworks for understanding information systems, and leave the interested reader to explore further:

- It prevents a truly sensitive approach to understanding the lifeworld or everyday experience because it forces us to take one sphere of meaning as our point of reference for all the others, thereby privileging it so that all others are reduced to it and not given their due. (Dooyeweerd, 1984, I, p. 15).

- By the same token, it leads to unmethodical treatment of the coherence between the normative aspects (Dooyeweerd, 1984, II, p. 49), making a genuinely interdisciplinary research extremely difficult.

- It is the source of many "-isms" (Dooyeweerd, 1984, I, p. 46).

- Meaning is distinguished from reality in immanence-philosophy (Dooyeweerd, 1984, II, pp. 25, 26). But in everyday experience, especially in IS use, meaning and reality are intertwined.

- Immanence-philosophy has never posed the problem of the relationship between different spheres of meaning of our experience (1984, II, p. 49). So any frameworks for understanding IS would be forced to rely on arbitrary speculations. (Arguably, systems theory is now attempting this via emergence, but it is not yet clear that it can escape speculation.)

- Immanence-philosophy is incapable of positing the problem of concept formation correctly because of "the disturbing influence on the formation of concepts exercised by the form-matter scheme, or by the disruption of the integral empirical reality into a *noumenon* and a *phenomenon* and by the reduction of this reality to a merely 'physico-psychical' world" (Dooyeweerd, 1984, II, p. 50). (For example, this leads to problems in Object-orientation.) So any concepts included in a framework are likely to be out of kilter with everyday life at some point.

Dooyeweerd argued that such problems are inherent in the very nature of the immanence-standpoint itself, and cannot be overcome from within that standpoint by merely shifting to a different immanence-philosophy or to a different pole of the current ground-motive. Rather, a new standpoint must be explored.

2.3.4 Transcendence-Standpoint

Dooyeweerd adopted a transcendence-standpoint, which locates the basic Principle (which Clouser calls the Divine) outwith temporal reality. One (perhaps the only) ground-motive that makes this presupposition is the Creation-Fall-Redemption ground-motive. He was not alone in arguing for this transcendence presupposition: Martin Buber and Michael Polanyi shared it but did not work it out in the way Dooyeweerd did.

What Dooyeweerd did was to work out the philosophical, not theological, implications of CFR, in terms of what it allows us to "see" of reality. (Example: if the cosmos is created, then we find it easier to acknowledge irreducible diversity that coheres.) He discovered that it is an important tool in helping him "clear away" various impediments to understanding the structure of reality as it presents itself to us in everyday experience. As a consequence, he was able to offer a positive philosophy that has a very different feel from most others. It does justice to all the branches of philosophy mentioned in Chapter I—ontology, epistemology, philosophical ethics, methodology, anthropology, critical philosophy—and enables us to at least tackle all the issues relevant to "the whole story that is IS". This is a sixth reason for interest in Dooyeweerd.

Some examples of the kinds of philosophical implications of presupposing CFR compared with those of immanence-philosophy are set out in Table 2-1 (Dooyeweerd's own longer and philosophically-oriented comparison is found in 1984, I,

Table 2-1. Philosophical implications of CFR

CFR	Immanence Philosophy
The whole cosmos has dignity	Half is Good, the other half Evil or inferior
Nothing in cosmos is absolute (including Reason)	One aspect is absolute, self-dependent
Diversity coheres. Coherence is diverse.	Focus either on diversity or coherence
Self and world cohere with each other	Kantian gulf between Self and world
The foundation of all in cosmos is cosmic meaning-and-law	Foundation is either (a) Being / process (b) Autonomous ego
World reveals itself	Can never know 'Ding an sich'
We can let everyday experience speak to us	Everyday experience is reduced to a theory
Theory has religious root	Theoretical thought is absolute
Hope and Destiny	'Towards death' (Heidegger)

p.502 ff). If they are valid, then these may be taken as reasons for IS researchers and practitioners to take the CFR ground-motive seriously; some will be referred to in formulating frameworks for understanding IS.

The author has found some of these implications of CFR helpful in stimulating practical discourse in both research and practice. By keeping in mind the alternative possible ways of seeing things under CFR, it is possible to pose questions that lead participants to question the usual (immanence-standpoint) assumptions.

The positive side of Dooyeweerd's philosophy is sometimes called *Cosmonomic Philosophy* (law of cosmos). He also sometimes referred to his philosophy as a "Christian philosophy". This should not be misunderstood as being driven by Christian theology or anti-secular reaction, nor as requiring personal Christian commitment in those who adopt it; see "Religious Root" later.

2.4 The Different Flavour of Dooyeweerd's Approach

No theoretical thinking is without presuppositions. Dooyeweerd himself openly began from the CFR ground-motive, with its transcendence-standpoint, and sought to work out a number of fundamental implications thereof. Dooyeweerd's positive philosophy has several major parts and several secondary parts which he derived from them.

Part I of Volume II of Dooyeweerd (1984) is devoted to "The general theory of the modal spheres". This shows how modal spheres (or "aspects") account for diversity and coherence, being and doing, normativity, and so forth. The discussion, which is always related to the works of other thinkers from the early Greeks, all the way through to Husserl and Heidegger, ranges widely through many spheres of human experience, from mathematics, physics, biology, through psychology, social science, to art, law, ethics, and theology. There is a lengthy discourse on how the spheres anticipate each other, and on their analogical coherence, wrapped round a theory of history. Dooyeweerd's radically interesting non-Cartesian subject-object relationship is expounded, with an introduction to the problem of individuality.

In Part II of Volume II, Dooyeweerd discussed the problem of knowledge and epistemology. First he argued that immanence-philosophy has not properly posed the problem of knowledge. He discusses theoretical knowledge, intuition, the horizon of experience, and "truth". He offers considerable critiques of Kant and Heidegger especially.

Volume III comprises a comprehensive theory of entities. "As far as I know," concluded Dooyeweerd (1984, III, p. 53), after a review of attempts to understand thingness from Greek metaphysics through to the 20th century, "immanence-philosophy, including phenomenology, has never analysed the structure of a thing *as given in naïve experience*." He therefore formulated a notion of "thing" as multi-layered, complex, in relationship with other "things". He then applied this theory to understand things as widely different as the linden tree in front of his window, a sculpture, music, utensils, planks of wood, social institutions, and the State.

Volume IV is an encyclopaedic index to the other three volumes.

His ideas are outlined in Chapter III. But before this, it is useful to understand something of the different flavour of Dooyeweerd's positive thought compared with other strands.

2.4.1 Starting Point 1. Religious Root and Destiny of All— Including Information Systems

It has been long assumed that philosophy and science should be carried out without reference to God; this has especially been so in information systems. This is not to deny God's existence, so much as to declare God as unnecessary to philosophy because theoretical thinking has been assumed to be autonomous. But this is now being questioned. The eminent systems thinker, C. West Churchman, once said, "'Does God exist?' is the most important question in systems thinking" (1987, p. 139).

Dooyeweerd believed it to be important to all philosophy, because if God exists, in the sense of his Existing having an impact on the content of our philosophy, then that should be taken into account. He rejected treating the Divine as an object about which to philosophise, such as is done by treating God as a First Cause, or by formulating Scholastic (NGGM) arguments for God's existence. Rather he sought to work out the philosophical implications of what it means that the cosmos is Created rather than "just is". Clouser clarifies Dooyeweerd's position (2005, p. 342):

Our claim is not that all theories are produced or forced on us by some divinity belief ... The claim is that the nature of a theory's postulates is always interpreted in the light of what is presupposed as divine.

The nature of frameworks for understanding is thus influenced by our religious presuppositions and this affects, in turn, our research and practice. One, albeit rather humorous, example of the religious root of thought is the "holy wars" in which groups of computer scientists engage regarding what is the best operating system or text editor.

Dooyeweerd's own religious background was that of Dutch Calvinism, and this author detects also some ideas from Celtic Christianity, both of which entered that area of Europe at different points in history. Both streams emphasise the importance of the "secular", "everyday" side of life in God's sight, and were thus more in line with Creation, Fall, and Redemption rather than Nature-Grace. [Note: Dooyeweerd's creation presupposition should not be confused with fundamentalist Creationism. Clouser (2005) shows how fundamentalism is incompatible with Dooyeweerd's thought.]

But Dooyeweerd distinguished philosophy from theology and tried to avoid using the latter's way of thinking, so that his philosophical arguments could be accepted by people of any faith but without denying the importance of one's faith stance. For example, Witte (1986) cites P. B. Cliteur, professor of philosophy at the Technical University of Delft and president of the Dutch Humanist League, and thus not in agreement with Dooyeweerd's own Christian principles, as saying:

Herman Dooyeweerd is undoubtedly the most formidable Dutch philosopher of the 20th century ... As a humanist I have always looked at my own tradition in search for similar examples. They simply don't exist. Of course, humanists too wrote important books, but in the case of Herman Dooyeweerd we are justified in speaking about a philosopher of international repute.

But, for formulating frameworks for understanding, the issue of which religious stance is held must be separated from that of the religious root of all we do, which, Dooyeweerd argued, cannot be escaped and affects all at a deep level. Kuhn (1996, p. 175) gave the word "paradigm" two meanings: a concrete puzzle-solution that can act as a model in science, and "the entire constellation of beliefs, values, techniques, and so on shared by the members of a given community." The second especially is of a religious nature, when "religious" is given Dooyeweerd's earlier definition. While Kuhn discussed its importance, Dooyeweerd's insights enable us to differentiate four ways of addressing the religious root of IS directly:

- The faith aspect of what we do and are (see theory of aspects) helps account not only for traditional religious experience but also for the tenacity with which we commit to, and defend, positions.

- Life-and-world-views (*Weltanschauungen*) and perspectives to which a group of people are committed are often centred on an aspect; this helps us understand and analyse tenaciously-held perspectives in IS as well as elsewhere.

- Presuppositions that are religious in nature, such as ground-motives and the immanence presupposition, which underlie the way we assume reality to be, can explain why some of the debates in IS have taken the shape they have.

- The orientation of the human self toward the true Absolute, or a pretend one, can help us understand religious dysfunction, in terms of absolutization, and its possible remedy.

These are explained in more detail later, and used especially in Chapters VI and VIII.

Religious root implies Destiny. That is, if all in the cosmos is religious (in whatever way we allow, whether Christian, Hindu, Humanist, Islamic, etc.) then it has a Destiny as well as an Origin. More specifically, humanity has a Destiny, a mandate to which it is responsible. And all it generates or produces as part of that mandate, such as IT/IS, has its own Destiny. Destiny implies meaningfulness of all, and this is reflected in the primacy of Meaning (see below). Dooyeweerd, under CFR, held the Destiny to be Completeness in Jesus Christ, and this pervaded all he wrote. Though this author shares Dooyeweerd's belief, he has tried to allow for other possible destinies. Nevertheless, the general notion of Destiny (and meaningfulness) will pervade all in this work, and will be specifically referred to again in Chapter VIII.

(Some readers might detect an echo of this in Aristotle's notion of final cause, but Dooyeweerd's notion of Destiny is richer and allows for immense diversity, freedom and an open future, as will become plain later.)

Dooyeweerd's positive philosophy may be seen as an attempt to work out the philosophical, rather than theological, implications of such a stance. He particularly wanted to avoid theology dominating philosophy or imposing on philosophy constraints that philosophy itself would not recognise as valid. This was one more reason he undertook a transcendental critique of theoretical thought.

A seventh, practical reason for considering Dooyeweerd today is that it provides a basis for addressing the relationship between Islam, Christianity, Hinduism, Humanism, and other religious stances as they meet in the areas of research and practice in information systems.

2.4.2 Starting Point 2. Everyday Experience

Since the birth of philosophy, the everyday, pre-theoretical (or "naïve") attitude has been deemed inferior to the theoretical or scientific attitude in one way or another. Sometimes it has been seen as something to escape, as expressed by Moran (2000, p.146): "Given the difficulty of doing philosophy (i.e., escaping from the natural attitude which constantly seeks to reassert itself) ... ." Often it has been equated with (reduced to) sensory functioning, as in Bertrand Russell. Where it has been deemed a topic worthy of reflection, it has often been seen through the lens of a theory or theoretical framework, such as in Adam et al. (2006), who expressly state, "we draw upon the theoretical constructs of the gender and technology literature to theorise the relationship between gender and technical skill ... ."

But to Dooyeweerd, the "natural attitude" (also called "everyday", "naïve", and "pre-theoretical") is not only important and worthy of philosophical attention, but a vital starting point. Everyday life should be reflected upon in a way that does not theorise it but listens sensitively to it as it presents itself to us—though not uncritically. So he opened Volume I of his *magnum opus* (1984) with:

If I consider reality as it is given in the naïve pre-theoretical experience, ...

Dooyeweerd believed that philosophy (including phenomenology) has fundamentally misunderstood the nature of everyday experience. The lifeworld is not the generator of all meaning, as Husserl (1970) assumed (see Chapter I), but it itself operates within a cosmic meaning-framework, which cannot be separated from reality. (This is not to say that no meaning-attribution occurs; see the following.) Having shown that immanence-philosophy drives meaning and reality apart from each other (see earlier), he turned to the transcendence standpoint, which allows us to accept the possibility that both diversity and coherence transcend us in everyday experience. It is for this reason he continued his opening (1984, I, p. 3) with:

... and then confront it with a theoretical analysis through which reality appears to split up into various modal aspects, then the first thing that strikes me, is the original indissoluble interrelation *among these aspects ..."*

By this he was not making a dogmatic statement, but rather inviting the reader to similarly recognise a coherence in the lifeworld even when we split it up theoretically.

Throughout his work, in developing both his critical and his positive arguments, Dooyeweerd kept on returning to everyday experience not as a source of individual empirical facts but rather as a corrective that keeps on resetting our direction of thought. He paid attention to the structure or nature of naïve experience more than to its content, because the latter is contingent and varies with personal and cultural history. For example:

To all of these speculative misunderstandings [made by philosophers] naïve experience implicitly takes exception by persisting in its pre-theoretical conception of things, events and social relationships. (1984, III, p. 28)

This relates to the second reason for interest in Dooyeweerd. Not only does this ensure his philosophy is sensitive to the lifeworld, but what Dooyeweerd says about everyday experience and attitude makes an important contribution to extant discussion of the lifeworld. This is outlined later in Chapter III, where it is also compared with the phenomenological notion of lifeworld.

2.4.3 Being as Meaning

Echoing the Greek thinkers, Gaines (1997) says, "The most fundamental properties which we impute to any system are its *existence* and *persistence* over time." But this causes difficulties when considering such things as the mouse pointer or virtual reality (see Chapter III). Most Western philosophical thought has presupposed Being as the "most universal", as indefinable, and as self-evident for 2,500 years. But Heidegger (1962, pp. 22-3) tried to question and define Being, because he did not presuppose Being. He tried to understand Being in terms of situatedness in a context among other things (*Dasein*).

But Dooyeweerd went further, and held that Being must be understood as Meaning (1984, I, p. 4):

Meaning *is the* being *of all that has been* created *and the nature even of our self-hood. It has a* religious root *and a* divine origin.

Things "exist" by virtue of their meaningfulness in certain different ways. Things do not *have* meaning, as a kind of property, they *are* meaning. For example, a mouse pointer "is" by virtue of pointing and has no other existence as mouse pointer apart from that. A similar thing can be said about anything, including rocks through organisms, feelings, beliefs, utterances, social institutions; see §3.2, which also reviews problems with extant presuppositions about Being.

But what is Meaning? Dooyeweerd criticised Husserl for promiscuously mixing meaning and signification, and reviewed Hoffman's subjectivism (Dooyeweerd, 1984, II, p. 25 ff.). He then gave a careful explication of his own view (p. 30). A clearer explanation might perhaps be found in the Appendix of Dooyeweerd (1999):

Meaning—Dooyeweerd uses the word "meaning" in an unusual sense. By it he means the referential, non-self-sufficient character of created reality in that it points beyond itself to God as Origin. Dooyeweerd stresses that reality is *meaning in this sense and that, therefore, it does not* have *meaning. ... "Meaning" becomes almost a synonym for "reality".*

Meaning is "referring beyond" (Dooyeweerd, 1984, I, p. 110). This use of the word "meaning" is akin to when we speak of "the meaning of life": a referring to something that transcends us, something higher. If everything is meaning, then every single thing (or event) in the cosmos is interconnected, and nothing can be understood "in itself", and the whole interconnected cosmos itself is meaning, referring ultimately to its Origin. This is a philosophical implication of CFR. Heidegger's *Dasein* might, of course, be seen as a kind of "referring beyond", but he makes that depend on, and presuppose, relatedness to the world, whereas to Dooyeweerd it was the other way round, as will be explained later: relatedness and being itself depend on, and presuppose, referring beyond.

Unfortunately, the word "meaning" has many meanings. In this work in order to differentiate them the following terms will be used (not terms that Dooyeweerd used, but necessary in formulating frameworks):

- **Cosmic meaning:** This is Dooyeweerd's use, by which the very existence and occurrence of all in the cosmos is meaning and is made possible by meaning. Dooyeweerd held that cosmic meaning transcends us and makes all concrete reality possible, including the types of meaning that follow.

- **Attributed or ascribed meaning:** We subjectively attribute concrete meaning to things; it might not be conceptualised. This is how Descartes, Weber, Hoffman, and so forth, used it, and how it is used in subjectivism and much phenomenology.

- **Signified meaning:** This refers to the meaning-content of symbols in discourse (involving speech, writing, graphics, gestures, etc.). It is how thinkers influenced by the linguistic turn, such as Habermas and Foucault use "meaning".

- **Conceptualised meaning:** We form a concept of meaning, for example, that of justice (cosmic meaning). Whenever we are aware of meaning we have already conceptualised it. Lack of awareness of meaning does not imply meaninglessness, it is simply that it has not been conceptualised.

When the word "meaning" is used on its own it will usually be clear from the context which is meant, and if not, it will refer to cosmic meaning. Attributed, lingual, and conceptualised meaning always echo cosmic meaning because they are made possible by it and function within it.

Dooyeweerd's emphasis on meaning is an eighth reason why his thought is of interest.

2.4.4 Law and Subject Sides

This cosmic meaning, which transcends us and applies to all, has the character of law. In contrast to Peirce (1898), who said "the first germ of law was an entity," Dooyeweerd held that entity emerges from law, and is subject to a transcendent, *a priori* law-framework.

This cosmic law must be sharply distinguished from concrete rules, social norms, laws, and the like. Cosmic law does not constrain so much as enable, and it has the character of promise rather than authoritarian demand: contrast "Keep to the syntax of the language you use!" with "If you keep to the syntax of the language you use, then people will understand you better." For this reason, we will often refer to this kind of law as "law-promise". This transcendence of cosmic law will be especially useful in understanding impacts of IS in use.

Law-promise is so shaped as to direct us (everything in the cosmos) toward what is meaningful. Thus cosmic law is cosmic meaning is cosmic law. Therefore the phrases "cosmic meaning", "cosmic law", "meaning-and-law", "law-promise", and (see later) "aspectual law" and "aspectual meaning" will be used interchangeably, depending on what is to be emphasised at the time.

Law presupposes subjects (things that are subject to it). What is subject to cosmic law is the cosmos and all in it: all of concrete temporal reality. Temporal reality thus has two sides: the law side and the subject side.

- The **subject side**, also called entity side or fact side, comprises all that exists or occurs in the cosmos, as concrete reality, including concrete meanings that are ascriptions we make, and includes all our experience, past, present, future and potential.

- The **law side** comprises the framework that enables the entire subject-side cosmos to exist or occur. Dooyeweerd differentiated two ways in which things are subject to law, firstly, in their functioning and, secondly, in their structure:

  - *Functional law:* What is called *aspectual law* in this work includes the law of gravity, the laws of economics, and so on, and guides functioning (or activity) of things and is expounded in the theory of modal spheres and of knowing in Chapter III.

  - *Structural law:* What Clouser (2005) calls *type laws* guide the structure of things and is expounded in the theory of individuality structures (things) in Chapter III.

Figure 2-2. Law and subject sides

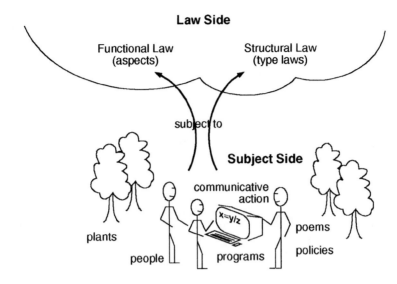

This is shown in Figure 2-2. This author has always been more interested in aspectual law and less in type laws, and in this way his view of Dooyeweerd might be one-sided. Even so, type laws depend on and presuppose aspectual law. Note that law and subject side are not two worlds but two sides of one world (Dooyeweerd, 1984, II, p. 44), but it will be found very useful to separate them conceptually in formulating frameworks for understanding IS. Philosophically, it helps reconcile universality and individuality, and explain why there might be a reality that transcends us but which itself cannot ever be fully known.

The novel way in which Dooyeweerd understood law and subject helps untangle many knotty issues in IS and leads to four more reasons why he is of interest.

2.4.5 Escaping Descartes and Kant

Wolters suggests (1985, p. 17), "the law-subject correlation ... bears new and important philosophical fruit, pointing a way which can break through such dilemmas as natural law versus historicism and substance versus function." Here we look at some philosophical fruit relating to IS.

Western thought has been deeply influenced by Descartes' view of subject and object, either accepting it uncritically, or reacting against it and thereby mistakenly acknowledging its presumed right to set the agenda for debate (as does, e.g., existentialist or feminist discourse in IS). But Dooyeweerd offers a radical alternative to all these views, in the form of a fundamentally different notion of subject and object. While Descartes had a human subject thinking about (or acting upon) an external object-in-itself, Dooyeweerd has a subject responding to aspectual law-promise and something else responds to that subject-functioning as object. See Figure 2-3.

- **Subject:** To be an active subject (an agent) is constituted in being subject to law; there is no other way in which we can be active agents than in being subject to law because it is the spheres of meaning-and-law that enable this. Dooyeweerd brings together the two meanings of the English word, subject.

- **Object:** To function as object is to be involved in some entity's subject-functioning.

This non-Cartesian notion of the subject-object relationship can be quite complex, though its complex versions seldom concern us (a multi-object example is found in §5.3.1.4). What is important is that it offers a subtle dignity to things, in that subject does not necessarily imply human (e.g., a plant functions as subject in the biotic aspect) and because objects are not to be seen as passively acted-upon. Free-

Figure 2-3. The relationships between subject, law (promise), and object

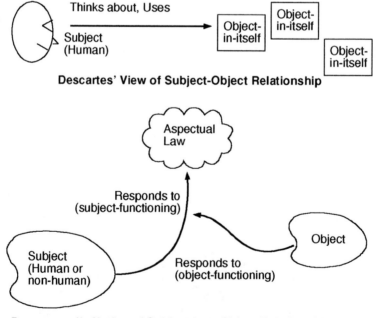

ing us from the Cartesian view, at least as it has been understood in IS, has three implications. The Cartesian view drives subject and object apart into a "distal" relationship, which is blamed for making computers unusable (see §4.6.3), but the Dooyeweerdian subject-object relationship is a "proximal" one of close engagement: computer becomes usable. The Cartesian view makes too sharp a distinction between human and non-human. In order to maintain "symmetry" between them as "actants", Latour (1987) had to dissolve this difference, but many believed he went too far. To Dooyeweerd the difference between human and non-human lies not in the subject-object relationship (see Chapter III), but in responses to different spheres of law, so computer and human may be seen as both similar and different though in different spheres; see Chapter V. The Cartesian approach presupposes the primacy of Being over Meaning and thus seeks to understand the behaviour and characteristics of an object in terms of something inherent in that object. This makes it difficult to understand the nature of computers (Chapter V). But to Dooyeweerd the nature of computers is to be understood by reference to a law side, which throws fresh light on artificial intelligence issues. Dooyeweerd's non-Cartesian subject-object relationship is a ninth reason why he is of interest.

Whereas Heidegger tried to overcome the Cartesian divide by destroying the distinction between subject and object, as discussed in Chapter IV, this makes it difficult to critique social structures and the status quo. Dooyeweerd did it by redefining both subject and object in relation to law and can maintain the possibility of critique while allowing many of Heidegger's ideas (such as *Dasein*, "worlding", being-in-the-world, etc.) to flourish. But Heidegger, "mov[ing] in the paths of immanence philosophy" (Dooyeweerd, 1984, IV, p. 88), as mentioned earlier, could not conceive of a distinction between law and subject sides, and seems to have been unable to complete the quest that he began in his *Being and Time* (Heidegger, 1962). It may be that Dooyeweerd offers a way to achieve what Heidegger sought, including a new notion of Time (which portion of Dooyeweerd's philosophy is not examined here, because it is not needed for IS). This is a 10th reason for interest in Dooyeweerd.

Kant's "Copernican revolution" which put the human mind at the centre, made it impossible for the world to be known if the human knower is free (a shift from the nature to the freedom pole of NFGM). This drove thought and thing apart, and the various turns of philosophy since then have been trying to solve the problems this has caused. Ontology and epistemology can no longer live together. Kant's insight into human freedom is important, but the thought-thing rupture is particularly problematic in understanding the everyday experience in ISD (Chapter VI) and in the shaping of technological resources to match the diversity of reality (Chapter VII) because both presuppose that thought has at least some true access to thing. Dooyeweerd's view of knowing (§3.3) provides a way to ameliorate these problems without denying Kant's insight, by reinterpreting the knower-known relationship in the light of the CFR ground-motive. Ontology and epistemology join hands again. With Kant,

Dooyeweerd believed we are part of the reality we experience but, unlike Kant, who was forced under NFGM to give supremacy to Freedom, Dooyeweerd could allow both knower and known to be subject to the same framework of law-and-meaning. It underpins not only the non-neutrality of knowledge and non-absoluteness of reason but also a knowledge-friendly cosmos, which may be truly known even if never perfectly. This is an 11th reason for interest in Dooyeweerd.

"Is" and "Ought" (being and normativity) were also driven apart by Kant, having been separated explicitly by Hume. This is problematic because in the everyday attitude, built into the question of what an information system (or anything else) is, is the question of what *is* a "good" or "true" one, and what it *ought* to be. That this was so was recognised by Plato and Aristotle, but NGGM and NFGM both separate being from normativity. But they are brought back together by Dooyeweerd because the cosmic meaning that gives rise to "Is" is exactly the cosmic law that gives rise to "Ought". This provides a philosophical basis for restoring the normative implications to information technology, and is a twelfth reason for interest in Dooyeweerd.

As the human and social aspects of IS have been recognised, various post-Kantian "turns" in philosophy have been referred to because they seem to address issues that present themselves in the everyday experience of IS. The author believes that Dooyeweerd's approach might make these turns less necessary to us because Dooyeweerd provides a different way of tackling the same issues. No longer is a phenomenological turn necessary to tackle the relationship between theory and practice or science and lifeworld, no longer is an existential turn necessary to understand situatedness of being, no longer is a linguistic turn necessary in order to understand intersubjective meaning and social life, no longer is a critical turn necessary in order to reinstate normativity. Dooyeweerd himself discussed the first two turns in depth, but not the latter two. (A substantial engagement between Dooyeweerd and the latter two has yet to be fully explored.) The author's belief about these latter two is based on his investigations into how these turns have been referred to in IS and how Dooyeweerd has helped him formulate alternative frameworks for understanding IS.

How it manages to do this will occupy the discussion in the next chapter.

2.5 Conclusion

We have encountered at least twelve reasons why Dooyeweerd should be of interest when critiquing or formulating FFUs for IS.

1. There is reason to believe that the "lens" Dooyeweerd used provides a clearer view of the structure of reality than others might. This promises a clearer view especially of the "whole story".

2. He prepared tools suited to thinking about everyday experience.

3. He provides a basis for examining where the foundations of philosophical thinking in IS might be shaky, so that deep problems and incommensurabilities in extant FFUs may be explored.

4. Many of his insights into the nature of religious roots of IS are particularly useful, for example in helping us to understand perspectives and conflicts of various kinds.

5. Dooyeweerd proposed a deeper understanding in which both control and freedom can co-exist and various fractures can be healed. This is especially relevant to use of IS, to ISD and to management in general.

6. He explored the philosophical implications of the transcendence-standpoint as embodied in the CFR ground-motive. As a consequence, he was able to offer a positive philosophy that has a very different feel from most others, which does justice to all the branches of philosophy mentioned in Chapter I—ontology, epistemology, philosophical ethics, methodology, anthropology, critical philosophy—and might enable us to tackle the issues relevant to "the whole story that is IS" outlined in Chapter I.

7. A practical reason for considering Dooyeweerd today is that it provides a basis for addressing the relationship between Islam, Christianity, Hinduism, Humanism, and other religious stances as they meet in the areas of research and practice in information systems.

8. Dooyeweerd emphasised Meaning over Being. Many of the issues faced in all areas of IS, and also in management, are issues of meaning.

9. Dooyeweerd's non-Cartesian subject-object relationship overcomes a number of deep problems in various fields of IS.

10. Dooyeweerd's way of affirming the engagement and situatedness that Heidegger sought for, and yet allowing for critique of social structures helps us understand how IS may be situated in organisations and yet stimulate important changes.

11. Dooyeweerd's notion of a knowledge-friendly cosmos gives those working in both ISD and the shaping of technological resources some confidence in their work.

12. Dooyeweerd restores the normative implications to information technology alongside an understanding of their nature.

Those are the "why". What Dooyeweerd actually proposed, and how he might be able to meet these challenges, are explored in the next chapter.

References

Adam, A., Griffiths, M., Keogh, C., Moore, K., Richardson, H., & Tattersall, A. (2006). Being an "it" in IT: Gendered identities in IT work. *European Journal of Information Systems, 15*, 368-378.

Basden, A. (1999). Engines of dialectic. *Philosophia Reformata, 64*(1), 15-36.

Chesterton, G. K. (1908). *Orthodoxy*. London: The Bodley Head.

Choi, Y.-J. (2000). *Dialogue and antithesis: A philosophical study of the significance of Herman Dooyeweerd's transcendental critique.* Unpublished doctoral thesis, Potchefstroomse Universiteit, South Africa.

Churchman, C. W. (1987). Systems profile: Discoveries in an exploration into systems thinking. *Systems Research, 4*(22), 139-146.

Clouser, R. (2005). *The myth of religious neutrality: An essay on the hidden role of religious belief in theories* (2nd ed.). Notre Dame, IN: University of Notre Dame Press.

Dooyeweerd, H. (1935). *De Wijsbegeerte der Wetsidee (The philosophy of the law-idea)* (3 vols.). Amsterdam: H. J. Paris.

Dooyeweerd, H. (1979). *Roots of Western culture: Pagan, secular and Christian options* (J. Kraay, Trans.). Toronto, Ontario, Canada: Wedge. (Original work published 1963).

Dooyeweerd, H. (1984). *A new critique of theoretical thought* (Vols. 1-4). Jordan Station, Ontario, Canada: Paideia Press. (Original work published 1953-1958)

Dooyeweerd, H. (1986). *A Christian theory of social institutions* (M. Verbrugge, Trans.). La Jolla, CA: Herman Dooyeweerd Foundation.

Dooyeweerd, H. (1999). *In the twilight of Western thought: Studies in the pretended autonomy of philosophical thought.* Lewiston, NY: Edwin Mellen Press.

Friesen, J. G. (2003). *The mystical Dooyeweerd: The relation of his thought to Franz von Baader.* Ars Disputandi, 3.

Gaines, B. R. (1997). Knowledge management in societies of intelligent adaptive agents. *Journal of Intelligent Information Systems, 9*(3), 277-298.

Geertsema, H. G. (2000). Dooyeweerd's transcendental critique: Transforming it hermeneutically. In D. G. M. Strauss & M. Botting (Eds.), *Contemporary reflections on the philosophy of Herman Dooyeweerd* (pp. 83-108). Lewiston, NY: Edwin Mellen Press.

Habermas, J. (1972). *Knowledge and human interests* (J. J. Shapiro, Trans.). London: Heinemann.

Habermas, J. (1992). *Postmetaphysical thinking: Philosophical essays* (W. M. Hohengarten, Trans.). Cambridge, UK: Polity Press.

Habermas, J. (2002). *Religion and rationality: Essays on reason, God and modernity* (E. Mendieta, Ed.). Cambridge, England: Polity Press.

Heidegger, M. (1962). *Being and time* (J. Macquarrie, & E. Robinson, Trans.). Oxford, UK: Blackwell. (Original work published 1927)

Heidegger, M. (1971). *Poetry, language and thought* (A. Hofstadter, Trans.). New York: Harper Collins.

Husserl, E. (1970). *The crisis of European sciences and transcendental phenomenology* (D. Carr, Trans.). Evanston, IL: Northwestern University Press.

Kalsbeek, L. (1975). *Contours of a Christian philosophy*. Toronto, Ontario, Canada: Wedge.

Kuhn, T. S. (1996). *The structure of scientific revolutions* (3rd ed.). Chicago: University of Chicago Press.

Latour, B. (1987). *Science in action*. Cambridge, MA: Harvard University Press.

Moran, D. (2000). *Introduction to phenomenology*. London: Routledge.

Peirce, C. S. (1898). *Reasoning and the logic of things*. Cambridge, MA: Harvard University Press.

Polanyi, M. (1962). *Personal knowledge: Towards a post-critical philosophy*. Chicago: University of Chicago Press.

Vollenhoven, D. H. Th. (1950). *Geschiedenis der Wijsbegeerte I: Inleiding en geschiedenis der Griekse wijsbegeerte voor Plato en Aristotles (History of philosophy I: Introduction and history of Greek philosophy before Plato and Aristotle)*. Franeker, Netherlands: Wever.

Witte, J., Jr. (1986). *Introduction. In H. Dooyeweerd, A Christian theory of social institutions* (M. Verbrugge, Trans.). Jordan Station, Ontario, Canada: Paideia Press.

Wolters, A. M. (1985). The intellectual milieu of Herman Dooyeweerd. In C. T. McIntire (Ed.), *The legacy of Herman Dooyeweerd: Reflections on critical philosophy in the Christian tradition* (pp. 1-19). Lanham, NY: University Press of America.

Endnote

[1]   Jim Skillen, president of the Herman Dooyeweerd Foundation, said this at a meeting in Leeds, UK, March 2007.

Chapter III

Some Portions of Dooyeweerd's Positive Philosophy

This chapter explains some portions of Dooyeweerd's positive philosophy which the author has found useful in understanding IS. It covers Dooyeweerd's theory of modal spheres (aspects), his theory of things, his theory of knowing, experience, and assumptions, and, on the basis of these, it draws together his approach to everyday life. What it does not cover is his notion of cosmic time and the relationship between the self and the Divine, nor his extensive discussion of the State. It also reviews some criticisms of Dooyeweerd.

Some ideas are derived from Dooyeweerd rather than used by Dooyeweerd himself; they will be differentiated where necessary by the use of the adjective "Dooyeweerdian" rather than "Dooyeweerd's".

3.1 Dooyeweerd's Theory of Modal Aspects

Consider a bunch of keys (I am indebted to Gareth Jones for this example). They exhibit several aspects—a spatial aspect is important in that each key must be a certain shape, and a physical, because the metal must not wear or bend. There is a lingual aspect, in that each key bears its maker's name. But the most important aspect is the legal one, in that each key is designed to protect against theft or trespass. A situation likewise exhibits aspects, for example if you were to reflect on reading this book, you would most probably agree that there is a lingual aspect to what you are doing. You would probably agree there is also a biotic aspect—you are breathing, digesting food, and so forth. Generosity in your reading, overlooking the mistakes in style, grammar or spelling you might encounter might be yet another aspect.

What aspects might there be? Is there an infinite number of possible aspects? If not, what are they, and why is there only a limited set? How should they be identified? Are the aspects of the keys also aspects of reading the book, or of someone else reading it? Can we agree on sets of aspects? And why does it matter; what does the notion of aspects do for us? These are the kinds of questions that Dooyeweerd reflected upon, and which his theory of modal aspects tries to address.

Regardless of what aspects might be delineated, what are aspects as such? Webster's Dictionary (1971) defines aspect, in roughly the sense meant here, as "a particular status or phase in which anything appears or may be regarded" and Clouser (2005, p. 267) defines aspect as "a basic kind of properties and laws." But Dooyeweerd himself introduced aspects without definition and in a way that appeals to our intuition, only later gradually exposing their nature. We will follow Dooyeweerd here, introducing Dooyeweerd's suite of aspects before outlining more precisely what aspects are and what they do for us philosophically.

3.1.1 Dooyeweerd's Suite of Aspects

Dooyeweerd delineated 15 aspects of our everyday experience. The first page of his (Dooyeweerd, 1984, I) has:

A indissoluble inner coherence binds the numerical to the spatial aspect, the latter to the aspect of mathematical movement, the aspect of movement to that of physical energy, which itself is the necessary basis of the aspect of organic life. The aspect of organic life has an inner connection with that of psychical feeling, the latter refers in its logical anticipation (the feeling of logical correctness or incorrectness) to the analytical-logical aspect. This in turn is connected with the historical, the linguistic, the aspect of social intercourse, the economic, the aesthetic, the jural, the moral aspects and that of faith.

Though Dooyeweerd defended this list, he was usually cautious about presenting it systematically, for reasons discussed later. But we will do so, for clarity and for later reference.

Dooyeweerd was not consistent in the names he gave the 15 aspects in his suite, so the following list gives the names used throughout this work. Each aspect is a sphere of meaning, centred on a kernel meaning:

- Pistic aspect, of faith, commitment, and vision
- Ethical aspect, of self-giving love, generosity, care
- Juridical aspect, of "what is due", rights, responsibilities
- Aesthetic aspect, of harmony, surprise, and fun
- Economic aspect, of frugality, skilled use of limited resources
- Social aspect, of respect, social interaction, relationships, and institutions
- Lingual aspect, of symbolic signification
- Formative aspect, of formative power and shaping, in history, culture, creativity, achievement, and technology
- Analytical aspect, of distinction, conceptualizing, and inferring
- Sensitive (or psychic) aspect, of sense, feeling, and emotion
- Biotic (or organic) aspect, of life functions, integrity of organism
- Physical aspect, of energy and mass
- Kinematic aspect, of flowing movement
- Spatial aspect, of continuous extension
- Quantitative aspect, of discrete amount

This list of aspects and their kernel meanings have been compiled by taking account of what is said about them in many places in Dooyeweerd's writings and that of others. Note that they are listed in the reverse order from the earlier list, from what Dooyeweerd called the latest (pistic) to the earliest (quantitative), in order to avoid any connotation that the quantitative aspect can be treated as a self-sufficient foundation for the rest.

The aspects cannot be directly observed, but only as they are expressed in things, events, situations, and so forth, as ways these can be meaningful. Table 3-1 gives, alongside the kernel meaning, issues in running a chemical-producing business that are meaningful in each aspect.

Surrounding the kernel of each aspect is in fact a complex constellation of meaning, of properties, relationships, things, processes, events, norms, social roles and institutions, and the like. All things within our experience—whether dynamic or static—make sense by reference to one or more of the aspects. It is important not to confuse the concrete things or events that are meaningful by reference to an aspect [what Dooyeweerd (1984, I, p. 3) called the "what"] with the aspect itself, which is a way (a "how") in which the things might be meaningful.

Many things are of multiple aspects. For example digits, which express figures, are lingual signs with strong quantitative meaning. For example, an idiom is of the lingual aspect, but it has a strong social aspect because its meaning depends on the writer and reader sharing social or cultural assumptions.

Table 3-1. Aspects of running a chemical production business

Aspect	Kernel	Management situations or issues
Quantitative	Amount	Figures, Accounting
Spatial	Extension	Spatial layout of site and buildings
Kinematic	Movement	Movement around site and building; movement of product
Physical	Energy	Physical integrity of buildings and plant; Chemical reactions
Biotic/Organic	Life	Food hygiene in canteen; Product safety
Psychic/Sensitive	Feeling	Use of eyes, ears in production; How employees feel
Analytic	Distinction	Clarity and logicality of instructions, of goals, of vision
Formative	Shaping	Planning Designing Manufacturing
Lingual	Signification	Signage; Records; Trademarks
Social	Respect	Team building; Structure of organisation
Economic	Frugality	Budgets, deadlines; Waste, efficiency; Resources
Aesthetic	Harmony	Harmony of organisation; Fun; Decoration
Juridical	Due	Legal matters, contracts; Responsibility
Ethical	Self-giving	Generosity, Ethicality of organisation
Pistic	Faith	Vision and mission; Loyalty; Religious faiths

Some concepts in common use have several meanings. For example, the word "debt" is usually given economic meaning, but it may also be used juridically, when something remains due to another. One is a kernel meaning, the other analogical; see later.

3.1.2 Other Recognition of Aspects

It is very natural to think aspectually, whenever we delineate a set of things that should be taken into account separately from each other and not reduced to each other. Usually we do so informally, as in Adam (1998, p. 180): "The way that a number of aspects of knowing are not reducible to propositional knowledge, but rely instead on some notion of embodied skill, points to the role of the body in the making of knowledge." While many use the word "aspects", others use other words. Maslow (1943) offered his hierarchy of needs. Dahlbom and Mathiassen (2002, p. 135) distinguished three types of quality: functional, aesthetic, and symbolic. y spoke of regional ontologies. Even Foucault's regimes of truth might be centred on aspects.

Some try to identify aspects more formally. Husserl suggested there are three aspects: material, psychological, and social—which is a very common set. To Husserl (1970, pp. 233-4) it was important to distinguish the psychological aspect from the physical and believed that Brentano was prevented from doing so by his "prejudices". Hartmann believed there are four "strata": inorganic, organic, animal-psychic, and supraindividual-cultural, and possibly the historical is a fifth. Bunge (1979) omits the psychological, splits the material into physical, chemical, and biological, and adds a technical. Habermas (1986) identified five action types. Table 3-2 allows comparison of several sets of aspects with Dooyeweerd's. To a large extent, they accord reasonably well with Dooyeweerd's, as a subset, and, if any order is given, this is usually approximately the same order as Dooyeweerd's. Dooyeweerd warned against claiming any absolute truth for any suite, and later the extent to which his suite may be trusted (and hence adopted and used) is discussed.

Table 3-2. Some suites of aspects

Aspect	Maslow	Husserl	Hartmann	Bunge	Habermas
Quant'ive					
Spatial					
Kinematic					
Physical		Material	Inorganic	Physical, Chemical	
Biotic	Biological	Material	Organic	Biological	
Psychic	Safety	Psychological	Psychic		
Analytic	Enquiry				
Formative			Historical	Technical	Instrumental, Strategic
Lingual	Expression				Communicative
Social	Affiliation, Esteem	Social	Supra-individual	Social	(Strategic)
Economic					
Aesthetic	Aesthetic				Dramaturgical
Juridical					Normatively regulated
Ethical					
Pistic	Transcendence, Self-actualization				

3.1.3 More than Categories

Most thinkers treat aspects as little more than categories that are irreducible to each other. Hartmann's (1952) discussion in his "new ontology" was somewhat more sophisticated than most, in particular discussing the bearer-borne relationship between aspects ("strata") and a normative concept of "perfection" in each. A comparison between Hartmann and Dooyeweerd may be found in Seerveld (1985).

Probably because of his different ground-motive, Dooyeweerd went further than most in exploring what aspects are. According to Henderson (1994, pp. 37-8), Dooyeweerd's way of working out his philosophy had no predecessors, and he recounted shortly before his death in 1977 how the shape given to his idea of aspects occurred to him:

*I enjoyed going for walks in the dunes in the evening. During one of these walks in the dunes I received an insight (*ingeving*) that the diverse modes of experience, which were dependent upon the various aspects of reality, had a modal character and that there had to be a structure of the modal aspects in which their coherence is reflected. The discovery of what I called 'the modal aspects of our experience horizon' was the point of connection.*

Thus aspects, though separate, cohere and do much more than categorise.

To Dooyeweerd aspects are spheres of meaning and law that constitute the law side (§2.4.4). They form an enabling framework that enables the entire cosmos to Be and Occur, meaningfully and "good" (that is, to be sought rather than avoided), diversely and coherently. The cosmos includes not just physical things and occurrences, but conceptual, social, moral, and so on. Poems, programs, people, and policies, for example, are part of this meaning-and-law-enabled cosmos.

Each aspect is some kind of origin, not the absolute Origin, but that which enables being, doing, knowing, and the like. Heidegger (1971) seems to have understood something similar:

Origin here means that from and by which something is what it is *and as it is. What something is, as it is, we call its essence or nature. The origin of something is the source of its nature.* (p. 17)

Heidegger continued, concerning art:

The question concerning the origin of the work of art asks about the source of its nature. On the usual view, the work arises out of and by means of the activity of the artist. But by what and whence is the artist what he is? By the work; for to say that the work does credit to the master means that it is the work that first lets the artist emerge as a master of his art. The artist is the origin of the work. The work is the origin of the artist. Neither is without the other. Nevertheless, neither is the sole support of the other. In themselves and in their interrelations artist and work are each of them by virtue of a third thing, which is prior to both, namely that which also gives artist and work of art their names—art.

To Dooyeweerd, the aesthetic aspect is that "third thing" which is prior to both: a source of the nature of works of art and also that by which the artist is artist.

But Heidegger, trapped in the immanence-standpoint (Dooyeweerd, 1984, I, p. 112), did not allow himself to conceive of a law side, so he could not develop this theme as far as Dooyeweerd did. In Dooyeweerd's extensive treatment in Volume II several philosophical characteristics of aspects and several ways in which aspects fulfil this role of "origin" (philosophical roles) may be discerned which will prove useful in formulating frameworks for understanding IS.

3.1.4 Characteristics of Aspects

What characteristics thinkers believe aspects to have is to a large extent determined by religious presuppositions. Dooyeweerd developed his view presupposing CFR, and, though he did not set these out systematically, we can find throughout his writings the following characteristics of aspects.

3.1.4.1 Transcendence of Aspects

The aspects, as a framework of meaning-and-law that enables the cosmos, transcends the entire cosmos. This means the aspects pertain, across all situations, all cultures, all times, whether we acknowledge or understand them, or not. This is especially important, for example, in understanding unexpected impacts of computer use (Chapter IV).

3.1.4.2 Irreducibility

Aspects are irreducibly distinct in respect of their meaning. For example, the psychic aspect cannot be reduced to the physical nor the juridical to the social, lingual, or ethical: justice (meaningful in the juridical aspect) cannot be reduced to discourse (lingual-social) or even to morality (ethical). It is in everyday experience that this

irreducibility is most important, even though we might not be aware of the distinctness of aspects therein.

Clouser (2005) explains Dooyeweerd's notion well. Irreducibility, usually called
sphere sovereignty by Dooyeweerd, is a stronger notion in Dooyeweerd than in most
thinkers. It means that no aspect can be eliminated in favour of another, neither by
declaring it to be a figment nor by treating it as essentially the same as another, and
no aspect "causes" another.

Reducing one aspect to another gives rise, over a period, to major problems in
philosophy, such as antinomy (Greek "against-law"). In research and practice,
aspectual irreducibility has other implications, providing philosophical grounds
for understanding diversity, delineating distinct horizons of meaning, helping us
to avoid overlooking important factors, and fostering and guiding interdisciplinary
thinking. In everyday experience, every aspect is important: none can be dismissed
as less meaningful, less interesting, or deserving of less of our attention.

3.1.4.3 Harmony of Aspects

"Being too ethical jeopardises the success of business!" This would only be true in
general if the laws of the ethical aspect were in conflict with those of the economic
aspect. But Dooyeweerd contended (1984, II, p. 3) that there is no such conflict.
Philosophically, if aspects are irreducible to each other the laws of one *cannot* be in
conflict (antinomy) with those of another. If the Origin of all is loving, as Dooyeweerd
believed, they *will not* be. Any apparent disharmony occurs, argued Dooyeweerd
(1984, II, p. 334, ff.), because of the immanence-standpoint. In IS, for example, the
late Enid Mumford seemed to believe in aspectual harmony, and so do the more
recent business authors like Collins and Porras (see §4.4.3).

3.1.4.4 Non-Absoluteness of Aspects

Though the aspects constitute the enabling framework for the temporal cosmos, no
aspect is absolute, in the sense that no aspect can be the foundation for all the others, and no aspect has its full meaning within itself. Rather each refers beyond itself
(which is the nature of meaning discussed in Chapter II). Dooyeweerd mentioned
three ways in which an aspect refers beyond itself (1984, III, p. 632):

*The idea of meaning-modality points above itself to the temporal coherence of all the
modal spheres [aspects] and to the fulness of meaning in the transcendent religious
root and to the Origin of the creation.*

The first is of most interest here, in that it implies that each aspect refers to, or relates to each of the others, either before or after it; see the following.

Reductionism absolutizes an aspect (e.g., rationalism absolutizes the analytic). If we absolutize an aspect we treat that aspect as of overriding importance, as the only aspect that should be considered, as able in principle to achieve all we want, and as worthy of all attempts to defend it. Absolutization fundamentally affects the research questions we pose and seek to answer in that the research community is driven to see everything as an extension of the meaning of that aspect, and to try to justify everything in terms of that aspect. It prevents a true interdisciplinary attitude. Most immanence-philosophy has tended towards absolutization of one aspect or another.

3.1.4.5 Anticipation and Retrocipation

But giving too much emphasis to the irreducibility of aspects (sphere sovereignty) can lead to fragmented views. Dooyeweerd was clear (1997, p. 154):

Sphere-sovereignty does not yield a watertight compartment or mechanical division among the areas of life. It is, as we have seen, an organically most deeply cohering principle, for it begins with the religious root-unity of the life-spheres.

One of the very few diagrams that Dooyeweerd devised, reproduced in Figure 3-1, indicates how he conceived of the relationship between aspects. It shows a kernel with a constellation of meaning, half of which anticipates later aspects and half of which retrocipates earlier ones. Dooyeweerd called these directions anticipatory [or "transcendental" (Dooyeweerd, 1984, III, p. 109)], and foundational or substratum.

The lingual aspect, for example anticipates the social (in that it would be largely meaningless without social intercourse, restricted to private note-taking) and retrocipates the analytic (in that all symbolic signification involves concepts) and the formative (in that it involves their structuring). To speak of "later" and "earlier" aspects presupposes a linear order among them, which Dooyeweerd called cosmic

Figure 3-1. Spheres of meaning in sequence

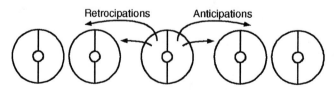

time (Dooyeweerd's theory of Time is not discussed here). Immanence-philosophy had never posed the problem of this order (II, p. 49) (though some systems theory has now perhaps begun to do so). The quantitative and pistic aspects, as terminal aspects, are special (Dooyeweerd, 1984, II, pp. 52-54). This order of aspects offers a basis for labelling groups of aspects. "Pre-" or "post-" meaning the aspects before or after an aspect. Other labels that will be referred to include:

- **Social aspects:** Social to pistic
- **Human aspects:** Analytic to pistic
- **Physical aspects:** Physical to psychic
- **Mathematical aspects:** Quantitative to kinematic (pre-physical)

Anticipation and retrocipation will be important in designing IT resources, discussed in Chapter VII, and in understanding technological progress, discussed in §3.4.6 and Chapter VIII. There are three main types of inter-aspect relationship: dependency, analogy, and what we will call "reaching out", which have both anticipatory and retrocipatory directions.

3.1.4.6 Inter-Aspect Dependency

Each aspect depends on earlier aspects for its facilitation and on later aspects for the opening of its full meaning (Dooyeweerd, 1984, III, p. 91), such as exemplified earlier by the lingual aspect's retrocipation of the formative and anticipation of the social.

Foundational dependency (a more common term for retrocipatory dependency) is important in IS because it is concerned with implementation, and helps us understand the levelled nature of computers and information (Chapter V), where it is explained more fully. (Dooyeweerd's notion is not unlike Hartmann's notion of lower strata "bearing" higher, but places more emphasis on anticipation than Hartmann did.)

Anticipatory dependency in the field of IS is concerned with application, and hence is important when anticipating use in information systems design (Chapter VI). Anticipatory dependency is open-ended and so is relevant when considering building future-proof computer system architectures (discussed in Chapter VII).

In Dooyeweerd dependency does not imply reducibility, so our frameworks for understanding do not need to be based on the system-theoretic notion of supervenience. Further discussion of dependency is found in §7.4.1 and §7.4.2.

3.1.4.7 Inter-Aspect Analogy

In each aspect there are echoes of all the others. For example causality is physical but there is something that resembles it in all aspects, such as logical implication and formative repercussions; see Geertsema's (2002) discussion of aspectual "causality" and Dooyeweerd's own discussion of many analogies (1984, II, p. 118ff.). Inter-aspect analogy is a major component of what Dooyeweerd called sphere universality (1984, II, p. 331), which is an antidote to too much sphere sovereignty.

While dependency has a certain necessity about it, inter-aspect analogy does not. For example the "movement" from premise to conclusion works well without any kinematic movement, while physical processes, which depend on the kinematic, cannot. Inter-aspect analogy is not to be confused with concrete analogies that we detect or create between things, such as metaphors; rather, it enables metaphors to "work". Inter-aspect analogy is important in computer system architectures and is explored further in Chapter VII.

3.1.4.8 Inter-Aspect "Reaching Out"

There seems to be a third inter-aspect relationship, which must be differentiated from both dependency and analogy, where each aspect "reaches out" to the meaning of all the others. Figure 3-2 shows two of these. The analytic aspect reaches out, in that we make all kinds of distinction: between amounts, shapes, feelings, and so forth, and the lingual reaches out in that it enables us to speak of all kinds of things. Reaching-out is clearly differentiated from dependency and analogy by the following example of lingual-aesthetic relationships:

- **Dependency (anticipatory):** Verse is a form of lingual structure and use that is particularly meaningful in the aesthetic aspect, but which would be a mere speculative curiosity without reference to that aspect.

- **Analogy:** The lingual notion phrase or sentence is used analogically to refer to a short section of music.

- **Aspectual reaching-out (to signify):** Words like "harmony", "beauty", "music", "art" signify aesthetic meaning.

Whereas the two directions of inter-aspect dependency differ markedly, aspectual "reaching-out" feels the same whether directed towards earlier or later aspects.

The author finds that a sound understanding of this relationship is important to formulating frameworks for understanding IS. The lingual reaching-out relationship is important to knowledge representation in Chapter VII, and the analytic and juridical,

Figure 3-2. Aspectual reaching-out

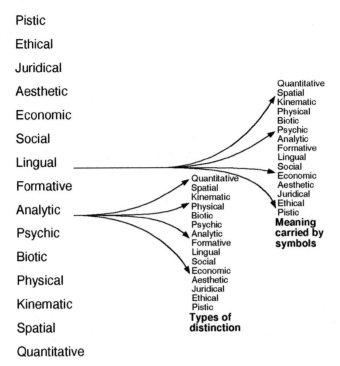

in ISD in Chapter VI. But Dooyeweerd did not seem to discuss this relationship as explicitly as the other two but rather tended to assume it as part of law-subject functioning (and did not use the term "reaching out"), except for a special form of it, the *Gegenstand* relationship. Whereas in everyday life, reaching-out involves intimate engagement, in the theoretical attitude it takes a distal form of "standing over against" the target aspect; it is discussed later.

3.1.4.9 Aspectual Normativity

Aspects are spheres of law that establish a variety of norms for the cosmos (cosmic norms). The earlier aspects (especially quantitative to physical) are determinative while the later aspects (especially from the analytic aspect onwards) allow some freedom. For example, we have freedom to go against the lingual law that it is better to abide by the syntax of the language we are using, but seven people in a room do not have the freedom to be four (a law of the quantitative aspect). Some hold that normativity begins with the analytic aspect, while others, such as de Raadt (1991), who employs them in business analysis, suggest a gradual increase in normativity along the aspects.

Though Dooyeweerd did not develop it clearly, there seems to be a difference between normativity and freedom (non-determinativity). Freedom means that the future is open (not determined, e.g., we can choose different syntactic structures), while normativity distinguishes what is "right" or beneficial from what is "wrong" or detrimental (e.g., refusing to abide by syntax altogether). (Note: though we have freedom to go against laws of normative aspects, we are never free from repercussions of so doing, because of aspectual transcendence.)

That determinativity, freedom, and normativity are brought together in such a way frees Dooyeweerdian thought from the dialectics of the NFGM presupposition and especially that observed in IS in the relationship between positivism, interpretivism, and criticalism. Aspectual normativity and non-determinativity will be found useful in considering IS success (Chapter IV), the difference between humans and computers (Chapter V), and the tension between guidance and freedom in ISD (Chapter VI).

3.1.4.10 Grasped by Intuition

Aspectual meaning is grasped by the intuition, but not by theoretical thought; for example, we know what justice is but find it impossible to define (§3.3.7 explains why). As a result, categories based on aspects tend to be easy to understand, and they inform rather than mislead in aspectual analysis used in Chapters IV to VIII. Intuition is not, however, absolute and is subject to cultural, experiential and religious modification (Dooyeweerd, 1984, III, p. 29), but can still help us in intersubjectivity, and even trans-cultural understanding such as is necessary to the Internet. See §3.3.5.

3.1.5 What Aspects Enable

Aspects, as spheres of meaning and law, enable the cosmos to be and occur; they are a transcendental condition for it. In the context of formulating frameworks for understanding IS, the following philosophical roles that aspects fulfil are ones the author has found important. (Dooyeweerd himself never set these out as systematically as is done here, so some of his views might be unwittingly misrepresented.)

3.1.5.1 Distinct Categories of Meaning

As spheres of meaning, aspects provide distinct ways in which things can be meaningful, especially to us who attribute meaning. This enables us, for example, to see a Web site as meaningful from the point of view of the economic aspect (e.g., how much money it makes), the juridical aspect (will anyone sue it?), the aesthetic aspect (its artistic merit), and to discuss it in such terms. This validates diverse ways of understanding a thing, such as a computer (Chapter V) and will be found helpful in accounting for the perspectives that people can take of things (Chapter VI, Chapter VIII). As spheres of meaning, aspects provide a basis for intersubjectivity. Though Dooyeweerd (1984, II, p. 50) stressed that aspects are not the same as Kant's categories of thought, aspects may be seen as a foundation for them.

3.1.5.2 Distinct Rationalities

"In science, for example," suggested Winch (1958), "it would be illogical to refuse to be bound by the results of a properly carried out experiment; in religion it would be illogical to suppose that one could pit one's strength against God's." As a sphere of meaning, each aspect provides a different way in which we exclaim "That makes sense!" or "That does not make sense!" Analytic rationality differs from pistic—and from mathematical, physical, juridical, and so on. The radical irreducibility of aspects means that we can never reason our way from (concepts of) one aspect to (those of) another by logic alone. The notion of distinct rationalities is important in understanding conflicts (Chapter VI) and in researching inference mechanisms in Chapter VII.

3.1.5.3 Distinct Modes of Being

Each aspect provides a distinct way in which a thing can exist. It is in this way that an aspect is an "origin" of both work of art and artist, as Heidegger struggled with (mentioned earlier). But Dooyeweerd saw this clearly, and moreover allowed things to be multi-aspectual. Praxiteles' sculpture of *Hermes and Dionysus*, for example, is both a block of marble (physical aspect) and a work of art (aesthetic) (Dooyeweerd, 1984, III, pp. 110-127). Likewise, a computer might have at least six modes of being, defined by physical to lingual aspects (Chapter V). A business is likewise multi-aspectual. As discussed next (§3.2), Dooyeweerd held that Being cannot be understood any other way, especially if we approach it with an everyday attitude. It is in this way that cosmic meaning is prior to, and a transcendental condition for, Being.

3.1.5.4 Distinct Ways of Functioning

As sphere of law, each aspect enables concrete entities to concretely function (occur, happen, behave):

Subject-side responds to Law-side --> Occurrence.

Figure 3-3. Aspectual functioning, with repercussions

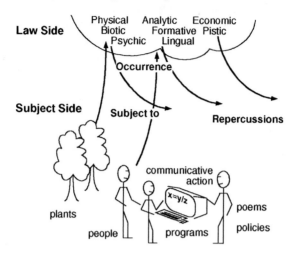

For example, as I write this, I function in the lingual aspect of signifying, in the formative aspect of structuring, in the analytic aspect of conceptualizing, in the psychic aspect of sensory-motor activity, and so forth. Such functionings are not to be seen as parts of a whole nor as forming a temporal sequence, but to be seen as different ways in which the whole experience of writing is meaningful. See Figure 3-3. Trees, for example, function in the biotic and physical aspects.

That aspects enable diverse ways of functioning is useful in understanding use of computers (Chapter IV) and ISD (Chapter VI). [In the normative aspects, we may be aware of responding to aspectual law, but in determinative aspects, "respond" takes on an unusual meaning, and we experience it more like something happening to us, for example the physical law of gravity, or having a property or attribute, such as being 6 feet tall (spatial aspect).]

Things may function in each aspect as either subject or object. For example, as I write, I function as subject in the lingual aspect and the book functions as object in the same aspect. However both book and I function as subject in the physical aspect as I leave ink on paper or exert force on plastic keys of computer. This will inform our understanding of the nature of computers in Chapter V and of HCI in Chapter IV. Several things arise from aspectual functioning, including:

• distinct **basic kinds of property** (e.g., physical mass, aesthetic rhythmic scheme)

- distinct **ways of relating** (e.g., physical causality, social friendliness, ethical love)
- different **kinds of possibility** [e.g., that the plant will thrive (biotic), that we will achieve our plans (formative)]

These will be useful when designing technologies (Chapter VII).

3.1.5.5 Distinct Types of Repercussion

Because aspectual law has the characteristic of promise there are inevitable repercussions to each type of functioning, a different type for each aspect. Some are exemplified in Table 3-3, which is referred to later when impacts of IS use are discussed in Chapter IV. Aspectual repercussion may be seen as an analogical echo of causality. To some extent, the time-response of repercussion lengthens with the aspects, from almost immediate in the earliest aspects to centuries in the pistic aspect [cf. Lonergan's (1992) "longer cycles", of decline or creation and healing] though these must be expected to vary in each case. Aspectual law pertains (§3.1.4.1)—and

Table 3-3. Aspectual repercussions

Aspect	Example Repercussion	Typical Time
Physical	Electric conductance	Pico-seconds
Biotic	Hormone release	Sub-seconds
Sensitive	Reflex response; nerve activity	Sub-seconds
Analytical	Switch attention	Second
Formative	Making a choice, given the info	10 seconds
Lingual	Write, utter or understand message	Minutes
Social	Initiate friendship or respect	Hours
Economic	Effect the saving of a resource	Week
Aesthetic	Make a thing fashionable	Months
Juridical	Bring to justice	Years
Ethical	Build attitude of forgiveness (S.A.)	Decades
Pistic	'Longer cycles'; see text	Centuries

repercussions occur—whether we are aware of it or not. Chapter IV shows how this provides an understanding of the complex impacts that IS use has.

3.1.5.6 Distinct Kinds of Normativity

In general, beneficial or positive repercussions come from functioning in line with the laws of aspects and detrimental or negative repercussions come from going against the laws of aspects. Each aspect yields a distinct type of Good and Evil, such as:

- **Biotic aspect:** Vitality, health vs. disease, threat to life
- **Sensitive aspect:** Sensitivity vs. sensory overload or deprivation
- **Analytical aspect:** Clarity vs. confusion, illogicality
- **Formative aspect:** Forming, creating, achieving vs. destroying
- **Lingual aspect:** Conveying truth, understanding vs. deceit and misunderstanding
- **Social aspect:** Friendship, respect vs. enmity, disrespect
- **Economic aspect:** Care, frugality vs. waste, squandering resources
- **Aesthetic aspect:** Harmony, fun vs. disharmony, boredom
- **Juridical aspect:** Justice, giving due vs. injustice, denial of what is due
- **Ethical aspect:** Generosity, giving, sacrifice, hospitality vs. selfishness, taking advantage of others, competition
- **Pistic aspect:** Loyalty, trust, orientation to true God vs. disloyalty, cowardice, idolatry.

This view of normativity is useful in understanding success and failure in IS use (Chapter IV), in formulating guidelines for ISD (Chapter VI), in defining constraints (Chapter VII), and in understanding impacts on society (Chapter VIII).

3.1.5.7 Distinct Ways of Knowing

Each aspect provides a distinct way of knowing: theoretical, social, and cognitive knowing, for example, are all different. Full knowing, which is everyday experience, involves every aspectual way of knowing in harmony. See more on knowing in §3.3.

3.1.6 How Aspects May be Delineated

A suite of aspects is one of the primary tools that will be used in formulating frameworks for understanding IS. Dooyeweerd's suite will be used here, but some readers might wish to modify it, use another, or even create their own. What guidance can be given? The author has found three principles useful.

First, while suites of aspects are offered, no suite is "given" in the sense of a claim to absolute truth being forced upon us. Dooyeweerd was clear (1984, II, p. 556):

In fact the system of the law-spheres designed by us can never lay claim to material completion. A more penetrating examination may at any time bring new modal aspects of reality to the light not yet perceived before. And the discovery of new law-spheres will always require a revision and further development of our modal analyses. Theoretical thought has never finished its task. Any one who thinks he has devised a philosophical system that can be adopted unchanged by all later generations, shows his absolute lack of insight into the dependence of all theoretical thought on historical development.

This is because the very activity of delineating and characterising aspects, composing a coherent suite, and then expressing it in words relies on subject-functioning in the analytic, formative, and lingual aspects at least, and all are non-absolute. But see also "Intuition" later.

Second, a suite is "taken" by us, when we adopt it for purposes of differentiating categories or any other task—even if in the next breath we question it.

Third, delineation of aspects is not invention but discovery. While our beliefs about, knowledge of, and expression of, aspects, and the construction of suites, may be socially constructed, their transcendent impact is not. Indeed, aspects are what enables social construction.

However, this is not a scientific but a philosophical form of discovery, so requires a different method, one that enables us to take a lifeworld rather than theoretical attitude towards the diversity of our experience. (The difference between science and philosophy is discussed later.)

From these principles, one can offer the following practical guidelines:

•   Be sensitive to our intuition of kernel meanings, reflecting on life experience, one's own and that recorded or written by others, especially those of distant cultures. Poetry is good.

- But always recognise that intuition is culturally shaped, especially to ignore certain aspects, and discourse is strongly influenced by prevailing world-views and ground-motives.
- This can yield an initial suite, for more precise consideration.
- Tentatively take account of what the sciences have so far discovered about their aspect (see "science").
- The method of antinomy (Dooyeweerd, 1984, II, pp. 37-41). Conflation of aspects can be detected by examining paradoxes; Dooyeweerd's examination of Xeno's Paradox (race in which hare can never overtake tortoise) led him to the conclusion that the kinematic aspect cannot be reduced to the spatial.

Most thinkers have been informed by only some of these. Dooyeweerd has been informed by all. This, coupled with his philosophical examination of the characteristics and roles of aspects, suggests that Dooyeweerd's suite is at least of comparable quality with the others, and thus we may place some trust in it. Thus it will be "taken" as the suite to inform our formulation of frameworks for understanding throughout this work.

3.2 Things

In everyday experience, we encounter things and and types of things. What is a thing, such as the bunch of keys, organisational culture, or computer? What is "thingness", Being, Existence? The answers given to, or assumed for, these questions deeply affects the frameworks for understanding we formulate.

As already mentioned earlier, Dooyeweerd believed that all temporal Being is founded in Meaning, and this differs radically from most other views. His theory of things presupposes his theory of modal aspects. Though he sketched out how the being of things relates to aspectual meaning in volume II of (1984), it was in III, 'The Structures of Individuality of Temporal Reality', that he most fully developed his ideas. Dooyeweerd was particularly interested in typology and in social institutions like the state.

3.2.1 Everyday Experience of Things

Dooyeweerd believed that a theory of thingness should be able to account for all kinds that we might experience with the everyday attitude. He rejected naïve realism, which he held to be a theory about what constitutes thingness, namely that a thing is nothing more than as we directly experience it through our senses.

But Dooyeweerd also rejected the Kantian gulf between thought and thing, which holds itself to be superior to naïve realism by declaring that we cannot know the *Ding an sich* (thing in itself). Dooyeweerd argued that this too is a theoretical abstraction which is rejected by everyday experience and, if accepted, distorts it. We engage with things, and they engage with us. This is a conundrum that phenomenology and existentialism tried to tackle. But they, taking the immanence-standpoint (see Chapter II), could not inquire into the nature of thingness. Dooyeweerd concluded (1984, III, p. 53), having made his comprehensive survey of 2,500 years of Western thinking:

As far as I know, immanence philosophy, including phenomenology, has never analysed the structure of a thing as given in naïve experience.

Failing to do this causes problems in attempts to understand the nature of things like computers, businesses, and so forth.

3.2.2 Some Problems with Extant Understanding of Things

Dooyeweerd's theory of things was motivated by deep problems at the root of our conventional view of things. What is the "being" of the keys mentioned earlier? Conventionally, we might say each *is* a piece of metal with a certain spatial shape as a *property* and that its juridical meaning is something it *has* to us. Dooyeweerd puts it the other way round because they have no existence as *keys* (as opposed to lumps of metal) without that juridical meaning.

The conventional view seems reasonable until we try to understand the Being of some other types of thing. This is not merely an issue of rival philosophical stances; this very problem has been encountered by the knowledge representation community as it had to come to terms with the diversity encountered in everyday life (especially via "natural language" utterances). Hirst (1991) reviewed a number of problems that arise from "existence assumptions in knowledge representation".

The first type of problem Hirst discussed is the distinction between different types of existence, such as between concrete (physical) and abstract (non-physical) existence (he cited the number 27), and to this might be added the mind, beliefs held, messages sent, symphonies, and the like. He gave it little consideration because he believed Quine and others had dealt with it. Hirst discussed the problems thrown up by Kant's proposal that existence cannot be treated as a predicate like colour or height, including when trying to treat existence itself as an object (as in "The existence of Pluto was predicted by mathematics and confirmed by observation").

But what occupied most of Hirst's discussion was "to account for the fact that in ordinary, everyday language we can *talk about* certain things without believing that they exist." A particular question was how logical inference, which lies at the root of knowledge representation, could cope with non-existent things—such as voids ("There are too many holes in this cheese," "A complete lack of money led to the downfall of the company"), things that aren't there ("There's no one in the bathroom"), events that don't occur, actions that are not taken ("The (threatened) strike was averted by last-minute negotiations"), fictional or imaginary objects ("Dragons like baklava," "Sherlock Holmes lived in London"), and unreal things ("Round squares make me seasick—especially the green ones"). His solution was to suggest that, despite Kant, existence should be treated as a predicate.

Yet another set of problems that Hirst discussed briefly is concerned with change, including existence at other times ("Alan Turing was a brilliant mathematician") and continued existence of some kind ("Alan Turing is a celebrated mathematician", "Alan Turing is dead").

A further problem is that it is easy to misunderstand the part-whole relationship, and thereby make the category error of assigning agency to the part rather than the whole, such as that the brain rather than the person understands (Boden, 1990).

All these serve to show that the notion of Being is problematic and thus that philosophy should not just account for types of things, but inquire into the nature of Being as such. Aristotle did so using various notions of substance. Heidegger tried to but, as already mentioned, never completed his quest.

3.2.3 Dooyeweerd's Approach

Dooyeweerd, rejecting the immanence-standpoint, presupposed Creation, Fall and Redemption as a ground-motive (§2.3.1.2), and thus presupposed that cosmic Meaning-and-Law are more fundamental than Being (§2.4.3). As has already been quoted:

Meaning *is the* being *of all that has been* created ... (Dooyeweerd, 1984, I, p. 4, his emphasis)

It is not sufficient to say that things "have" meaning (as a lump of metal "has" juridical meaning); cosmic meaning is the very nature of Existence itself. Meaning is the first thing we should say about anything, before talking about its existence. Though this might seem a radical departure in some areas of research, it is not so unusual in others. For example, Smircich (1983, p. 353) suggested that the important task of researchers of organisational culture is not to ask, "What is organisational culture?," so much as "How is organisation accomplished, and what does it mean to be organised?"

This applies not only to conceptual-social things like organisational culture: according to string theory physical Being is showing characteristics of being rooted in meaning-and-law.

As will be seen in Chapter V, many have tried to understand—and debate—the nature of computers by presupposing some self-dependent "substance" that is the essence of computer (e.g., Searle's contention that the computer works by physical causality while humans work by biological). But, of things in our everyday experience, Dooyeweerd said (1984, III, p. 108):

For it is really impossible to ascribe their typical nature to an independent "substance". Their very nature is meaning, realized in a structural subject-object relation. A bird's nest is not a "thing in itself", which has a specific meaning in the bird's life. It has as such no existence apart from this meaning.

So a key, as such, has no existence apart from the juridical meaning. "Organisational culture", as such, has no existence apart from social meaning. A business, as such, has no existence apart from economic meaning. Plants are because they mean (biotically). Likewise, rocks are rocks by virtue of the physical aspect.

To say something exists is to say it means, and to say this we must refer to at least one of the spheres of meaning. Though only one aspect is mentioned for each thing, things exhibit several ways of meaning, from several aspects. For example, a poem exists *qua* poem because of the aesthetic sphere of cosmic meaning, and without (that is, without reference to) aesthetic meaning, it does not exist as a poem, but it also exists as a piece of writing by virtue of the lingual sphere of cosmic meaning. The following "exist", as the things named, by virtue of their meaning in the stated aspects (among others):

- **The sculpture:** Aesthetic and physical
- **The poem:** Aesthetic and lingual
- **The bunch of keys:** Physical and juridical
- **The landslide:** Physical and spatial
- **The bird's nest:** Physical, biotic, and sensitive
- **The kennel:** Physical, biotic, sensitive, and formative
- **The greyhound racing track:** Spatial, kinematic, and aesthetic
- **The skeleton in a museum:** Biotic, physical, and formative (historical)

- **The ring:** Social and physical
- **The kiss:** Ethical and social
- **The speed limit:** Juridical and kinematic

We can also treat events or processes as things:

- **The race:** Kinematic, social, aesthetic
- **The act of programming:** Lingual, formative
- **The war:** Juridical, social, formative, pistic

Chapter V will develop this to understand the nature of computers:

- **The computer (running software):** Physical, psychic, analytic, formative, lingual (and other aspects)
- **The mouse pointer:** Sensitive, analytic

Thus things do not exist without the spheres of law-and-meaning. Their very being is enabled by, and constituted in them. Dooyeweerd was interested in "wholes" rather than parts, and here the meaning-nature of things will be emphasised by referring to things as "meaningful wholes". In an ingenious way, Dooyeweerd brought together an account of both physical, conceptual, and social things into a single framework, while providing a basis for differentiating them.

3.2.4 Becoming and Change

My hammer has had two new heads and five new handles—is it still the same hammer? When does a computer, organisation, or project change or stay the same? Dooyeweerd began Volume III with the question of how do we account for changes in a thing, through some of which it is the same thing, while others destroy it as that thing. The book with pages torn or scribbled on is still the same book, but throw it in the fire, so that it burns up, and it ceases to be the same thing. In a short section headed "Reality as a continuous process of realization" Dooyeweerd said (1984, III, p.109), poetically:

For the reality of a thing is indeed dynamic; it is a continuous realization in the transcendental temporal (i.e., anticipatory aspectual) direction. The inner restless-ness of meaning, as the mode of being of created reality, reveals itself in the whole temporal world. To seek a fixed point in the latter is to seek it in a "fata morgana", a mirage, a supposed thing-reality, lacking meaning as the mode of being which ever points beyond and above itself. There is indeed nothing in temporal reality in which our heart can rest, because this reality does not rest in itself.

Platonic ideal types cannot help because they presuppose all change is a departure from the ideal. (Even more illogical, perhaps, is the modern assumption of progress, that all change is good, especially if it means more advanced technology.) As discussed in §5.1.3 in considering the nature of computers, neither can Aristotle's substance concepts, nor is an answer that confuses naïve experience with sensory experiences, nor naïve realism, nor several other views.

Dooyeweerd's proposal was that things come into being, exist, change, and cease to be, by aspectual functioning—whether subject- or object-functioning. Perhaps the proverbial hammer can be understood in this way. From the physical aspect it is a different thing. But from the formative aspect (its meaning as a tool) and the juridical aspect (my ownership of it), it remains the same thing. Likewise, a book, an organisation, a project. Dooyeweerd's meaning-based approach is nuanced enough to allow us to see a thing as both same and different without contradiction in different aspects. This will inform ISD in Chapter VI.

Pacé Plato, it is not change itself that is either good or evil. Whether the change is good or evil depends on whether the functioning in each aspect is with or against its laws.

3.2.5 Types of Things

The question of whether a thing such as a hammer or an IS remains that same thing presupposes a notion of type. In discussing the duration of a thing (1984, III, pp. 76-79), Dooyeweerd concluded that:

In general we can establish that the factual temporal duration of a thing as an individual and identical whole is dependent on the preservation of its structure of individuality.

"Structure of individuality" refers to the ways a thing is meaningful in the various aspects and it is what enables it to be that particular individual despite changes. It is also referred to as "internal structural principle" and as type law by Clouser (2005).

Dooyeweerd identified three levels of type: radical types, geno- or primary types, and pheno- or variability types (Dooyeweerd, 1984, III, p. 93). Radical types are

Table 3-4. Types of thing by qualifying aspect

Aspect	Example things
Quant'tive	Amount, proportion
Spatial	Shape, Distance, Angle, Direction
Kinematic	Path or route, Flow
Physical	Energy, Waves, Particles, Material, Fields, Forces, Rock
Biotic	**Plants**, Organism, Organ, Tissue, Cell
Sensitive	**Animals**, Sound, Colour, Feeling, Emotion, Excitation
Analytical	Concept, Distinction, Deduction, Awareness
Formative	Goal, Achievement, Forming, Will, Tool, Skill
Lingual	Word or sentence, Book, Writing, Utterance, Diagram, Index
Social	Friendship, Institution, Status, Respect
Economic	Resource, Limit, Production+consumption, Money, Management
Aesthetic	Music, Sculpture, Cuisine, Humour, Fun, Sport, Nuance
Juridical	Responsibility+rights, Reward+punishment, Laws
Ethical	Self-giving love, Generosity, Sacrifice
Pistic	Faith, Trust, Loyalty, Worship, Commitment, Ritual

the first division, differentiated by the aspect that is most important in defining the main meaning or destination of the thing, which is known as the qualifying aspect; Table 3-4 illustrates the notion of radical types.

It is the qualifying aspect that determines the unity of a thing as of that type of thing. All its other aspects serve the qualifying. For example, the qualifying aspect of a book is the lingual, and the formative aspect of its structure, and the psychic aspect of the marks on its page, serve the lingual function of being-read. Dooyeweerd did acknowledge that the qualifying aspect of a thing might change; an antique shawl might become a wall hanging (III, p. 146). The notion of qualifying aspect is discussed more in Chapter IV, where it helps differentiate types of human activity, and Chapter V where it is seen how all aspects of a computer serve its lingual qualification.

Geno-types are usually defined by a second reference to aspects; for example while all social institutions are qualified by the social aspect, a business, church, state are led by the economic, pistic, and juridical aspects, respectively. Pheno-types are subtypes that arise from interaction with other things of a different geno-type. Geno- and pheno-types are of little concern here, though they could be useful in guiding the construction of robust domain ontologies (Chapter VI).

Dooyeweerd explored typology among natural things, and also in social institutions, but not much of other types of thing (an exercise he left to us). He made reference to some things that cannot be fully understood in such ways, including semi-manufactured products, which are explained in Chapter VII, and *Umwelt*, explained in Chapter VIII.

Any superficial resemblance between Dooyeweerd's notion of type laws and Plato's notion of ideal types dissolves when the inherent dynamicity of the former is acknowledged, which allow enormous latitude of variability without any hint of departure from perfection. Dooyeweerd dealt with the tension between individuality and universality, not by positing a realm of ideal perfect types, which denigrates individuals, nor by denying universality as philosophical nominalism does, but by placing universality within the law side and individuality in the subject side. A true universal can never arise purely from study of the subject side. [Strauss (2000, p. 21, footnote 2) however, who dislikes nominalism, criticises Dooyeweerd for being too nominalistic.]

3.2.6 Relationships

"In veritable naïve experience," said Dooyeweerd (1984, III, p. 54), "things are not experienced as completely separate entities." In this, Dooyeweerd was signalling his agreement with existentialism that we can never understand a thing without reference to its context, but he offered a fresh approach and new conceptual tools for understanding relationships a thing has with this context.

First, Dooyeweerd differentiated two contexts—law and subject side—while existentialism largely conflates them. Relationships with the law side take the form of law-subject relationships, in which the subject responds to law-and-meaning that enables. Subject-side relationships are of several types, some functional and some structural.

3.2.6.1 Functional Relationships

Functional relationships that Dooyeweerd discussed include the subject-object, subject-subject, and *Gegenstand* relationships. All functional relationships are enabled by agents functioning in the aspects, subject to aspectual law. As has already been mentioned, Dooyeweerd's subject-object relationship is conceived very differently from the conventional, Cartesian one.

What is perhaps even more interesting is that his notion of law spheres gives grounds for understanding subject-subject relationships, where both entities function in the same aspect as subject. If I understand Dooyeweerd aright, then it is in subject-subject relationships that genuine interaction between entities occurs. For example, for communication to take place, both speaker and hearer must function as subjects in the lingual aspect. Subject-subject and subject-object relationships are useful in considering human-computer interaction in Chapter IV and the nature of computers in Chapter V.

Gegenstand relationships are similar to subject-object relationships but, while a subject-object relationship involves intimate engagement between subject and object, the *Gegenstand* relationship involves distance between them because it is not the object as such with which the subject relates but an abstracted aspect of the object. The most common *Gegenstand* Dooyeweerd discussed was that of the analytic aspect (see the following), but he hinted (1984, II, p. 275) that there are *Gegenstand* relationships involving other aspects. This will be useful in understanding distal HCI in Chapter IV.

3.2.6.2 Structural Relationships

A structural relationship is one that contributes to the Being of a thing as that thing. There is a degree of necessity in structural relationships that is absent from functional relationships. The best-known structural relationship is the part-whole relationship (or system-subsystem), but Dooyeweerd gave it a new twist because he viewed it from the point of view of cosmic meaning rather than from one of structure (Being) alone.

Things function as meaningful wholes, not as parts. It is not the brain that thinks or feels but the person. The brain only functions in aspects in which it can meaningfully be seen to be a whole in its own right, such as the physical. The psychic aspect is not one of these, so the brain cannot feel. This is referred to in Chapter V.

In Praxiteles' sculpture, *Hermes and Dionysus* (Dooyeweerd, 1984, III, p. 110), a part-whole relationship exists between the torso, head, limbs, and so forth, and the body, and between eyes and head, but it does not seem right to say that calcium carbonate molecules of which the marble is composed are parts of the torso of Hermes, nor that the limbs are part of the piece of marble. Likewise, while the computer, motherboard and memory chips and components form a true part-whole hierarchy, and so do data structures, modules and program, we cannot say that data structures are part of the motherboard.

Why not? Dooyeweerd accounts for this intuition by defining parts as having the same qualifying aspect as their whole. The part has no meaning as a part without reference to the whole (Clouser, 2005, p. 287). For example, my pancreas is meaningless if separated from me, even though its physical functioning can be made to continue in a laboratory. Aggregation of parts into larger parts and eventually a whole only takes place within an aspect. This will be important in separating out bits and bytes from characters and words in Chapter V.

What, then, is the relationship between the block of marble and the torso, head, eyes, and so forth? Dooyeweerd borrowed the word "enkapsis" from biology for these, and extended its meaning. In an enkaptic relationship, two or more wholes are joined in a structural relationship in which both are necessary. He identified five types of enkapsis:

- **Foundational enkapsis** is that which occurs between meaningful wholes and the same thing viewed from a particular aspect, such as the sculpture and the block of marble from which it is made. As will be seen in Chapter V, a piece of data and its bit pattern are also related by foundational enkapsis.
- **Subject-object enkapsis** is exhibited by a hermit crab and its shell.
- **Symbiotic enkapsis** is exhibited by clover and its nitrogen-fixing bacteria.
- **Correlative enkapsis** is the relationship that exists between an *Umwelt* (environment, such as a forest) and its denizens. It is useful in understanding our technological ecology (Chapter VIII).
- **Territorial enkapsis** is the relationship between, for example, a city and its university, orchestra, or football team.

It may be that there are other types of enkaptic relationship that Dooyeweerd did not conceive of, especially to be found in information systems.

Enkapsis is an insight unique to Dooyeweerd, and is useful for formulating frameworks for understanding in almost every area of research and practice in IS.

3.3 Experience, Knowledge, and Assumptions

In the light of Dooyeweerd's approach, the whole epistemological problem of knowledge, experience and assumptions had to be re-examined, including even the questions that traditional philosophies had asked about it. He wrote in the context of an age-old assumption that theoretical thought is superior to everyday experiencing as a route to knowledge, and so had first to spend considerable time showing how the problems of epistemology had been approached in the wrong way.

In Volume I of Dooyeweerd (1984) he argued, by means of immanent critique, that immanence-philosophy has always presupposed the autonomy of theoretical thought. Part II of Volume II, entitled "The Epistemological Problem in the Light of the Cosmonomic Idea" presents a transcendental critique of what makes a theoretical attitude of thought possible, which is briefly outlined next and how it relates to intuition and truth. But the problem of knowing may also be approached via Dooyeweerd's positive philosophy, as a type of aspectual functioning.

Coming to know things is important in knowledge elicitation in IS development (Chapter VI) and, in Chapter VII, for encapsulating the diversity of the world in computers. The supposed opposition of objectivism and subjectivism in much recent IS research makes these difficult.

Here we consider the roots of this problem, how Dooyeweerd addresses it by reference to the status of the knower among what they know, intuition, theoretical thinking (including science and philosophy) and presuppositions.

3.3.1 Objectivism and Relativism/Subjectivism

In IS the Burrell-Morgan model of sociological paradigms has had considerable influence (see Chapter VI for discussion thereof). It opposes what it calls objectivism and subjectivism, the apparently incommensurable beliefs that there is a world "out there" which we can know versus that we can never reliably know the world but merely have subjective interpretations, or social constructions, of it.

This view is rooted in Kant's "Copernican revolution", in which he drove apart thought and thing, phenomenon and noumenon. Kant's insight was that absolutely reliable knowledge of the world is impossible if humanity is part of a determined universe but equally impossible if humanity is completely free. Knowledge requires interpretation of the immense variety of sense impressions we receive. In the first case, interpretation is merely a determined physical process in our brains, so we can never be sure that this yields valid knowledge. In the second case, interpretation is free and thus might completely mislead us about the world (if there even is a world), as do visual illusions. Western thought lives in the shade of the Kantian gulf between thought and thing. As a result, the individual is infinitely lonely and isolated in their attempts to know the world, and in science and elsewhere is dominated by what Tarnas (1991, p. 366) calls a "secular skepticism".

This stance is obviously at variance with everyday experience which, though often denigrated as misleading, nevertheless continues to appeal quietly to our spirits. As a result, several attempts have been made to overcome the Kantian gulf.

One tries to ameliorate the infinite loneliness by the notion of intersubjectivity—that there is a shared knowledge of the world which is somehow more "true", or at least legitimate, than our individual interpretations. As outlined in Chapter I, Husserl (1970) introduced the notion of lifeworld as that which gives meaning to the terms that scientists use in their attempt to know the world, and tried to find a "transcendental method" to penetrate to the essence of things—but many question whether he succeeded in this. As suggested earlier, that lifeworld is the generator of meaning is merely a presupposition.

Bernstein (1983) tries to move "beyond objectivism and relativism" by a slightly different route. He suggests that we need to let "the things themselves" "speak to us", recruits prejudgements to this task, and ends (p. 231) with the call to "dedicate ourselves to the practical task of furthering the type of solidarity, participation and mutual recognition that is founded in dialogical communities." Unfortunately, Bernstein's call is purely speculative and is not worked out.

Yet another attempt is made by critical realism (CR), such as by Bhaskar, of which Mingers (2004) is a prime exponent in the IS field. CR begins by asking the transcendental question, what the world must be like for accepted occurrences to occur and be intelligible, especially to science. It answers this by positing that the three domains of causal laws, of events they generate, and of particular experiences of those events (e.g., experimental observations used in science) may be conceptually separated. For scientific knowledge there must be some regularity in our experiences and this regularity must not be coincidence nor generated by the nature of our experience, so CR posits also enduring entities, physical, social, and conceptual in nature, which have tendencies to act in particular ways. Bhaskar uses the notion of emergence to account for the stratification of types of entity.

CR has been criticised by Klein (2004) on a number of points, the main ones of which are that CR does not adequately acknowledge either subjectivity or normativity. Dooyeweerd briefly discussed CR (1984, III, pp. 44-47) but it was Riehl's version rather than Bhaskar's. To this he admitted some similarity (in denying the Kantian gulf) but argued that they differ in that CR ends up theorizing everyday experience. While CR accepts both ontology and epistemology, more recent versions of CR privilege the latter and end up with impoverished ontologies especially relating to the human and social aspects of everyday experience. These criticisms may be taken in addition to Klein's criticisms. Thus CR is at best of only limited value in the formulation of frameworks for understanding IS. A full immanent critique of CR and comparison with Dooyeweerd must wait for another occasion.

It might also be noted that much of the debate from Husserl onwards, and especially in Bhaskar's CR, has been about how scientific knowledge can be reliable, rather than about how in the everyday attitude we can take the world to be real and develop our intuition. To formulate FFUs for IS that have an everyday orientation requires a rather different approach to bridging the Kantian gulf.

A proper analysis of these attempts to overcome the Kantian gulf needs more rigorous treatment than can be given here. One gets the impression however, not least from the rejection of all such "grand narratives" by postmodernism, that all these attempts ultimately fail.

Dooyeweerd took a completely different route, and argued that the Kantian gulf cannot be resolved under the Nature-Freedom ground motive because it is the meridian expression of the NFGM itself. Under NFGM there is no escape from the driving apart of thought and thing. So his approach to knowledge is based on the different ground-motive of CFR.

3.3.2 The Knower-Known Relationship

Geertsema (2000, p. 96) suggests that one of the major contributions of Dooye- weerd's approach to knowledge and theory has been to put the human knower at the centre:

A long tradition in Western philosophy suggests that the subject of philosophy and science is or should be human reason alone. In our century this view has led to a strong emphasis on method. ... Over against this tradition Dooyeweerd emphasized that it is the human person who thinks, does scientific work, and theorizes.

But Dooyeweerd stressed that the knower is part of what is known, rather than an autonomous (free) ego taking the role of a detached observer. He did not seek to mend the Kantian gulf; rather, he held the gulf to be falsehood from the very start. Geertsema (2000, p. 101) said, of Dooyeweerd's view:

Knowledge and understanding do not start with the subject as if knowledge has to bridge an original gulf between the two. ... To do so we have to ignore that in actual life we experience ourselves in coherence with the world around us. There is no original gap that needs to be bridged. Knowledge presupposes that we are in a relationship already.

Dooyeweerd's view is akin to Polanyi's (1967) that all knowledge is "dwelling" but he makes it more specific.

3.3.3 Knowledge as a Law-Subject-Object Relationship

This may be understood aspectually. To know (or experience, think, reflect) is to function as subject in the various aspects of knowing, and to be known is to function as object in those same aspects (§2.4.4). Table 3-5 gives examples of various aspects of knowing or, as we might say, different ways of knowing; these will be referred to in Chapters VI and VIII. For example, the psychic aspect enables cognitive memory, the analytic enables concepts, critical distance and doubt, the formative, skill, and the pistic, certitude (it is sometimes called the certitudinal aspect).

Table 3-5. Aspects or ways of knowing

Aspect	Ways / aspects of knowing
Physical	Physical knowing is persistent change of physical state resulting from some functioning in the physical aspect. This is the physical 'implementation' of all other types of knowing (e.g. computer memory chips have a persistent electric charge).
Biotic / Organic	The way things have grown, etc. e.g. plant bent towards light 'knows' where the light is. Also the growth of nerve connections.
Psychic / Sensitive	a) Memory. Receiving stimuli and holding a memory of them in the nervous system. b) Recognition of a pattern (seen or heard) c) Instinct (of the animal kind).
Analytic	a) Making distinctions between things. b) Conceptualizing. c) Making inferences from those distinctions; reflection; what is deducible from what I already know. d) Theorizing.
Formative	a) Knowledge of structure; 'knowing my way around'. b) Skills: knowing how to achieve things.
Lingual	a) Discourse, debate that sharpens and disseminates. b) Stuff set down in symbolic form, e.g. 'knowledge' stored in books, libraries, records, archives, web sites.
Social	a) Buber's 'I-Thou' encounter, but see Ethical aspect. b) Shared cultural knowledge, assumptions. c) Networks of knowledge.
Economic	Managing limits on knowledge (personal and communal memories, etc.). assumptions.
Aesthetic	Harmonizing what we know with what else is known, and with what we experience in life. That what we know 'fits comfortably'. That insight. Example: Habermas' triples all harmonize. How an artist helps us understand reality. and communal memories, etc.).
Juridical	Giving due weight to various pieces of knowledge and to the whole; proportion and a sense of 'perspective', an informed sense of the essence of things.
Ethical	A complete 'entering in' to the other person, in Bergson's sense, is only possible with complete self-giving. Hebrew in Genesis 4:1 the word "he-knew" for 'have intercourse with'. Buber's I-Thou relationship contains at least an element of self-giving.
Pistic	Certainty. Committing to a belief, both the little commitments in everyday living and the large commitments for which we might lay down our lives. Also prejudice etc.

It should be noted that we are not, here, talking about the diversity of what we can know about but about knowing itself. The first three aspects are missing because knowing implies irreversibility, which enters the meaning-scheme only with the physical aspect. Some ways of knowing are what we often call experiencing.

In Descartes and Kant, the knower-known relationship is one of distance, based on *Gegenstand* (even though it is conventionally called "subject-object") and the known thing is "passive". But in Dooyeweerd, the knower-known relationship is one of multi-aspectual engagement, and the known thing lets itself be known, not helpless against being-known. This affords the known thing more dignity in the relationship, which can be important in knowledge elicitation (Chapter VI). The long-held assumption of a gulf between knower and known has been questioned and the world is now seen as knowledge-friendly, tending to reveal rather than hide itself.

It overcomes the dialectic between objectivism and relativism without denying the insights offered by other thinkers. Taking for example Bernstein's points, it can be seen that Dooyeweerd both affirms their intent but goes further and gives reasons for doing so:

- Objectivism is rejected because of the subject-functioning in the normative aspects of knowing.

- Subjectivism-relativism is rejected because of the object-functioning of the known thing, and because the thinking ego is not autonomous.

- Our prejudgements, which Bernstein presupposes but does not critically examine, are revealed by Dooyeweerd as constituted in the knower's subject-functioning.

- The "things themselves" "speak to us" because of their object-functioning.

- While Bernstein can only point speculatively in the direction of "dialogical communities" and "practical discourse", Dooyeweerd can provide more specific understanding, not least in his insight that knowing is multi-aspectual human functioning in which the known-things let themselves be known (i.e., "speak to us").

3.3.4 Non-Neutrality of Knowledge: Life-and-World-Views

Around the 1970s many were arguing that scientific knowledge is never neutral, ranging from historians like Kuhn to philosophers like Habermas and Foucault. Dooyeweerd also argued this—but 20 years earlier, and even before World War II, which might be why his ideas are not widely known. But Dooyeweerd still has something to contribute because he comes from a different direction. This has important consequences for what we take to be knowledge, appropriate ways to obtain it, and appropriate ways to use it. Even more importantly for those who seek to understand everyday experience and living, it has important consequences for how we live and experience things, such as in management and the practice of IS.

As mentioned in Chapter II, Clouser (2005) credits Dooyeweerd with being the first to take seriously self-performative coherence. Whereas thinkers like Habermas and Foucault were influenced by Nietzsche to assume it is power that lies at the root of non-neutrality, Dooyeweerd held that the non-neutrality is religious because it is the human being who theorises and human thinking is governed by our fundamental view of the nature of reality, especially as exhibited in our ground-motives. In Chapter II it was suggested that among four types of religious function that might lie at the root of research and practice in a community or area, one is the aspectual life-and-world-view (LWV).

LWVs were discussed at length by Dooyeweerd (1984, I, pp. 114-164). He found them to be often centred on an aspect, giving examples of intellectualism, aestheticism, mysticism, and so forth. Such LWVs then exercise a strong guiding influence on intellectual thought of a community because they embody the deeper assumptions, aspirations, quality criteria, and so forth. Frequently the editorial policy of a journal espouses one LWV or another, and this influences the discourse of the community so that certain aspects are privileged while others become ignored, exacerbating the imbalance in the discourse and research of that community.

In management, for example, people's roles, what they emphasise and the perspectives they take during discussion, tend to centre around aspects, as suggested in Table 3-6.

But perspectives and especially LWVs are usually more subtle, involving several aspects and yielding aspectual profiles. Figure 3-4 illustrates some feasible aspectual profiles of various LWVs expressed by the given statements.

Table 3-6. Aspectually-centred perspectives and roles

Aspect	Emphasis (e.g.)	Role (e.g.)
Quant'ive	Number	Accountant
Spatial	Distance, layout	Geographer, draftsman
Kinematic	Movement	Transport planning
Physical	Energy, forces	Energy analyst
Biotic	Health	Nurse, Doctor
Psychic	Emotion, sensory-motor function	Psychologist
Analytic	Logic, Analysis	Analyst
Formative	Achievement, Power, History	Planner
Lingual	Communication, documentation	Communicator, External relations
Social	Social activity, Relationships	Host, Group therapist
Economic	Resources, frugality	Manager, Economist
Aesthetic	Harmony, Fun	Coordinator, Clown
Juridical	Due, Contracts	Lawyer
Ethical	Generosity	Charity worker
Pistic	Vision, Loyalty, Identity Religious activity	'Champion', Padre

Figure 3-4. Aspectual profiles of life-and-world-views

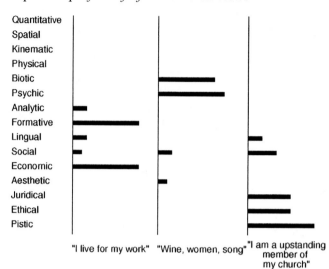

Understanding the aspectual profile, whether of everyday life or of a professional research community or journal can help in formulating frameworks for understanding the area concerned because it provides a way to understand the values and assumptions that drive both research and practice in the area. That LWVs are religious in nature implies pistic commitment, which is useful in understanding conflicts, in Chapter VI.

But LWVs are not static. Individual profiles and those of small groups can be modified because of pistic functioning, by life experiences or environment (the impact of IT as an environment is considered in Chapter VIII). But also, when an LWV is strongly reductionist, which many scientific ones are, it can become suddenly overthrown because all aspects pertain, including those ignored by it (§3.1.4.1). One example is the sudden shift from behaviourist to cognitive psychology in the 1970s. Another is the various dialectical swings in environmentalism (Basden, 1999).

3.3.5 Intuition

Intuition is in the ascendency in IS, so it is important to understand what it is. Dooyeweerd discussed the relationship between intuitive and theoretical or analytical thinking at some length (Dooyeweerd, 1984, II, pp.472-85 and elsewhere). He commented on ideas by Kant, Husserl, Fichte, Schelling, Volkelt, Bergson, Riehl, and Riemann, and also discussed the recognition given to intuition by Greek, Hindu, and Scholastic thinkers. Bergson, for example, contrasted intuition to analytical thought, whereas Dooyeweerd maintained that they cannot be divorced from each other, Bergson characterised intuition as "entering into" objects and persons, identifying with them, and it is by intuition rather than scientific knowing by which we know our deepest selves and our duration. But Dooyeweerd's account is richer. Dooyeweerd can offer us three accounts of what are often called intuition (though he himself did not set these out as systematically as here):

•   Multi-aspectual ways of knowing the subject-side
•   An intuitive grasp of law-side aspectual meaning
•   The immediate experience of self

Intuition as multi-aspectual knowing of subject-side things is constituted of functioning in every aspect—in a coherence of all the aspects of knowing mentioned earlier. Notice how it includes, for example, pistic certitude, which is necessary for making assumptions, which are so characteristic of intuition. But it also includes analytic knowing, and thus in Dooyeweerd intuition and everyday thinking can involve critical distance.

Intuitive grasp of aspectual meaning is very different. It refers to our inexplicable knowledge of what, for example, justice, amount, and signification are (as kernel meanings of the juridical, quantitative, and lingual aspects). Such grasp always exceeds any attempt to think about them analytically, such as to define them. Contrary to what many believe, this kind of intuition cannot even be reduced to feeling. It is not the engagement of the knowing-subject with known-object but the engagement of the human self with law-meaning.

Those two types of intuition are important in ISD (Chapter VI). Note, however, that intuition is not infallible. Though it may be richer and more reliable than theoretical thought (§3.3.7), intuition is influenced by culture, LWVs, and so forth.

3.3.6 The Human Self

Awareness of the self plays a small part in formulating frameworks to understand IS, in Chapters V and VIII, but, to Dooyeweerd, it is radically different from other types of knowing, and he would not usually call it intuition (though conventional thinking might). Dooyeweerd believed the self—he also called it ego or (human) heart—to be:

•   Trans-aspectual, in being a subject that can respond to all aspects
•   Supra-temporal, in that we have "eternity in our hearts"
•   Religious, in that we orientate ourselves towards an absolute

(In this he was informed by generations of reflective thought on the topic which do not concern us here.) Because of the first, the human self cannot be seen through the lens of aspects and in particular it cannot be penetrated by analytical or theoretical thought (because to do so it must function as object therein, which is not possible). This means there can be no theory, definition nor even conceptualisation, of the human self and that any attempt at such will always be wholly speculative rather than critical. Because of the second, we differentiate awareness of self from the intuition of aspectual meaning, because such intuition is temporal. All we can have is an immediate experience of our own selves.

Because of the third, the human self religiously orientates itself either to the True Absolute (the Divine, the Origin, God) or to some pretend, false absolute. Orientation towards a false absolute distorts all our aspectual functioning. (Does this sound like a re-emergence of the mediaeval spirit-body or Greek soul-body dualism? Dooyeweerd argued it was not the case, but as his argument is complex and not relevant to our task, we need not rehearse it here.)

3.3.7 Analytical and Theoretical Knowing

Dooyeweerd did not react against theoretical thinking and knowledge in a romanticist or anti-rationalist manner, but acknowledged its validity and its special place in Western culture.

But even in theoretical thought the central place of the knower as a human subject is maintained. As Geertsema (2000, p. 96) put it, Dooyeweerd "did not deny the importance of method ... but argued convincingly that theoretical thought cannot be accounted for without considering the 'ego' as the hidden player on the instrument of theoretical thought." This means that, as Geertsema (2000, p. 89) put it, "the object of theoretical thought is not reality as it is given in human experience, but the result of an abstraction and therefore intentional not ontic." This reaffirms that what we reason about is not "objective" reality but our normative conceptualisations.

As indicated previously, a Dooyeweerdian understanding of this important way of knowing may be approached via his positive philosophy, as a way of knowing focused on the analytic aspect. But it can also be approached via his critical philosophy, by means of the two transcendental critiques mentioned in Chapter II. Dooyeweerd tried to understand theoretical thought in terms of the fundamental (transcendental) conditions that make it possible, which does not require prior acceptance of his positive philosophy. The main thing one must accept is the givenness of everyday experience, and a determination not to distort it by making a speculative theory about it first.

Clouser (2005) has provided a particularly clear rendering of Dooyeweerd's second way of critique, and what follows is a summary of this. Clouser sometimes interprets

Dooyeweerd's ideas more narrowly, and he short-circuits some of Dooyeweerd's more tortuous arguments, but his rendering is nevertheless very useful especially in understanding IS development (Chapter VI).

After a discussion of what a theory is (an attempt to explain, either a concrete situation (entity theory) or a general state of affairs (perspectival theory), on the latter of which he focuses), he draws attention to the importance of abstraction, which he defines (p. 64) as "to extract or remove something (mentally) from some wider background." There are three degrees of abstraction, by which the thinker becomes aware of distinct aspects:

- **Everyday thinking:** We distinguish things or events but are not aware of aspects of those things separately; rather, we respond to aspectual laws tacitly in full engagement with reality. This is the multi-aspectual functioning that is intuition.
- **Low-level abstraction:** We abstract certain properties of things or events, such as the cost, colour, or weight of a car, but these properties are still of the thing.
- **High-level abstraction:** We abstract the property from the thing—colour as such, cost as such—and this enables us to formulate laws, of some generality, about how such properties relate to each other, e.g., momentum = mass times velocity.

High-level abstraction involves the analytic *Gegenstand* relationship, though Clouser avoids using that term. *Gegenstand* pulls aspects apart from each other. But since the aspects are interconnected they "resist" being pulled apart, and this, suggested Dooyeweerd, is why theoretical thinking is difficult. This is why aspectual meaning is grasped by the intuition rather than by theoretical thought.

Clouser then went on to argue that this presupposes a religious stance about what is assumed to be self-dependent.

3.3.8 Science and Philosophy

Higher abstraction of properties is necessary for science. To Dooyeweerd (and to Clouser) the role of science is to study the laws of aspects: universality, which science seeks to know, is of the law side. Each aspect is the centre for a distinct arena of science—such as physics, psychology, sociology—because each science works within the horizon of meaning given by that aspect. Since aspects are different spheres of meaning, with different ways of being, different rationalities, and so forth (§3.1.5), each aspectually-centred science has its own distinct idea of what it

Table 3-7. Sciences centred on aspects

Aspect	Science	Research methods
Quant'ive	Arithmetic, Algebra, statistics	Deduction, theorem proof
Spatial	Geometry, Topology	Geometric proofs
Kinematic	Kinematics, Fluid dynamics	Infinitesimal calculus
Physical	Quantum physics, physics, chemistry, mechanics, materials science	Laboratory experiment, with physical reasoning
Biotic / Organic	Physiology, life sciences, biology	Greenhouse experiments, field studies, taxonomic analysis
Psychic / Sensitive	Psychology, Sensory sciences Cognitive sciences	Stimulus-response trials, Control groups
Analytic	Logic, Analysis	Logical proof, Some as for cognitive science
Formative	'Sciences of the Artificial' [Simon] History	Game playing, puzzle-solving, model building, forensics
Lingual	Linguistics, semiotics	Hermeneutics, cognitive studies, model building, theorizing
Social	Social science	Surveys, questionnaires, interviews, model building
Economic	Economics, management science	Statistics, model building, As social science
Aesthetic	Aesthetics	As social science
Juridical	Juridics, Legal science	Review of cases and histories, Reflection, legal deduction
Ethical	Ethics	Attitude surveys
Pistic	Theology, Some anthropology	Reference to sacred writings, hermeneutics, theorizing, anthropological studies, dogmatics

is meaningful to research, research methods, criteria for truth, and so on. Table 3-7 shows various sciences corresponding to each aspect. It indicates that each requires a distinct research method: the positivist assumption that methods from physical sciences may be applied in all is untenable.

What is important in formulating frameworks for understanding IS is that "The theoretical object of scientific thought can never be the full or integral scope of reality" (Dooyeweerd, 1999, p. 93), so scientific research finds it very difficult to be interdisciplinary in the sense of embracing several aspects. It is, rather, the role of philosophy to reflect on this integral scope. It is the role of philosophy, not science, to address diversity and coherence, origin and destiny, being and norms, good and evil, the nature of science itself, the diverse nature of knowledge and truth, of the relationship between theory and practice. This is why philosophy is found useful in addressing the interdisciplinary areas of IT. Philosophy like science depends on higher abstraction, but it is not centred on individual aspects. Rather, it concerns itself with the connections between aspects, and thus requires an Archimedean Point, a viewpoint outwith the aspects from which to view them.

This, incidentally, also provides a Dooyeweerdian justification for the choice in Chapter I of what this work takes philosophy to be about. Some other views of what philosophy is base it on one or other of the basic sciences, such as logic or mathematics, but "The very notion that philosophy is founded upon self-sufficient basic sciences is rooted in the immanence standpoint" (Dooyeweerd, 1984, I, p. 544). He argued that most streams of philosophy failed in these points (and listed those that did).

3.3.9 Limits to Knowing

Neither science nor philosophy are neutral, but both are strongly influenced by world views and religious presuppositions. So neither of them should be treated as the avenue to absolute truth.

We can understand this non-neutrality aspectually. Though science and philosophy are primarily analytical processes, they are nevertheless multi-aspectual human activity. Neither would progress well without, for example, recording and dissemination (lingual functioning), debate and social groups, resources (economic), the desire to do justice to the aspectual laws (juridical), and (pistic) vision, among other aspects. What it means to be human is discussed next. None of our aspectual functioning is absolute (see §3.1.4.4), including none of our ways of knowing. We cannot, and should not, ever hope for "truth" in the way objectivism has conceived it.

Dooyeweerd very explicitly stated that, even in the ideal, "There is no truth in itself" (Dooyeweerd, 1984, III, p. 577). Whereas subjectivism questions the notion of truth, Dooyeweerd questions "in itself"; i.e., self-dependent and able to stand as truth without reference to anything else. It is not that there is no truth, but there is no truth "in itself", because all meaning refers beyond itself to its Origin. "Hypostatized 'truth' is a lie," he said (Dooyeweerd, 1984, II, p. 561), "there is no selfsufficient partial truth," and so "$2 \times 2 = 4$ becomes an untruth if it is absolutized into a truth in itself" (Dooyeweerd, 1984, II, p. 572). This means that "our insight is fallible" (p. 574). Non-absoluteness of our knowledge is not an imperfection, but rather a stimulation to humility.

But, as mentioned earlier, we should not despair of thought altogether. Dooyeweerd said (1984, II, p. 556) that thought "remains bound to a modal horizon which has a constant determining character as to all the changing concrete facts." In other words, thought is "friendly" to the Cosmos rather than inimical to it. Reality, he believed, has a tendency to reveal rather than hide itself, to inform us rather than mislead us. This belief sounds unfamiliar to Western ears that are used to what Tarnas (1991, p. 366) has called "secular skepticism" by which we have long believed there is absolutely no connection between knowing and known, and our capacity to be misled is, consequently, infinite. But that is a presupposition as to what the relationship is between thought and thing, and Dooyeweerd happens to have made a different presupposition. These led him to acknowledge the ability of intuition to grasp cosmic meaning more fully than analytic thinking can.

This cautious confidence has implications for knowledge elicitation (Chapter VI).

3.4 Human Life

If the doing of science and philosophy involve, in themselves, part of human life, then it is important to understand this. In Dooyeweerd, "human being" may be understood in two ways: the human person (where we are concerned with characteristics of human living), and the human self or "I". The ontic status of the human self, as trans-aspectual, supra-temporal, and religious, was briefly outlined earlier, but it is an understanding of the living human person that is more relevant to formulating frameworks for information systems. Human living is seen in Dooyeweerd as multi-aspectual agency, active as subject in every aspect, and this has a number of important implications for all areas of research and practice in IT/IS.

3.4.1 Multi-Aspectual Human Functioning

First, it means that human behaviour involves functioning in a variety of aspects (usually all of them). Aspectual functioning does not refer to different parts of such behaviour, but to different ways in which it occurs meaningfully. Table 3-8 shows (some of) the functioning in all the aspects while creating a book like this, and may be used as an example when multi-aspectual functioning is important.

Multi-aspectual functioning is not just a bundle of aspectual functionings; there is a coherence of meaning in it, made possible by the inter-aspect relationships and the inter-aspect harmony (§3.1.4.3). It is not only richer than a uni-aspectual view of functioning such as those offered by psychology, linguistics, or economics, but also more "true", in that everything is interconnected, and the meaning of any aspect of our functioning cannot be discerned properly without reference to all the other aspects. This multi-aspectual richness of meaning is important in understanding everyday life (see the following), and this includes especially human use of computers and IS development (Chapters IV, VI).

Table 3-8. Multi-aspectual functioning in writing book

Aspect	Functioning in writing book (examples)
Quant'tive	Word count
Spatial	Size of diagrams
Kinematic	Movement while drawing, writing
Physical	Paper jams in printer!
Biotic	Stuffy air hinders thinking
Sensitive	Moving my fingers. Seeing what I've written.
Analytical	Differentiating between ideas that seem similar
Formative	Structuring my sentences and diagrams
Lingual	Writing and drawing to convey what I intend
Social	Obtaining support and critique
Economic	Size limit on document, Keep backups. Manage time carefully
Aesthetic	Ensure the work integrates; Pleasant style of writing
Juridical	Do justice to topic, readers, publishers.
Ethical	Write generous prose that gives more than is due
Pistic	Ways I see myself.

Note that, despite possible implications to the contrary given earlier, multi-aspectual functioning is not the integration or synthesis of originally-separate aspectual functionings. Rather, it is a whole that is meaningful in a variety of aspectual ways. It is originally integral but may be conceptually separated by reference to aspects. Dooyeweerd used the term "systasis" to denote such integrality-prior-to-separation-into-aspects.

3.4.2 The Human Person

The multi-aspectual approach could inform our ideas of what it means to be "fully human". Table 3-9, which was created by applying the normativity of each aspect to human activity, gives examples, to be referred to when the issue of full humanity is important, such as in IS development (Chapter VI), IS use (Chapter IV) and the gender issue (Chapter VIII). More generally, it may be useful as a checklist for managers, or for assessing the quality of the human process of research in each science.

Table 3-9. Examples of aspects of being fully human

Aspect	.. of being fully human
Biotic / Organic	Healthy
Psychic / Sensitive	Emotionally stable Good sensory-motor functioning
Analytic	Clear thinking, logical Good critique
Formative	Planning, achieving
Lingual	Articulate, multi-lingual; Recording and communicating
Social	Friendly, sociable
Economic	Careful, frugal
Aesthetic	Delightful, enjoying life, interesting and interested
Juridical	Just, proportionate, Good citizen
Ethical	Generous, self-giving or selfish
Pistic	Loyal, of good morale; Vision Orientation to True Absolute.

This multi-aspectuality of the human person precludes a science of human behaviour as such (Dooyeweerd, 1984, II, p. 55). Rather, the various sciences allow us to study various aspects of the human person. This emphatically means that the study of the human person cannot and must not be reduced, as it has often been in the past, to such sciences as psychology, linguistics, or sociology. Anthropology is, instead, a branch of philosophy because it is philosophy which enables us to consider multiple aspects.

Unfortunately, each science not only rightly focuses on a single aspect, but also tends to overlook other aspects, especially the later ones that it anticipates. For example psychology, especially the forms that are most frequently referred to in the HCI community, tends to assume that such diverse human functionings as love, religious adherence, friendship and artistic appreciation can all be viewed as merely mental phenomena. Likewise social science tends to conflate post-social aspects into the social. Each science tends to arrogantly claim the whole of human life to itself. This impoverishes research in IS because the full richness of everyday life is hidden from view.

But if the multi-aspectual view of the human person is kept in mind, this invites the psychologist or sociologist to be always aware that there are later spheres of meaning and law, which retrocipate their central aspect in different ways and whose meaning may be found to impact on that of their aspect by dependency and analogy (§3.1.4). Thus they can begin to differentiate, for example, different types of mental phenomena or social relationships without feeling compelled to account for these differences from within their own special science.

3.4.3 The Shalom Principle: A Useful Approach to Ethics

Human living is multi-aspectual. All but the earliest aspects are normative. Therefore human activity exhibits diverse normativity.

Thus far we have focused on the positive, but the negative meets us at every corner. Our functioning and repercussions in one aspect might be Good or beneficial while in another they might be Evil or harmful. Examples given in Chapter IV show how adoption of IT might improve efficiency yet damage other aspects such as trust (pistic) or social relationships (social). Or vice versa. Analysis of the quality of human situations will be enriched by considering positive and negative functioning in all aspects separately. This is especially useful in understanding success and failure in IS (Chapter IV).

If, as Dooyeweerd believed, the aspects are in harmony (§3.1.4.3), then it is possible in principle to achieve what Van der Kooy (1974, pp. 40-41) calls "simultaneous realization of norms". This leads to what might be called the shalom principle: that if we function well in every aspect then things will go well, but if we function poorly in any aspect, then our success will be jeopardised.

The shalom principle may be, like Habermas' ideal dialogue, a counterfactual ideal, but it is useful as a conceptual tool for addressing the normativity of different areas of research and practice because it provides a basis for understanding the diverse normativity we encounter in everyday life. It has been employed, for example, by Brandon and Lombardi (2005) to address the diverse aspects of urban sustainability. It also prevents false justification of actions that bring benefits in one aspect (e.g., the economic) while bringing harm in others (e.g., the social). The danger that stressing the irreducibility of aspects can lead to fragmentation, expressed earlier, can be ameliorated by the shalom principle.

"The purpose of norms" said the Dooyeweerdian economist Goudzwaard (1979, p. 243), "is to bring us to life in its fullness by pointing us to paths which safely lead us there. Norms are not straitjackets which squeeze the life out of us. I stated as my conviction that the created world is attuned to those norms; it is designed for our willingness to respond to God and each other. If man and society ignore genuine norms, such as justice and restitution of rights, respect for life, love of neighbour, and stewardship, they are bound to experience the destructive effects of such neglect. This is not, therefore, a mysterious fate which strikes us; rather, it is a judgment which men and society bring upon themselves."

Aspectual laws are not primarily a set of social or religious do's and don'ts, but rather a framework within which we live and prosper and bring blessing to the whole Creation.

3.4.3.1 A Practical Device

The shalom principle has been very useful to this author in practical situations. Look at Table 3-4, of aspectual profiles. If the holders of these LWVs were to live precisely in line with their LWVs, then the first would function very well in the formative and economic aspect but very poorly in the aesthetic to pistic, the holder of the second might function well in feeling and eating, but poorly in reasoning, frugality, justice, faith, and the holder of the third might function well in responsibility, ethics, and faith but might downplay both bodily functioning, rationality, and economics.

The aspectual profile, whether drawn or merely held in the mind, can give a very useful indication if any aspects are over- or under-emphasised. Frequently, certain aspects are taken for granted, overlooked, or even denigrated. The shalom principle motivates us to seek to restore some harmony among aspects, down-playing those which are over-emphasised, and drawing attention to, and activating, those which are under-emphasised. That the aspects are grasped by intuition (§3.1.4.10) makes such analysis very natural and acceptable. This is very useful not only when reflecting and providing consultancy advice but also in the very midst of discussion. A number of examples of aspectual analysis occur throughout the framework chapters.

3.4.4 Brief Comparison with Extant Views of Ethics

Compared with most approaches to ethics, Dooyeweerd's allows us to distinguish different types of dysfunction or evil, such as in Table 3-10, which might be useful reference when considering detrimental impacts of IS (Chapter IV).

A number of distinct types of ethics have been assumed in IS but, there is a tendency to absolutely separate each from others, often denigrating some types in the process. But a Dooyeweerdian approach would rather recognise the insight in each and integrate them. For example, MacIntyre's (1985) notion of *eudaimonia* is not unlike shalom, but Dooyeweerd provides a basis for understanding its diversity and coherence. The idea that good is constituted in adherence to aspectual law echoes deontological ethics. But deontological ethics usually refers to humanly-created,

Table 3-10. Dysfunction or evil in each aspect

Aspect	Dysfunction
Biotic / Organic	Disease
Psychic / Sensitive	Emotional instability
Analytic	Confusion
Formative	Laziness
Lingual	Lying
Social	Hatred, disrespect
Economic	Waste, squandering
Aesthetic	Boredom, Ugliness
Juridical	Injustice
Ethical	Selfishness, self-centredness
Pistic	Idolatry, disloyalty

socially-constructed rules and norms while Dooyeweerd refers to deep spheres of law that make such norms meaningful and fruitful. The difference between aspectual law and human rules must be clarified: the latter arise from aspectual functioning in the first place, to obedience to aspectual law, and thus cannot be absolute. Specifically, for example, norms will be distinguished and conceptualised (analytic functioning), developed (formative), expressed and discussed (lingual), and agreed on (social). It is in this sense that Dooyeweerd affirms the concern that partly motivates social constructionism, namely to question the absoluteness of human rules and ideas. The notion of aspectual repercussions echoes functional ethics. The way aspects define what is good and meaningful echoes axiological approaches to ethics. The notion that it is a human being who is subject to all aspects and tends to respond in certain ways echoes virtue ethics. Emancipatory ethics, as pursued by critical social theory, may be seen as shalom (see §4.4.3). And the long timescales of later post-social aspects, especially the pistic, echoes Lonergan's (1992) long-term ethics of longer cycles of decline or creation and healing; these can also be seen in terms of ground-motives. The possibility of such integration of the various approaches to ethics, exploration of which cannot be pursued here, is due in no small measure to Dooyeweerd's notion of law and subject (§2.4.4).

A fundamental philosophical requirement for realizability of normativity is that the basis on which we make normative judgement must be the same as that which enables and empowers us to take corrective action. The Humean and Kantian divorcing of Is from Ought effectively brought much Western thinking into the unfortunate position of breaking this link. But Dooyeweerdian aspects are both what enable us to function and also define our norms. So Dooyeweerdian ethics are, in principle, realizable—though a false orientation of the self in practice prevents this (see Chapter VIII).

To liberal Western individualism, law is seen as constraint or oppressor, but to Dooyeweerd, aspectual law is an enabler, without which nothing would be possible. However such a view is not unknown in mainstream thought; see for example Giddens (1993, p. 129ff/B2).

3.4.5 Everyday Life: The Lifeworld

Perhaps the most attractive feature of this Dooyeweerdian approach is that it enables us to consider the rich normativity of the lifeworld. We are continually responding to the norms of all the aspects all the time, but not always aware of them as norms. For example, when I write my computer programs, I am continually sensitive to such things as:

- "I'd better make this easier to understand." (lingual norm)
- "If I do this, I'll make it more efficient." (economic norm)
- "I'm tempted to take a short cut here, but that would not do justice to what I'm trying to express here." (juridical norm)

—though I might be unaware of them as norms and sometimes go against them.

The need to understand and be sensitive to the everyday lifeworld of research and practice in each area has been emphasised throughout. Some characteristics of the everyday lifeworld were outlined in Chapter I, with implications for lifeworld-oriented frameworks for understanding. How compatible is Dooyeweerd with these and how may Dooyeweerd contribute to such lifeworld-oriented frameworks for understanding? Here we draw together some implications of his thinking to enrich the characteristics outlined earlier.

Though Dooyeweerd did not explicitly set out a theory of the lifeworld, and never actually used that term, preferring instead the words "naïve", "everyday", or "pre-theoretical", he referred to it continually throughout (1984). In almost all places at which he referred to this, it was to its structure rather than its content. (Lifeworld content is actual assumptions prevalent in a culture; structure is nature and characteristics of lifeworld; see Chapter I.) This gives his deliberations a cross-cultural relevance that is second to none.

Dooyeweerd's thought can affirm most of the characteristics of the lifeworld outlined in Chapter I, and can clarify and enrich them. A full discussion of this must wait for another occasion, but the following is a brief summary (which contains references to much of the material previously mentioned):

- The lifeworld as intersubjective stock of shared experiences is affirmed by the Dooyeweerdian notion of multi-aspectual ways of knowing, and clarified in differentiating between meanings we attribute and cosmic meaning, of which we have intuitive grasp.

- The lifeworld as "pre-given reality" is affirmed by Dooyeweerd in that the human ego is not autonomous but we are part of the reality that we intersubjectively experience, and may be clarified in differentiating two types of "givenness": subject-side and law-side givenness. The subject-side is contingent and open, but it is "given" in the sense that it is actualised in time, and our knowledge of it is constituted in a law-subject-object relationship. Dooyeweerd called it the "plastic horizon" of the naïve attitude (1984, III, p. 31). The law-side is "given" in the sense that it transcends us; Dooyeweerd called this the "modal" or "transcendental" horizon.

- That lifeworld is engagement is affirmed by Dooyeweerd's law-subject-object relationship, and enriched by his diversity of such relationships and by the possibility that critical thought can be part of it. Dooyeweerd stressed that "naïve experience ... does not know of a *Gegenstand*" (1984, II, p. 431).

- That lifeworld cannot be explicitly known is affirmed in Dooyeweerd by his picture of reality resisting being pulled apart by analytical thought, and may be clarified by differentiating between intuitive grasp of aspectual meaning and intuitive experience of things.

- The difference between the lifeworld and theoretical attitudes is affirmed by Dooyeweerd's exposition of theoretical thinking as *Gegenstand*, especially as explained by Clouser (2005), and enriched by the idea of lower abstraction.

- That lifeworld is characterised by meaning and normativity is affirmed by the idea that cosmic law-and-meaning pervades all and may be enriched by Dooyeweerd's drawing attention to the diversity of types of meaning-and-law.

But Dooyeweerd can also address the problems with extant discourse noted in Chapter I. Some concern was noted that most discourse has been about the lifeworld as shared experience rather than about the everyday world and living as such. The root of this emphasis probably lies in the Kantian gulf between knowledge and world, which tends to make us focus on one at the expense of the other; this gulf is itself rooted in the Nature-Freedom ground-motive. Because Dooyeweerd rejected this presupposition, he felt under no obligation to divorce knowledge from world, and thus his interest was wider, extending beyond everyday experience to the everyday attitude as such and the everyday world of which we are part. This re-integrates epistemology and ontology and Dooyeweerd had, in his notion of aspects, a basis on which the diversity and coherence of this world could be made visible as it presents itself to us. This, along with his transcendental critiques, enabled him to see more clearly the relationship between everyday and theoretical attitudes of thought. Finally, while much extant discourse has assumed that lifeworld meaning is generated by intersubjective activity, Dooyeweerd has shown how both are made possible by a transcending law side, and whereas phenomenology sought to remove the "lens" through which we view the world, Dooyeweerd argued that the "lens" cannot be removed and is religious in nature.

Finally, Dooyeweerd discussed whether modern life, which is so heavily invested with the results of theoretical scientific and technological activity, may still be seen as everyday experience? This is a crucial question for this work, of formulating lifeworld-oriented frameworks for understanding IT/IS, because it questions the validity of our whole exploration. Dooyeweerd concluded (1984, III, p. 31) that "The naïve attitude cannot be destroyed by scientific thought. Its plastic horizon can only be opened and enlarged by the practical results of scientific research." That is, the results of theoretical activity can themselves be used in an everyday manner. Everyday life can indeed involve IT/IS.

3.4.6 Time, Destiny, and Progress

Dooyeweerd's theory of time is complex and most of it does not concern us here. At one time he thought in terms of four "dimensions" of reality: the subject, law, time, and religious root. We have already seen (§2.4.4) that the subject dimension—subject-side actuality—always depends on the law dimension. The law dimension in turn depends on that of cosmic time, and all in turn depend on religious root. So "theory of time" is a misnomer, since we can never actually make a theory about it.

However, Dooyeweerd did discuss several features of cosmic time. One is that cosmic time expresses itself in every aspect. What Westerners habitually think of as time, clock time, is physical time. But two other types of time concern the discussion here, formative time, which is historical events that result from our formative functioning, and pistic time, which is concerned with the entire temporal being of the cosmos, from beginning to end.

Pistic time involves the Origin and Destiny of the whole cosmos (and makes sense especially under the CFR ground-motive). Aristotle's "final cause" might be seen as an echo of the Destiny, but Dooyeweerd's notion is richer and allows for dynamic response and an open future. What is relevant here is that humanity, and all else in the cosmos, has a mandate, a Destiny. This belief pervaded all Dooyeweerd's thought even where it is not explicit. If Dooyeweerd was right, then IT/IS also has a Destiny—and this has in fact pervaded three decades of the author's work in all these fields, as an intuition rather than as an explicit theory.

Pistic time is related to formative or historical time. History, Dooyeweerd believed, has a normative direction. The formative mandate, or "normative task", of humanity is to "open up", "unfold" or "disclose" the aspects and their potential especially in relation to other aspects.

In our current state of modernity humanity has "unfolded", among other aspects, the analytic aspect as, and via development of, science and critical thought, the formative aspect as technology and technique, and the social aspect as differentiated institutions, the economy as money and markets, and the juridical via democracy. It is likely that much has yet to be discovered and opened up in the later aspects.

What Habermas (1987) means by system may be seen in Dooyeweerd as functioning of institutions in the formative mode of getting things done, especially in the economic and juridical fields. Like Habermas, Dooyeweerd saw no fundamental incompatibility between modernity and life-world meaning and normativity, but the reason modern life has meaning and normativity differs. It is not primarily Habermasian critique of validity claims but that it, along with all else, is enabled by a law-framework that is Meaning. What critique of validity claims does is to contribute to concrete subject-side attributed meaning.

There is, however, a normative direction to progress and modernity as such (Dooyeweerd, 1984, II, p. 275):

The subjective individual dispositions and talents intended are not themselves to be viewed as the normative standard of the disclosed process of cultural development. They ought to be unfolded in accordance with the normative principles implied in the anticipatory structure (of the historical law-sphere).

That is, (1) what has actually happened might not be what should happen, and (2) progress must never be for its own sake, but for the sake of other aspects, especially later ones. Therefore IT/IS as it has developed is not necessarily how it should have developed; it should have been guided by the norms of, for example, justice and ethics, more than it has been. This will be discussed in Chapters VII, VIII.

Problems of modernity may be seen as arising from "closing down" rather than opening up aspects and their relationships. Habermas' (1987) notion of colonisation of lifeworld by system may be seen as the "closing down" of the lifeworld around the formative aspect (instrumental functioning). The formative aspect is elevated above others and shalom is prevented. This hinders both IT/IS and humanity in realizing their Destiny.

3.5 Critique of Dooyeweerd

3.5.1 Superficial Critiques

Sadly, most of the criticism levelled at Dooyeweerd has been from fellow Christians who mistook his philosophy as an attack on aspects of their theology or even on theology as a whole. With Dirk Vollenhoven who was also working on a similar approach to philosophy, Dooyeweerd was once accused on heresy and brought before a religious court. An interesting story, which reflects the radical status of Dooyeweerd's thinking, but most of the criticisms from the Christian community are of little relevance to this work.

There is also a tranche of criticism that reflects back more on the pre-theoretical stance of the criticiser than being an actual criticism of Dooyeweerd as such. The comment of one colleague, "It's too essentialist for me," is one such—he did not even attempt to understand that Dooyeweerd is no essentialist, but reacted against an immediate impression arising from his own adherence to the Freedom pole of NFGM than from any real attempt at critique. Such criticisms are of little value.

Several people have criticised Dooyeweerd for using abstruse or ambiguous terminology or otherwise being difficult to understand. Such criticisms are valid, but do not affect the substance of his thought.

Some criticisms reflect more on those who have used Dooyeweerd's thought than on the thought itself. Klapwijk sees Dooyeweerdian thought, and especially how it has been worked out in America, as too antithetical to other types of thought. But this seems to be the fault of those who adopted his thought; Dooyeweerd's intention was to engage. Choi points out (2000, §331) "Dooyeweerd is also open to positive fruits of non-Christian cultures seen as gifts of God's common grace. For instance, he considers the Roman *ius gentium* as such a gift. Thus it is not correct to say that Dooyeweerd is totally negative to non-Christian cultures. Rather he is very careful in his assessment of other cultural heritages."

A number of criticisms are of Dooyeweerd's failure to address certain issues. Nash (1962), for example, noted that Dooyeweerd had yet to engage with Anglo-Saxon thought in the USA and linguistic analysis; his thought has still not engaged with the linguistic turn in philosophy, though it did with historicism. A colleague has suggested that Dooyeweerd does not seem to address issues of democracy, participation, emancipation, and the like. In view of his social theory and juridical theory, one might have expected much more discussion of these things.

It is also unfortunate that Dooyeweerd did not engage directly with social constructionism, which is so important in many areas of information systems. Many Dooyeweerdian thinkers instinctively react against it (seeing it as a denial of reality, even the reality of God!). But, in this author's view, the two are largely compatible once one differentiates law from subject side. In practice, social constructionism seems to concern itself with subject-side occurrence, which Dooyeweerd himself stressed was open-ended and highly plastic, and as having an important social dimension. Moreover, the importance of discourse in social construction has its place in Dooyeweerd, in the lingual aspect. It is the law side which transcends us. Extreme social constructionists will no doubt dislike any whiff of that, but philosophically, if one challenges them how social construction is itself possible, one must fall back on something exhibiting similar characteristics to Dooyeweerd's law side. The contact between these two streams has yet to be seriously explored.

It has also been suggested that while Dooyeweerd's theory of internal structural principles is excellent, he did not give much attention to the individual that actually occurs as a response to that law. This would seem not just a lack but a real gap. This is what social constructionism concerns itself with.

Dooyeweerd's suite of 15 aspects may obviously be criticised (even though, as argued earlier, it might be the best available). See, for example, Seerveld (1985), De Raadt (1997), and Stafleu (2005). This is an on-going process which Dooyeweerd expressly welcomed (1984, II, p. 556).

3.5.2 Substantial Critiques

However, there have been a number of more substantial philosophical criticisms recorded from within the Dooyeweerdian community itself, which are here summarised. A discussion of a number of them may be found in Choi (2000), to which this author is indebted.

Dooyeweerd's transcendental critique of theoretical thought comes in for several criticisms. While Klapwijk (1987) applauds it for making "the structure of theoretical thought transparent," he suggests that his transcendental epistemology and cosmology form a vicious circle. This might reflect a misunderstanding on Klapwijk's part that arises from ambiguity in Dooyeweerd's writings; Clouser's (2005) clearer rendering of Dooyeweerd's transcendental critique does not seem to exhibit this vicious circle, though Klapwijk has yet to comment on whether this is so.

Choi (2000) criticised Dooyeweerd's attempt at cultural critique for being too abstract and theoretical and orientated mainly to the state and society than to culture as such, and not attracting the attention from other scholars that it deserved. Choi nevertheless believes (§333) "It gives marvellous insight into understanding the root and dilemma of Western culture throughout its history ... but also offers insight into the possibility of reforming or transforming it from a Christian perspective." Choi's criticism was thus only that Dooyeweerd did not work on this issue as fully as Choi had wished.

Klapwijk points out that Dooyeweerd and others (e.g., Vollenhoven) are adamantly opposed to "synthesis-philosophy" (NGGM) while, in another place, Dooyeweerd argues that synthesis-philosophy is impossible. How can one be opposed to what is impossible? But this apparent inconsistency might be explained if Dooyeweerd was working at two levels, the first as a human being with belief-commitments, the second as an analytical thinker.

Olthius (1985) criticises Dooyeweerd's notion of the supra-temporal self as being too like Plato's notion of an eternal realm of Forms, and as leading to dangerous duplication in his philosophy. It is not clear what response was made to this.

Dooyeweerd's view of time and progress is criticised by McIntire (1983) from several angles, the root of which is that he believed Dooyeweerd had conflated the problem of unity and diversity with that of time—but that might be seen as a philosophical insight of Dooyeweerd's rather than a deficiency. More relevant to us, Klapwijk (1987, p. 123) criticised Dooyeweerd's concept of culture and progress as the unfolding of aspects as "a speculative product of German idealist metaphysics of history" which has "romantic-organismic, progressivistic and universal-historical connotations," rather than as truly emerging from his main thought. "Dooyeweerd continued to espouse the basic idea of a universal-progressive process of disclosure that in one way or another eventuates, as it turns out, in modern *Western* culture." This is a little unfair because Dooyeweerd himself distanced himself from such a view, but the slope of this part of his thought seems always towards modernity.

But there has been very little, if any, criticism of the portions of Dooyeweerd's thought that are of most interest here—including the primacy of meaning, the non-Cartesian law-subject-object relationship, the exposing of religious roots, the notion of aspects as spheres of law and meaning, the approach to things, and the respect for everyday experience. As far as this author is aware, critique of these has yet to occur. Maybe the proposals in this work will stimulate application of Dooyeweerd's thought which will then lead to critique.

Dooyeweerd himself welcomed criticism, and the criticism levelled at his first *magnum opus* (1935) led to significant revision of his thought. But even after this he was relatively modest about his philosophy, acknowledging that much of it stood in need of critique and refinement, and he was rather disappointed that it received so little good quality attention. The reason for his modesty is to be found partly in his Christian faith, that sees humility as a virtue, but is also to be found in his philosophy itself, which stresses that philosophy is not some avenue to absolute truth (see §3.3.9).

There has, unfortunately, been very little if any substantive criticism of Dooyeweerd from the communities of mainstream philosophy, nor even any attempt to engage with Dooyeweerd despite its radical approach. This might be because people assume, wrongly, that because he spoke about a "Christian philosophy" his thought must be irrelevant. But it is probably also due to the characteristics of the current streams of philosophy themselves that limit their capacity to understand Dooyeweerd immanently. Analytic philosophy finds most of the issues he dealt with meaningless. Likewise positivism. Postmodernists will be put off by what they (mistakenly) dismiss as a "grand narrative" running though Dooyeweerd's work. Neo-Kantian thought will be upset by Dooyeweerd's severe criticism of it (despite his obvious admiration for Kant as such). Phenomenology and existentialism might empathise with much of what Dooyeweerd was talking about, but will not be able, from within their own thought, to truly understand his separation of law from subject side. Critical theory might like Dooyeweerd's recognition of transcending normativity and the possible enrichment of their notion of emancipation (see §4.4.3), but Habermas' (1986, 1987) theory of communicative action is so unquestioningly accepted now (almost an absolutization of the lingual aspect) that critical theorists are unlikely to recognise the validity of the other spheres that engaged Dooyeweerd.

The problem is that Dooyeweerd worked at a level deeper than the wars between Humanism and Scholasticism, between pre- and post-Kantian thought, between positivist, interpretivist, and criticalist thought, between modernism and postmodernism. Until that is understood, no serious critique from mainstream philosophy is possible. It will probably take someone of the calibre of Jürgen Habermas, who, like Dooyeweerd, has engaged in careful immanent critique of other thinkers and transcendental critique of issues, to sensitively yet powerfully get to grips with what Dooyeweerd was trying to do, understand it in its own terms, not get sidetracked by secondary issues with which Dooyeweerd's text is sprinkled, and on that basis make a real critique.

3.6 Conclusion

This chapter has explained and discussed some portions of Dooyeweerd's thought which will be used in formulating frameworks for understanding research and practice in IS. The next five chapters apply them along with a few notions introduced in Chapter II. It should be clear that all the main branches of philosophy are represented in Dooyeweerd:

- **Ontology:** His approach to things and at a deeper level his general theory of modal aspects help us understand the nature of computers (Chapter V) and the diversity of a world to be represented (Chapter VII).

- **Epistemology:** His approach to knowing, including his re-examination of the epistemological problem itself helps us understand how we know that diversity, and also how IS development can be enhanced (Chapter VI).

- **Philosophical Ethics:** The intrinsic normativity of the aspects as spheres of law, and the shalom principle, help us understand the difference between IS success and failure (Chapter IV) and the wider problems of the information society (Chapter VIII).

- **Methodology:** The normativity of the aspects as guidance for the future contributes to ISD (Chapter VI).

- **Philosophical anthropology:** Multi-aspectual human functioning as human living, plus Dooyeweerd's notion of self, ensure a full everyday understanding of IS use (Chapter IV), the human process of ISD (Chapter VI) and the relationship between humanity and technological ecology (Chapter VIII), and also sheds new light on the artificial intelligence question (Chapter V).

- **Critical philosophy:** His immanent critique of thinkers and his transcendental critiques to expose the nature of theoretical thought, including its religious root as presuppositions, give a basis for engaging with extant thought in all areas, so that extant ideas need not be completely uprooted but rather transplanted to a new ground-motive, resulting in their enrichment.

References

Adam, A. (1998). *Artificial knowing: Gender and the thinking machine.* London: Routledge.

Basden, A. (1999). Engines of dialectic. *Philosophia Reformata, 64,* 15-36.

Bernstein, R. J. (1983). *Beyond objectivism and relativism: Science, hermeneutics and praxis.* Philadelphia: University of Pennsylvania Press.

Boden, M. A. (1990). Escaping from the Chinese room. In M. A. Boden (Ed.), *The philosophy of artificial intelligence* (pp. 89-104). Oxford, UK: Oxford University Press.

Brandon, P. S., & Lombardi, P. (2005). *Evaluating sustainable development in the built environment.* Oxford, UK: Blackwell Science.

Bunge, M. (1979). *Treatise on basic philosophy, Vol. 4: Ontology 2: A world of systems.* Boston: Reidal.

Choi, Y.-J. (2000). *Dialogue and antithesis: A philosophical study of the significance of Herman Dooyeweerd's transcendental critique.* Unpublished doctoral thesis, Potchefstroomse Universiteit, South Africa. Subsequently published 2006 Norwalk, CA: Hermit Kingdom Press.

Clouser, R. (2005). *The myth of religious neutrality: An essay on the hidden role of religious belief in theories* (2nd ed.). Notre Dame, IN: University of Notre Dame Press.

Dahlbom, B., & Mathiassen, L. (2002). *Computers in context: The philosophy and practice of systems design.* Oxford, UK: Blackwell.

De Raadt, J. D. R. (1991). *Information and managerial wisdom.* Pocatello, ID: Paradigm Publications.

De Raadt, J. D. R. (1997). A sketch for humane operational research in a technological society. *Systems Practice, 10*(4), 421-441.

Dooyeweerd, H. (1935). *De Wijsbegeerte der Wetsidee* (The philosophy of the law-idea) (3 vols.). Amsterdam: H. J. Paris.

Dooyeweerd, H. (1984). *A new critique of theoretical thought* (Vols. 1-4). Jordan Station, Ontario, Canada: Paideia Press. (Original work published 1953-1958)

Dooyeweerd, H. (1997). *Essays in legal, social and political philosophy.* Lewiston, NY: Edwin Mellen Press.

Dooyeweerd, H. (1999). *In the twilight of Western thought: Studies in the pretended autonomy of philosophical thought.* Lewiston, NY: Edwin Mellen Press.

Geertsema, H. G. (2000). Dooyeweerd's transcendental critique: Transforming it hermeneutically. In D. G. M. Strauss & M. Botting (Eds.), *Contemporary reflections on the philosophy of Herman Dooyeweerd* (pp. 83-108). Lewiston, NY: Edwin Mellen Press.

Geertsema, H. G. (2002). Which causality? Whose explanations? *Philosophia Reformata, 67,* 173-185.

Giddens, A. (1993). *New rules of sociological method.* Cambridge, UK: Polity Press.

Goudzwaard, B. (1979). *Capitalism and progress, a diagnosis of western society* (J. Van Nuis Zylstra, Ed. & Trans.). Grand Rapids, MI: Eerdmans.

Habermas, J. (1986). *The theory of communicative action: Vol. 1. Reason and the rationalization of society* (T. McCarthy, Trans.). Cambridge, UK: Polity Press.

Habermas, J. (1987). *The theory of communicative action: Vol. 2. The critique of functionalist reason* (T. McCarthy, Trans.). Cambridge, UK: Polity Press.

Hartmann, N. (1952). *The new ways of ontology.* Chicago: Chicago University Press.

Heidegger, M. (1971). *Poetry, language and thought* (A. Hofstadter, Trans.). New York: Harper Collins.

Henderson, R. D. (1994). *Illuminating law: The construction of Herman Dooyeweerd's philosophy.* Amsterdam: Free University.

Hirst, G. (1991). Existence assumptions in knowledge representation. *Artificial Intelligence, 49,* 199-242.

Husserl, E. (1970). *The crisis of European sciences and transcendental phenomenology* (D. Carr, Trans.). Evanston, IL: Northwestern University Press.

Klapwijk, J. (1987). *Reformational philosophy on the boundary between the past and the future.* Philosophia Reformata 52.

Klein, H. K. (2004). Seeking the new and the critical in critical realism: Déjà vu? *Information and Organization, 14,* 123-144.

Lonergan, B. (1992). *Insight: A study of human understanding.* Toronto, Ontario, Canada: University of Toronto Press.

MacIntyre, A. (1985). *After virtue: A study in moral theory.* London: Duckworth.

Maslow, A. (1943). A theory of human motivation. *Psychological Review, 50,* 370-396.

McIntire, C. T. (1983). Dooyeweerd's philosophy of history. In C. T. McIntire (Ed.), *The legacy of Herman Dooyeweerd: Reflections on critical philosophy in the Christian tradition* (pp. 41-80). Lanham, NY: University Press of America.

Mingers, J. C. (2004). Re-establishing the real: Critical realism and information systems. In J. C. Mingers & L. Willcocks (Eds.), *Social theory and philosophy for information systems* (pp. 372-406). Chichester, UK: Wiley.

Nash, R. H. (1962). *Dooyeweerd and the Amsterdam philosophy.* Grand Rapids, MI: Zondervan.

Olthius, J. M. (1985). Dooyeweerd on religion and faith. In C. T. McIntire (Ed.), *The legacy of Herman Dooyeweerd: Reflections on critical philosophy in the Christian tradition* (pp. 41-80). Lanham, NY: University Press of America.

Polanyi, M. (1967). *The tacit dimension.* London: Routledge and Kegan Paul.

Seerveld, C. (1985). Dooyeweerd's legacy for aesthetics: Modal law theory. In C. T. McIntire (Ed.), *The legacy of Herman Dooyeweerd: Reflections on critical philosophy in the Christian tradition* (pp. 41-80). Lanham, NY: University Press of America.

Smircich, L. (1983). Concepts of culture and organizational analysis. *Administrative Science Quarterly, 28,* 339-358.

Stafleu, M. (2005). On the character of social communities: The state and the public domain. *Philosophia Reformata, 69*(2), 125-139.

Strauss, D. F. M. (2000). The order of modal aspects, In D. G. M. Strauss & M. Botting (Eds.), *Contemporary reflections on the philosophy of Herman Dooyeweerd* (pp. 1-29). Lewiston, NY: Edwin Mellen Press.

Tarnas, R. (1991). *The passion of the Western mind.* Pimlico, London: Random House.

Van der Kooy, T. P. (1974). De gereformeerde wereld en de sociologie (The Calvinist world and sociology). *Anti-Revolutionaire Staatkunde, 2,* 37-56.

Webster's Dictionary (1971). *Webster's third new international dictionary of the English language unabridged.* Springfield, MA: Merriam.

Winch, P. (1958). *The idea of a social science.* London: Routledge and Kegan Paul.

Section II
Frameworks for
Understanding

Chapter IV

A Framework for Understanding Human Use of Computers

"I know how to use it, and it's easy to use—but what the heck do I use it for?" Uttered by the manager cited in Vignette 2 in the Preface, this epitomises the difference between ease of use and usefulness and, in doing so, can serve to begin opening up the issue of the human use of computers. This chapter seeks to open it up further and explore how Dooyeweerd's philosophy can be used to understand human use of computers (HUC) by exposing some of the challenges and issues therein and enable us to formulate a framework for understanding it. The framework developed aims to be sensitive to everyday issues as they present themselves to us as users in both research and practice (see §2.4.2).

In this first exploration of a Dooyeweerdian framework, the framework will be developed step-by-step. After drawing out some issues from a few example cases, Dooyeweerd's notion of multi-aspectual human functioning (§3.4.1) is employed to ensure a broad, integrated focus on the everyday life of HUC. This reveals a number of such multi-aspectual functionings, three of which are distinguished:

- Human-computer interaction (HCI)
- Engagement with represented content (ERC)
- Human living with computers (HLC)

The structure of each is explored with the aid of Dooyeweerd's non-Cartesian notions of subject-object relationship and qualifying aspect. The normativity of each is explored by reference to the innate normativity of the aspects and, for HLC especially, this suggests a new way of understanding its complexity. A number of practical devices suggested by a Dooyeweerdian approach are presented, and the chapter ends by showing briefly how the Dooyeweerdian approach can engage with, underpin, and enrich other frameworks for understanding HUC. The framework for understanding HUC developed is not necessarily the only framework that could be developed, or even the best, but it seems to be a reasonable one. The validity of the framework is merely explained and illustrated, rather than proved.

4.1 Toward an Everyday Understanding of IS Use

In order to indicate the kinds of characteristics a framework for understanding HUC should have, four cases of IS in use will be examined.

4.1.1 A Major IS Failure

Mitev (1996, 2001) discussed the failure that was the early SNCF (the French national railways) Socrate rail ticketing system. The following paragraph from (2001) summarises her findings succinctly:

Technical malfunctions, political pressure, poor management, unions and user resistance led to an inadequate and to some extent chaotic implementation. Staff training was inadequate and did not prepare salespeople to face tariff inconsistencies and ticketing problems. The user interface was designed using the airlines logic and was not user-friendly. The new ticket proved unacceptable to customers. Public relations failed to prepare the public to such a dramatic change. The inadequate database information on timetable and routes of trains, inaccurate fare information, and unavailability of ticket exchange capabilities caused major problems for the SNCF sales force and customers alike. Impossible reservations on some trains, inappropriate prices and wrong train connections led to large queues of irate customers in all major stations. Booked tickets were for non-existent trains whilst other trains ran empty, railway unions went on strike, and passengers' associations sued SNCF. [Mitev's referencing removed]

Though brief, this analysis shows clearly the diversity of impacts an IS can have when in use—not just managerial or technical, but personal, social, and legal. The impacts were not just on the formal aspects of SNCF such as efficiency but on the everyday life of people—who were many and of wide diversity. This suggests that any framework for understanding HUC must provide a basis for understanding diversity of impacts and of types of people involved or affected (for whom the usual word in the HUC community, "stakeholder", will be used).

A strong normativity is clearly evident here, by which certain things (all, as it happens) are deemed negative and to be avoided while others are deemed positive and to be sought. More from the tone of her writing rather than from any explicit statements, it is clear that this normativity is to be treated, not as mere subjectively-experienced discomfort, but as something about which action should be taken. This suggests that a framework should provide a basis for understanding normativity without reducing it to mere description.

4.1.2 Unexpected Impacts

Eriksson (2006) discusses the case of a stock control system installed by a Swedish vegetable wholesaler in the hope of enhancing profitability via IT-assisted business process re-engineering. Specifically, a middle-man between wholesaler and retailer was replaced by a computer ordering system in the hope of increasing both flexibility and profits. To their dismay, the wholesaler found profits fell rather than rose. The reason was that, under the original regime, the middleman would offer both a friendly contact for the retailer and also nuanced advice about the quality of the day's stock and, based on a sympathetic awareness of the retailer's situation, proactively suggest alternatives. All this disappeared under the new regime.

Walsham (2001) shows unexpected impacts even more clearly. He discusses the problems of the EPS (electronic placing system, used in insurance market trading) (described later). Degree of use of the system was still low several years after its launch. The study found a number of reasons for this, but a key one was that "the EPS system ... undermined the Lloyd's rule of Utmost Good Faith. This rule essentially states that a broker must display all known relevant information about the client and the insurance risk to underwriters upon presentation of the risk to them" (p. 155). But the system did not facilitate this. "Trust comes out as a crucial element that must be successfully created and maintained throughout the negotiation process in the insurance chain" (p. 158). Use of the system hindered the building of such trust, so its use remained low.

In both cases, while the original objectives of the ISs may have been met, those objectives themselves did not adequately reflect vital aspects of the everyday life of its use, and so unexpected impacts occurred. These examples suggest that a framework should provide a basis for thinking about such complex issues as unexpected, indirect impacts of use as well as those we might anticipate.

4.1.3 The Case of *Elsie*: A Small IS Success?

Those three examples show failure. The *Elsie* knowledge based system (KBS) (Brandon, Basden, Hamilton, & Stockley, 1988) was a success, albeit a rather smaller IS. Perhaps because failures are more newsworthy than successes, good, penetrating discussion of successes is rare in the literature, and where it occurs, is rather narrow in focus. Seldom are all the everyday aspects of successes discussed because success is usually understood as fulfilling narrow objectives. The author was intimately involved in creating *Elsie* and can recall many of the everyday aspects, including some not published. So the account of *Elsie* will be rather longer, in order to provide sufficient material for later discussion.

4.1.3.1 Overview of Elsie

Elsie was developed to assist quantity surveyors in giving advice to their clients who were at an early (pre-architect) stage in considering the construction of new office developments. Many hundreds of copies were sold to surveyors in the UK and came into use. Its use was studied by Castell, Basden, Erdos, Barrows, and Brandon (1992). Of its four modules, the budget module was the most widely used, which would help the surveyor in setting an appropriate budget for a client who wished to consider building office space.

Elsie operated by asking the user a sequence of questions about the proposed office development (such as the number of staff, whether it was a head office or a regional administrative office, or characteristics of the site). From the answers received, it made expert inferences (about size, number of floors, number of lifts, quality of materials and fittings, type of foundations needed, and so on), and on the basis of these it would calculate what the building might cost to construct.

The budget module would ask around three dozen questions, but their number and order varied from session to session because the next question to put is always dynamically selected by the backward chaining (see Brandon et al., 1988) that is characteristic of KBS technology. This results in redundant or unnecessary questions being suppressed. Forward chaining, by which each answer is immediately propagated throughout the network of inferences to see if any inference goals have now been satisfied, give the KBS a responsive feel. Such technical features made *Elsie* feel very friendly (even though it was driven via keyboard).

At any time the user (usually a quantity surveyor) could obtain explanation of questions. To some questions a degree of uncertainty could be accepted in the user's answers. At the end, the budget module would present its estimate of what the building might cost to construct, and invite the user to explore and even critique this result. The user could request a breakdown of costs per major element of building (walls, foundations, services, fittings, etc.). If any element-cost was deemed excessive, the reason for this could be explored to ascertain on which information it depended, and this could then be varied to see what difference it made to the cost. The user could even override *Elsie*'s reasoning, for example, if the cost of bricks available differed from that assumed by *Elsie*, or the aesthetic-amenity quality factor needed to be raised or lowered.

4.1.3.2 Elsie *in Use*

In the survey of use (Castell et al., 1992), from which the following quotations come, it was found that features valued most by the users were not the technical ones like the reasoning algorithms employed, but such things as the perceived accuracy of the system, its flexibility (for example, though designed for commercial buildings it was used also for magistrate courts, hospitals, etc.), and the rather mundane feature that projects could be archived for later recall.

Such features led to benefits like reduced cognitive load, speed in obtaining an initial first estimate, enhanced communication with client and flexibility in responding to unanticipated change. As a result:

... not only is the process of generating a budget estimate significantly shortened ... but the process enhances the clarity of the customer's requirements. It should be noted, however, that this increases rather than decreases the chances of the customer changing requirements, but such changes can be readily accommodated.

By employing these features, the specification of the building was gradually refined to suit the client's needs, step-by-step, and at each step *Elsie* provided information to help the user make quite nuanced decisions about the specification of the building and the quality of materials. Every project a surveyor meets is unique, leading to a wide variety of types of situations in which *Elsie* was used.

To make valid estimates requires a clear picture of the customer's requirements, but it is often the case that such requirements are not clearly known in the initial stages. Customers often change their requirements when they see the implications in terms of costs or building specification.

Elsie changed what had been a single-stage process (of supplying the client with a detailed estimate taking about a week) to a two-stage process, in which an initial rough estimate was supplied immediately, then:

Following the initial stage, the customer and surveyor meet to revise, update and clarify elements of the original estimate, and a new negotiation cycle, based on the outcome, is started. During such cycles, assumptions are revealed and changes in requirements and their consequences are quickly analysed, with the customer present. This stage is supported directly by Elsie in that it is employed at the time of discussion and provides a powerful medium for expressing new decisions and clarifying assumptions.

Elsie not only made extant tasks more efficient (which had been anticipated), but even changed the very tasks themselves, which had not been anticipated but proved to be where the real benefit lay. This, in turn, "led to significant changes in the relationship between the quantity surveyor and customer, and a change in the role of each" which had two components:

... several factors combine to increase the customer's commitment to the project. ... direct involvement in the second (revision) stage of the process has meant that the customer has felt more in control of the whole process. Being encouraged to actively engage in the process by providing information and suggestions to be input directly into the system, and assessing the consequences of such new information in a variety of ways opens up the process and engages the client in a much more direct way. The system thus becomes a tool supporting interaction at the level of social relationships as well as that of tasks.

and

The relationship between surveyor and customer has traditionally been one of expert versus novice ... With the kind of shared problem solving behaviour described above, the participants in the negotiation cycle are now likely to be working towards a better articulated, shared goal.

As a result:

Thus the interaction with the computer becomes less differentiated in terms of expert versus novice and more co-operative in terms of reaching the common goal of an acceptable budget estimate and building specification. The change in relationship is towards empowerment of the customer ...

It is interesting, however, that the surveyors welcomed rather than resisted this apparent shift of power away from themselves because "they see themselves as able to provide a better and more attractive service to their customers." As discussed later, this might cast doubt on Foucault's elevation of power which Walsham (2001) has as one of his conceptual tools for understanding situations.

It is also interesting to note that, in terms of meeting its original objective—to make high-level surveying expertise available to less senior surveyors so they could share some of the workload and undertake budgeting tasks more cost-effectively—*Elsie* would be evaluated as a failure, yet it was judged as a success (and became the second most widely sold KBS of its time), because of the kinds of benefits discussed earlier. The latter—and especially the change in role—were completely unexpected impacts.

This account not only exhibits diversity, normativity, and unexpected impacts found in the other three cases, but clearly displays the network of interrelated factors in the use of such software in professional situations of decision making and advice giving, and the complexity that plagues our attempts to understand such use. The direct relationship with the computer enters the picture as various types of user-friendliness, transparency, and ease of use, followed by the ability of lay people to use it, the importance of technically mundane features, the changes in user tasks and processes, especially those brought about by the users themselves, changes in roles and social structures, and the difference between formal and actual criteria for judging success.

One thing that is clearly obvious in this case is that meaning is very important, and is diverse. It is also obvious in the Socrate case. Therefore, that Dooyeweerd takes cosmic meaning as a starting point, rather than being or process (§2.4.3), at least recommends his philosophy as a way to understand human use of computers.

4.2 Computer Use as Multi-Aspectual Human Functioning

Chapter III showed that Dooyeweerd's philosophy addresses issues of diversity and normativity in a human-centred way orientated to everyday experience. So it is no surprise that the author discovered that it provides excellent understanding of such everyday issues of use as are mentioned earlier. The argument that follows will gradually develop the author's framework for understanding HUC by means of aspectual analyses of the cases employing Dooyeweerd's suite of aspects, which fulfil a dual purpose. One is to introduce, one by one, new ideas that are important components of the framework for understanding being developed. The other is to expose the reader to the practical device of aspectual analysis in various guises.

Dooyeweerd invites us to see use of computers as multi-aspectual human functioning (§3.4.1). Many existing frameworks for understanding computer use elevate a single aspect thereof and largely ignore the others; for example, HUC seen from the viewpoint of psychology elevates the psychic aspect and HUC seen from the viewpoint of business elevates the economic aspect. But in a Dooyeweerdian understanding of HUC, all aspects are given due recognition and respect, yielding a framework that enables us to address the diversity that is HUC when viewed from a lifeworld perspective (see §3.4.5).

4.2.1 Aspectual Analysis of Computer Use

We can find almost every aspect in the earlier mentioned accounts. But what philosophical roles or characteristics of aspects are important here? Aspectual analysis in general relies on aspects being ways in which things may be meaningful (§3.1.5.1), the "things" in this case being situations of use. And, since computer use involves us doing things that result in impacts, aspectual analysis of computer use recognises that aspects enable both functioning and repercussions of that functioning (§3.1.5.4, §3.1.5.5).

One type of aspectual analysis is to interpret an extant text. The main aspectual meaning of each phrase is identified, and perhaps tabulated as exemplified in Table 4-1, which present an analysis of Mitev's report.

Aspectual interpretation of extant texts, as with Socrate, can indicate to what extent the writer is focusing on certain aspects at the expense of others. Mitev's text contains almost every aspect, hereby showing her sensitivity to the lifeworld to be high. But here it underlines how important every aspect can be in HUC.

Table 4-1. Aspects of Socrate use

Aspect	Phrases from Mitev's report
Pistic	unions and user resistance
Ethical	-
Juridical	political pressure, inadequate implementation, The user interface was designed using the airlines logic, inappropriate prices, Booked tickets, passengers' associations sued SNCF
Aesthetic	chaotic implementation, prepare the public to such a dramatic change
Economic	poor management, ticket exchange capabilities, customers, non-existent trains, trains ran empty, railway unions went on strike
Social	the public, passenger associations
Lingual	inaccurate fare information, Staff training, Public relations failed, sales force
Formative	Technical malfunctions, prepare salespeople, Impossible reservations
Analytic	tariff inconsistencies, airlines logic, inadequate database information
Psychic	not user-friendly, unacceptable to customers, irate
Biotic	-
Physical	-
Kinematic	timetable and routes of trains, wrong train connections
Spatial	queues
Quant'ive	Large, all

But an even richer picture can emerge from direct aspectual analysis of situations of use, rather than of texts. A quick analysis of *Elsie* is shown in Table 4-2. Such analyses involve two types of reflection. We can reflect on the situation as such and record whatever come to mind against whichever aspect(s) make them meaningful. We can also consider each aspect in turn; this is useful for surfacing things overlooked in the first way. Though in Table 4-1 the pistic aspect is first, it is usually not a good aspect to start on. It is usually better to begin with the quantitative aspect—though the author often finds it useful to begin in the middle (e.g., lingual) and work outwards.

Is analysing aspect by aspect merely "filling slots"? See the discussion in Chapter IX. However, it has a particular use in addition to surfacing the overlooked, in that it can reveal several multi-aspectual functionings.

Table 4-2. Aspects of Elsie

Aspect	Usage of Elsie
Quant'ive	Number of questions, Number of floors, Magnitude of cost.
Spatial	Shape, size of building, Determine whether building fits site.
Kinematic	Lifts, stairs.
Physical	Ground loadbearing capacity, Finger force on keyboard.
Biotic	Muscle exerts force; light activates retinal cells, Opening windows for fresh air:
Psychic	Users feel good about Elsie.
Analytic	Clarify client requirements, Separated building elements, Shapes on screen as significant, e.g. as letters, digits
Formative	Flexibility of layout, functional quality, Developing specification for the building, User formulates answer with computer.
Lingual	Communicate with client, Explanations of questions, User understand the questions and give answers,
Social	Cultural meaning of the questions, Enhance relationship with client., Circulation space.
Economic	Work out how to reduce costs, Final cost, Resources of the UI, e.g. screen area, time waited for computation. Time taken to answer questions.
Aesthetic	Aesthetic-amenity quality, Harmony between surveyor and client., Style of UI.
Juridical	Knowledge of local byelaws, building regs, Does the UI do justice to info, users?, Do justice to client's real needs.
Ethical	Surveyors happy to sacrifice power as experts, A 'generous' UI that gives more than strictly necessary.
Pistic	Role change to partners with shared goal.

4.2.2 Interwoven Multi-Aspectual Functionings

Table 4-2 feels rather heterogeneous. Many of the factors feel of a different sort from others: for example the number of questions presented to the user and the number of floors of the building are both quantitative but seem to be meaningful in different ways; the questions on screen and communication between surveyor and client feel different, though both are lingual. Likewise, in Mitev's account we detect a similar difference between two formative aspects: technical malfunctions and impossible reservations. Dooyeweerd urges us to listen cautiously to such intuitions.

Doing so can reveal the inner structure of things, and in this case it reveals several different ways of functioning in each aspect of HUC. Taking the formative aspect, for example, we find:

- The user functions in the formative aspect insofar as s/he forms what they see on screen (or hear via the speakers) into a structured conceptual whole, relating it to what they already know, and as they plan their next answer.

- The user also engages with the formative content represented in *Elsie*: the formative aspect of the building, such as its degree of flexibility and the quality of its functional fittings.

- The user also functions formatively with *Elsie>* in their life as surveyor, such as to develop a specification of the building that will be submitted to the architect.

These three formative functionings are different. It almost every aspect these three ways of functioning may be detected, so that at least three multi-aspectual human functionings may be distinguished that together constitute HUC:

- **Human-computer interaction (HCI):** What the users experience of the interaction between the user and computer.

- **Engaging with represented content (ERC):** What the users experience of the meaning that is represented in the IS: that is of the content of the model (in *Elsie*'s case, the knowledge base or, in the case of a computer game, a virtual world).

- **Human living with computers (HLC):** What the users experience when employing the computer in everyday living—aspects of living that might somehow be affected by, or affect, the use of the computer beneficially or detrimentally: in *Elsie*'s case, aspects of being a surveyor for which *Elsie* is used.

There may be others, but the discussion here will be restricted to these three. They cannot be seen as parts of a whole but rather as bound together by enkaptic relationships (§3.2.6.2).

Table 4-3, derived from Table 4-2 with some additions, shows examples of most aspects of HCI, ERC, and HLC in *Elsie*.

The HCI column contains aspectual functions of the human activity of interacting with a computer via a user interface (UI), without regard to what type of software is being used or the specific reason it is being used—*Elsie*, e-mailer, computer game, process control software, and so on. The UI can be mouse rather than keyboard, loudspeakers as well as screen, tactile feedback, microphone input, and so on.

The ERC column contains aspectual functions, properties, beings, and so forth, that are represented in the computer, in its data structures and its algorithms that make calculations or undertake activity. In *Elsie*, only a few aspects are missing (having

Table 4-3. Aspects of HCI, ERC, and HLC in Elsie

Aspect	HCI of Elsie	ERC in Elsie	Elsie HLC
Quant'ive	Number of questions	Number of floors.	Find out cost of building.
Spatial	Spatial layout of the keys.	Shape, size of building.	Determine whether building fits site.
Kinematic	Movement of fingers to press keys.	Lifts, stairs.	-
Physical	Force on key; light emitted by screen phosphor on retina; elecl-chem activity in brain.	Ground loadbearing capacity.	-
Biotic / Organic	Muscle protoplasm exerts force; retinal cells activate with light.	Fresh air: opening windows.	-
Psychic / Sensitive	Nerve signals from retina/cochlea to brain, to fingers, within brain to recognise and remember patterns.	-	Users feel good about Elsie
Analytic	I distinguish shapes on screen as significant, e.g. as letters, digits, conceptualising and categorizing them.	Separated building elements.	Clarify client requirements.
Formative	User forms letters into words, questions, makes inferences, formulates answers.	Flexibility of layout, functional quality.	Develop specification for the building.
Lingual	I understand the questions and give answers.	Explanations of questions.	Communicate with client.
Social	Cultural meaning of the questions; Standardization of UI.	Circulation space.	Enhance relationship with client.
Economic	Resources of the UI, e.g. screen area, time waited for computation.	Final cost.	Work out how to reduce costs.
Aesthetic	Style of UI.	Aesthetic-amenity quality.	Harmony between surveyor and client.
Juridical	Does the UI do justice to info, users?	Knowledge of local byelaws, building regs.	Do justice to client's real needs.
Ethical	A 'generous' UI that gives more than strictly necessary	-	Surveyors happy to sacrifice power as experts.
Pistic	My vision of who I am as I interact: controller of computer.	-	Role change to partners with shared goal.

none of their meaning explicitly represented in the knowledge base nor in the texts the user sees). This should be so for good knowledge based systems or knowledge management systems, especially those built for ill-structured applications, because what is represented should relate to the everyday lifeworld in the domain to which they have been applied. But for computer systems that have a very specific application it is often the case that only a few aspects are represented; for example a calculator might explicitly represent meaning only of the quantitative aspect.

The HLC column contains aspectual functions from the life of the user as they use *Elsie* for various purposes. Those mentioned here are ones that have been mentioned by surveyors or that the author recalls. (Note: At the time of developing *Elsie*, he knew nothing of Dooyeweerd but did take an everyday approach; see Vignette 2.) Some of the purposes might be overtly stated, others might be covert. In principle, any software might be used for a wide range of purposes; for example though normally we might play a computer game for fun (aesthetic aspect), we might sometimes play it as a social activity, sometimes to boost our image of ourselves (pistic), and so on.

Why is the biotic aspect of *Elsie* HLC missing? Are we not functioning biotically by breathing? The answer is that only those aspects of HL (human living) are included that relate to C (computer). But the other aspects will return in Chapter V.

Early research recognised only HCI. HLC has also now been recognised for some time; for example Davis' (1986) technology acceptance model speaks of ease of use and usefulness. But its treatment of these is narrow, and neither it nor most others even today recognise ERC. Yet the earlier analysis suggests that ERC should not be conflated with either. Some support for this comes from virtual reality (VR). The VR community has, for many years, talked about the experience of immersion, but recently has begun to differentiate "presence" from immersion. Immersion, an experience of the hardware and style of the UI, may be seen as HCI, while presence, with its quality criteria of "plausibility" and "believability", refers to ERC.

As a practical device, aspectual analysis after having distinguished interwoven multi-aspectual functionings is usually more fruitful than before having done so. Those two might be used for initial brain-storming, followed by a cautiously intuitive review to see if several multi-aspectual functionings are interwoven, then by a fuller aspectual analysis of each, with further cycles as necessary. This process not only separates out different issues more clearly but also helps to stimulate a more holistic style of thinking so that issues are not overlooked.

4.2.3 Validating the Intuition: Qualifying Aspects

Intuition must be cautious. Intuition's self-critique can be assisted by philosophy (not science; see §3.3.8). On what philosophical basis may HCI, ERC, and HLC be differentiated in the general case? And on what basis may it be decided whether there are other multi-aspectual functionings that, with these three, constitute HUC?

Dooyeweerd's answer is that, when seeking to differentiate one type of thing from another on anything other than a subjective basis, it is often helpful to identify the qualifying aspect (§3.2.5), because this indicates the main meaning of the thing, in the sense of its purpose or destination, to which all its other aspects contribute. Each

Table 4-4. Qualifying aspects for HCI, ERC, HLC of Elsie

	HCI of Elsie	ERC in Elsie	Elsie HLC
Qualifying aspect	Lingual	Economic	1. Economic 2. Analytic 3. Any

of the three multi-aspectual functionings are led by a different qualifying aspect. Table 4-4 lists these for *Elsie*.

4.2.3.1 The Qualifying Aspect of HCI: Usually the Lingual

The qualifying aspect of HCI is usually the lingual for all information systems. This is because, regardless of application, the main thing we experience is symbols on the screen (or heard from speakers) that signify something, and in the actions we make that signify what we want the computer to do. This is clearly so in the question-answer style of HCI dialogue found in *Elsie*, but it is also so in graphical UIs, and whether the sensory channel is visual, aural, or tactile-haptic.

Most of the aspects of HCI serve the lingual function of understanding what is presented via the UI and responding. It is obvious how the earlier aspects do so: because of inter-aspect dependency (§3.1.4.6). But so do the later aspects. The importance of the social aspect of HCI lies not in the social intercourse that occurs when driving the computer (such as children gathering round a game player, which is HLC), but in whether the user understands the cultural connotations of, or assumptions behind, what is shown on the screen (or heard through the speakers), and with the standardisation of things like user interface style. The importance of the economic aspect of HCI lies not in the cost of the building but in such things as the effect of limited screen area: only a certain amount of information is visible. The aesthetic aspect of HCI concerns the harmony and artistic style of the UI. The juridical aspect concerns whether the UI does justice to either the users or the represented content, and so on. See Table 4-3.

(Occasionally it could be argued, however, that in certain highly proximal software, the qualifying aspect is pre-lingual. For example, if we are using a photographic editing package to subtly change the shading of the sky to make it a deeper blue, not by means of commands or menus, but by sweeping it with the mouse, then this involves very little other than motor control of mouse combined with visual feedback. It could be argued that this interaction between computer and user is qualified by the psychic-sensitive aspect. Others might still maintain the lingual qualification by suggesting that the very movements may be seen as symbolic signification of psychically-qualified purpose—but that is a little awkward. Here we will leave these possibilities open and focus mainly on lingual qualification.)

Table 4-5. Qualifying aspects of extant software

Aspect	Example software
Quant'tive	Calculator, Statistics package
Spatial	Drawing packages, Geographic Information Systems, Computer-Aided Design
Kinematic	Animation packages, Fluid flow packages
Physical	Weather forecasting systems, Solid modelling systems
Biotic	Medical software, Genealogical software, Life games
Sensitive	Painting and photographic software
Analytical	Mind-mapping software, Deduction software
Formative	Planning software
Lingual	Word processors, KBS, Web browsers
Social	Email, IRC
Economic	ERP systems, Critical path analysis software
Aesthetic	Music composition software
Juridical	Will-writing, contract-writing software
Ethical	?
Pistic	?

4.2.3.2 Qualifying Aspects of ERC and HLC

While the qualifying aspect of HCI is fixed at the lingual (usually), those of ERC and HLC are variable, but in different ways.

The qualifying aspect of ERC is that of the represented content with which the user engages. All the pieces of information about the building in the ERC column of Table 4-3 serve the purpose of fixing a budget for the building, so the qualifying aspect of the content represented in *Elsie* is the economic.

In well-designed software, the qualifying aspect of ERC will usually reflect the purpose for which the IS was developed—its main meaning or destination as a piece of software—and will usually be that of the main result of any calculation or process it carries out. In the case of a calculator it is the quantitative, in the case of a word processor, the lingual, in the case of a game, the aesthetic, and so on; see Table 4-5. The notion of qualifying aspect is to be preferred to that of functionalistic purpose because much IS use cannot be seen in functionalistic ways (e.g., computer games designed for enjoyment).

As will be discussed in Chapter VI and VII the decisions about which features are designed into any IS can be guided by considering the qualifying aspect and how each of the other aspects serve it. If close attention is given to all aspects in this way then the IS will be flexible enough to support changes in use. This happened in *Elsie*, such that its qualifying aspect changed. It was originally to set a budget (economic) but this changed when it began to be used, to the analytic function of clarifying requirements. For a deeper understanding of changes in qualifying aspect, see Dooyeweerd's discussion of antique shawls that become wall hangings and castles that become tourist attractions (1984, III, p. 143).

The qualifying aspect of HLC is that of the purpose(s) for which the IS is being used in the everyday life or work of the user and other stakeholders. For example, for one surveyor, *Elsie* could be used to give good service to a client beyond the call of duty (led by ethical aspect), to another, to fulfil a contractual obligation (juridical), to a third, to reduce costs (economic), to yet another, to maintain good relationships with the client (social) and so on. In general, HLC has a variable qualifying aspect, because any IS in which some flexibility has been designed, may be used in a concrete situation for a variety of purposes. Playing a computer game (aesthetically qualified ERC) might be for educational purposes (lingual) in HLC.

4.2.3.3 On Identifying Qualifying Aspects

The assignment of a qualifying aspect to a thing is not given *a priori*, but must be chosen by a process of reflection on the thing as it presents itself to us in its everyday (pre-theoretical, naïve) lifeworld context (see §2.4.2). The author has found three principles useful:

- The qualifying aspect must reflect the thing itself rather than our variable, subjective use of it.
- The qualifying aspect is what primarily differentiates this type of thing from another type.
- For all other aspects, it should be clear and natural to see how they serve the needs of the qualifying aspect.

Preferably the thing should be given time for its relationships with its context to mature and for reflective discourse to develop about the nature of the thing. Such discourse should be taken into account but should not be allowed to unduly dominate.

For example, reflection on HCI during the 1960s would have revealed little more than control of a machine (formative aspect) or calculation (quantitative). Today, the vast majority of HCI is constituted in human understanding of the meaning of symbols presented and presenting other symbols in return; hence the lingual aspect is what qualifies HCI. In HCI the formative aspect serves the lingual by enabling a structured interaction, the social serves it insofar as cultural intersubjective meaning of symbols helps the user understand. Note: It may also seem that the lingual serves the social, as in e-mail, but that is HLC or ERC rather than HCI.

4.2.4 Benefits of Understanding Types of Multi-Aspectual Functioning

Being able to distinguish HCI, ERC, HLC as three types of multi-aspectual functioning in this way has several benefits. It can help us separate out different issues in HUC, but more specifically:

- It explains why something that is easy to use or technically advanced might still fail to bring benefits in use, and why supposedly old-fashioned software like legacy systems can bring real benefits.

- Recognition of ERC separate from HCI, HLC clarifies how to deal with virtuality. In particular this approach gives us a framework by which to understand the paradox of virtual objects like powerful spells from MUD (multi-user dungeon) games being bought and sold on eBay. Both ERC and HLC are real multi-aspectual functioning.

- We have a clear basis for differentiating what the software is intended for from what we actually use it for each time.

- It can be useful in helping us tease apart the issues that emerge in changes in use of software.

- It can be useful in practice in helping to guide IS development, and enriching it by employing aspectual analysis of all three of HCI, ERC, and HLC.

- It can be useful in the formulation of useful research questions and even whole research programmes in HUC, in that it urges us to consider all three, even though we might focus on one, lest our research becomes confused. It warns us that attempting to understand HCI in relative isolation from ERC and HLC might result in flawed, or at least narrowed, research that generates unsustainable results.

That the qualifying aspect of HCI is fixed, mainly, on the lingual provides a link with the area of research and practice that is the nature of computers, discussed in Chapter V. The qualifying aspect of ERC, which may be any aspect that expresses the purpose of the IS, links with the task of creating the IS and with knowledge elicitation in ISD (Chapter VI). That the qualifying aspect of HLC is highly variable provides a way of understanding cases in which IS are used creatively for reasons outwith their normal purpose, without giving either too much or too little attention to intended purpose. This links with the task of anticipating use in ISD (Chapter VI).

HCI, ERC, and HLC will now be explored in more depth, first the structure of each of the three and then their normative direction. Separating structure from normativity has the danger of perpetuating the myth that they are independent of each other, but the alternative rendering, of discussing structure and normativity of HCI, then of ERC, then of HLC, was rejected because it would weaken the sense that they are all part of the same activity that is human use of computers.

4.3 The Structure of Human-Computer Relationships

Dooyeweerd was interested in the structure of things. Here we consider the structure of the three relationships between human and computer in HCI, ERC, and HLC. Because the HCI relationship is the most direct and is required by the other two, it will be explored first and will receive the longest discussion.

4.3.1 Structure of HCI: Law-Subject-Object and *Gegenstand* Relationships

To understand the relationship between the human user and the computer, it is necessary to understand Dooyeweerd's notion of subject and object clearly. In place of Descartes' subject-object relationship, as an undifferentiated human ego observing or operating on a (more or less) non-human other, Dooyeweerd offers a diverse law-subject-object relationship in which both humans and other entities all play a part as subject and object in many different ways (§2.4.4).

4.3.1.1 Aspectual Subject-Subject and Subject-Object Relationships

In HCI, there is a subject-subject relationship between the human and computer in the physical aspect because both are subject to the laws of that aspect; it is this S-S relationship that enables true interaction. However, we cannot reduce HCI to the physical aspect, since, as HCI, it gains its real meaning from later aspects, usually the lingual, as discussed earlier. It is the psychic to lingual aspects that are given most attention in HCI. For example, when moving and clicking the mouse I do not think

Table 4-6. Aspects of user and computer

Aspect of HCI	of user (input; output)	Rel	of computer (output; input)
Physical	Physico-chemical energy, fields and forces	← S-S →	Physico-chemical energy, fields and forces
Organic/ Biotic	Muscle cells active; Retinal, ear, nerve cells stimulated	S-O	Currents from mouse; Voltages that control screen
Psychic	Make mouse gestures; See shapes, colours, spatial arrangements (esp. vert, horiz, topological), Hear sounds, etc.	S-O	Receive mouse gestures; Display shapes, fonts, colours at various positions, Emit sounds
Analytic	Choose what to ask computer to do; Differentiate parts of visual-aural field that carry information from backgnd	S-O	Recognise tokens of command; Display pieces of data
Formative	Structure commands to computer, plans for finding content; Intelligently respond to displayed structure (e.g. following appropriate links)	S-O	Take action appropriate to command (e.g. jump page); Display structure (e.g. hyperlinks as blue underlines)
Lingual	Find and understand content from the web pages	S-O	Offer content to user
Social	Understand the humour and cultural connotations on the page	S-O	Providing explanation and help to assist user's cultural understanding.
Economic	No need to scroll page, which appears quickly	S-O	The web page makes frugal use of screen area, and downloads rapidly
Aesthetic	Appreciate the colours, balance, etc. of the web page	S-O	-

(right margin, vertical text: Gives meaning to physical functioning)

primarily of "exerting sideways and downwards forces on the mouse" (description according to physical aspect) but in terms of "clicking, dragging the mouse" (psychic aspect), "selecting" (analytic), "forming" (formative), or "expressing" (lingual).

The physical is the latest aspect in which the computer can function as subject, so in all post-physical aspects, the relationship cannot be subject-subject. If the use is proximal, in which the user is hardly aware of the computer, giving it little attention, then the relationship in all these is subject-object, as depicted in Table 4-6 for Web-browsing. (The pre-physical aspects are not shown since their meaning here is completely bound up with the physical.)

The two sides of the relationship are described differently, except for the subject-subject physical relationship. In the subject-object relationships the functioning of the user is that of agent (subject) while that of the computer is of meaning in each aspect ascribed by us to the computer's physical functioning. Now is a good time to peruse the table, to sense the subject- and object-functioning in each aspect.

Later aspects depend foundationally on earlier. So, for example, if we describe the computer as receiving gestures from the mouse and displaying shapes on the screen, this is shorthand in the psychic aspect for saying that the computer functions in a certain way physically, which we (users and others) interpret as receiving mouse gestures and displaying shapes. If we describe the computer as "giving us the latest profit figures," this is lingual shorthand for the computer's physical functioning, which we interpret as:

- displaying shapes, which happen to be standard ones that are ...
- ... letters and digits rendered in a certain font, which we take to be words and numbers, which ...
- ... form a structure (e.g., word "profits" followed by "=" followed by a number), which is interpreted as ...
- ... the latest profits.

And we might add:

- ... which are seen as bad or good, depending on cultural knowledge of the sector in which the firm is engaged.

We then make use of this information in our lives (HLC).

The benefit of viewing HCI in this way here is that it gives a place to all aspects in HCI, especially the later ones which are often overlooked, and that it highlights the debt we owe the physical aspect, which is the latest in which the computer functions as subject. Being a determinative aspect, this explains why we can rely on the computer's calculations and searches. But the main benefit here is that it provides a basis for evaluating and designing HCI in which confusion between the various issues is minimised.

Some of the benefits of viewing HCI in this way are to our discourse about the nature of computers in Chapter V. It throws fresh light on why it is that the computer seems intelligent but why its behaviour is determined, and it makes it valid to speak of the computer "displaying", "recognising", and even "working out" or "thinking", implying neither subject-agency on one hand nor anthropomorphic metaphor on the other. This means that arguments and deductions made about what the computer is doing are sound (in contrast to those based on metaphor).

4.3.1.2 Distal and Proximal

Polanyi (1967) discussed how use of a tool is "distal" when we are unfamiliar with it but becomes "proximal" once it becomes second-nature to us. In a proximal relationship with a tool our attention is not on it but rather on the task for which we are using it, but in a distal relationship, some of our attention is distracted to attend to the tool as such. The tool is now at a distance from us experientially.

This view has been applied to understanding two types of HCI (Basden, Brown, Tetlow, & Hibberd, 1996). In the 1970s and 1980s the HCI relationship was assumed to be distal, often understood metaphorically as dialogue between two agents (human and computer). For example, nearly one third of Downton (1991) is devoted to "human-computer dialog design". But this distal relationship is problematic; Norman (1990) remarked:

The real problem with the interface is that it is an interface. Interfaces get in the way. I don't want to focus my energies on an interface. I want to focus on the job.

Shneiderman (1982) pioneered "direct manipulation", which enabled the "direct engagement" that Laurel (1986) called for. Proximal HCI became the norm (for example in using word processors).

How can the difference between distal and proximal HCI be understood? Winograd and Flores (1986) (discussed later) employed the philosophical dialectic between the Cartesian subject-object relationship and Heidegger's notion of "at-hand". But Dooyeweerd's non-Cartesian subject-object relationship is one of intimate, proximal engagement, not of distance. Dooyeweerd does recognise a distance relationship, *Gegenstand*, in which we "stand over against" aspects of the thing with which we are engaged. If we must give attention to the UI itself to work out what is presented (for example because of indistinct colours) then our aspectual functioning becomes unbalanced, with for example the psychic functioning of distinguishing visual patterns dominating all else for a time. This is a psychic *Gegenstand*. If the trouble we have is because we do not understand English well, then it is a *Gegenstand* related to the lingual aspect.

As will be discussed later, this account of both proximal and distal HCI is more nuanced than that by Winograd and Flores because the various aspects of the relationship can be differentiated to analyse more precisely what the problem is. "Direct engagement" is multi-aspectual and can be interrupted by any aspectual *Gegenstand*. Its multi-aspectuality can explain Ben Shneiderman's surprise (as the author heard him tell) when one of his students, who was using a text editor, contended "With the cursor over a word, I hit 'dw' to delete it; what could be more direct than that?" Shneiderman had until then focused only on the psychic aspect of direct manipulation.

4.3.2 The Structure of ERC: Aspectual Reaching-Out

Whereas in HCI the relationship is with the symbols of the user interface as such, the relationship the user has with the computer in ERC is with what they represent—but ERC goes beyond what is mediated via HCI.

In the case of *Elsie*, for example, HCI-mediated content includes such things as gross floor area, cost per unit area, type of office, and other things in the ERC column of Table 4-3. It can also include explanations of various things. In the case of a computer game, it might include characteristics of the player, the items carried, the spatially extended territory of the game all mediated through UI, sometimes in text, sometimes in graphics, sometimes in sound.

In general, the represented content is of all aspects. In the public domain game, *ZAngband*, for example, we find nearly all the aspects represented, as shown in Table 4-7. (This table will be referred to later to illustrate aspects of virtual worlds.)

Table 4-7. Aspects represented in ZAngband *virtual world*

Aspect	Content represented in ZAngband virtual world (game)
Quant'ive	Number of items of equipment carried; amount of gold.
Spatial	The maze of rooms and corridors laid out on screen
Kinematic	My 'speed' rating determines how fast I flee enemies.
Physical	Weight of armour. I can fall through trapdoors.
Biotic	I must look for food. Killing creatures or being killed
Psychic	I cannot see far in darkened rooms, but can smell monsters.
Analytic	I must identify types of objects, weapons, armour, creatures, etc.
Formative	Creatures attack me, 'intelligent' enough to find their way, so I must avoid them. ZAngband gives me quests.
Lingual	Creatures speak to me; I gather rumours.
Social	Creatures may be friendly / aggressive; I try not to annoy the friendly ones
Economic	I use my funds to buy, sell equipment. Limit on equipment carried.
Aesthetic	What creatures say to me is sometimes quite humorous.
Juridical	Some creatures are guilty of crimes, so I receive a reward for killing them.
Ethical	I meet beggars - but have no means of giving them money
Pistic	Players assigned a deity, which affects how we play

All this is made possible by aspectual "reaching-out" (§3.1.4.8) from the lingual aspect to all the other aspects. It is the function of the lingual aspect to enable all kinds of meaning to be represented.

But ERC can also involve other aspectual reach-out, such as when the user's imagination is active, which is formative functioning. There is represented content that is not mediated via the UI but is imagined or remembered. This can include:

- The "innards" of the computer, which is represented content the user believes to be inside the computer. In *Elsie* this included the costs database, the inter-variable inference relationships and the weights of the latter, and so on. In a computer game, in includes things not presently visible such as the whole territory, other players, non-player characters. (This is the issue of "innards" discussed in Chapter V.)

- The contents of documentation, information passed on from others about what is in the computer, and so on. (This communication might be a fourth multi-aspectual functioning alongside HCI, ERC, and HLC, which should be explored for a full picture of HUC.)

Whereas the user's knowledge of these will have, at some time, be mediated lingually, in the actual engagement with represented content it is active in the imagination and memory, which are of the psychic-sensitive aspect, perhaps assisted by deduction (analytic aspect). Whichever reaching-out occurs in ERC, meaning relevant to the application is represented as content and this could be of any or all aspects.

Not all experience of represented content is held in symbols as program variables, fields in a database, and so forth. Some is mediated via the activity of the IS. In *Elsie* this might be the sequence in which questions are asked, but the action as something to engage with is most important in games and physical simulations.

How representation comes about will be discussed in Chapter VI, and the techno-logical resources for it, in Chapter VII.

4.3.3 The Structure of HLC: Aspectual Repercussions

In HLC the computer system becomes almost invisible, being completely subsumed into the human user's subject-functioning. In HLC we are primarily concerned with how human beings function in the aspects that are their everyday living. As explained in §3.1.5.4, functioning in each aspect is a response to the laws of those aspects and it thereby incurs repercussions, as depicted in Figure 4-1.

Figure 4-1. Aspectual functioning and repercussions

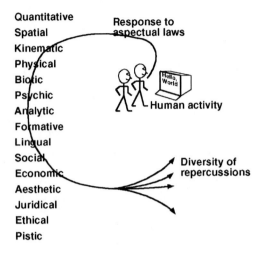

Because the later aspects are non-determinative, the repercussions therein cannot be predicted with certainty. But much analysis and useful prediction is still possible, including of troublesome issues like diverse, indirect and unexpected impacts.

4.3.3.1 Diversity of Repercussions

Each aspect gives rise to a distinct type of repercussion (see §3.1.5.5), so we can expect many types of repercussion in any single instance of HLC. None may be reduced to nor explained in terms of others. In Mitev's (2001) account cited earlier, of the Socrate rail ticketing system, the following aspectual repercussions can be identified (not including statements that express functioning that led to them):

- Led to an inadequate and to some extent chaotic implementation. – Juridical + Aesthetic
- Staff training was inadequate and did not prepare salespeople to face tariff inconsistencies and ticketing problems. – Formative
- The new ticket proved unacceptable to customers. – Formative (historical)
- Public relations failed to prepare the public to such a dramatic change. – Formative

- Caused major problems for the SNCF sales force and customers alike. – Formative

- Led to large queues of irate customers in all major stations. – Social + Sensitive

- Booked tickets were for non-existent trains. – Economic

- Other trains ran empty. – Economic

- Railway unions went on strike. – Economic

- Passengers' associations sued SNCF. – Juridical

All such repercussions (all negative as it happens) contributed to, and constituted, the failure of that IS. Table 3-3 and §3.1.5.5 give examples of general repercussions in each aspect.

Aspectual analysis of the functioning and repercussions in each aspect is useful either prospectively in prediction and design or retrospectively in evaluation. This can ensure that the moral and political issues (ethical, juridical aspects) are not overlooked, a criticism that Walsham (2001, p. 49) reports was made of Social Construction of Technology. It can also ensure that impacts on the environment are not overlooked, but usually these are indirect impacts.

4.3.3.2 Indirect, Long-Term Impacts

The most direct repercussions of HLC can be expected in the aspects in which the functioning occurs. In lingual functioning we would expect lingual repercussions; for example, if the text of a question the computer puts to the user is poorly worded, then we can expect the user to misunderstand the question. But the effects of this misunderstanding could then be passed on to others, affecting them. See Figure

Figure 4-2. Repercussions of IS use

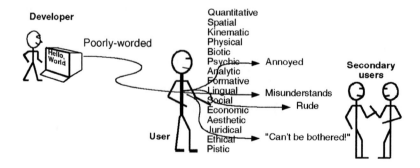

4-2. Ulrich (1994) calls such people "affecteds" and Eason (1988) and Friedman and Cornford (1989) called these "secondary users"; the human factors community calls them "stakeholders". Ferneley and Light (2006) suggest the notion has still not received adequate attention in the IS literature.

Whereas these authors merely draw attention to the problem, this Dooyeweerdian view can provide a basis for further analysis and research. The indirect impact is inevitably diverse because the human being responds in all aspects, even when stimulated by one. This could be called *aspectual crossover* in the human user (not a term Dooyeweerd used) since the repercussion crosses over to other aspects. These repercussions can then impact others—which makes the social aspect very important in HLC.

Many aspectual repercussions in post-social aspects occur because the aspectual functioning spreads throughout society and may thus take some time to materialise fully. As Table 3-3 showed, the latest aspects like the pistic can give rise to particularly *long-term repercussions*, often of an ecological nature.

We can perhaps see aspectual crossover in the Socrate case cited earlier, where the social and psychic repercussions "large queues of irate customers" resulted from "Impossible reservations on some trains, inappropriate prices and wrong train connections," which are analytic and juridical dysfunctions. But there is no mechanical link that determines that people will become "irate" in such circumstances (though they often will), because all the later aspects, at least from the analytic onwards, are normative rather than determinative. This precludes any attempt in Dooyeweerd to find a mechanistic methodology for predicting the impacts of information system use. But this Dooyeweerdian framework does at least provide a basis by which we may analyse and discuss it. Jones and Basden (2004), for example, shows how aspectual analysis can be useful in identifying a wide range of stakeholders, including those often overlooked, such as non-human ones.

4.3.3.3 Unexpected Impacts

As illustrated in the cases discussed by Eriksson (2006) and Walsham (2001), many important impacts are unexpected. The reason for unexpected repercussion, Eriksson argues, is because a number of important human aspects had been overlooked (e.g., the social relationship by which informal advice about the quality of vegetables was offered). Though Walsham does not speak of aspects in the same way, it is clear the same reason applies in his case of the insurance system: the entire pistic aspect of Utmost Good Faith was overlooked. The pistic aspect was also overlooked in the case of *Elsie* (changed view of one's role), but here it brought unexpected success.

The most serious unexpected impacts occur not because of lack of detailed knowledge but because aspects are overlooked during design, development, or use.

Table 4-8. Addressing complex HLC issues with Dooyeweerd

HLC Issue	Addressed by ...
Diversity of impacts	Multi-aspectual human functioning
Diversity of stakeholders	Social aspects
Unexpected impacts	Aspectual law pertains
Indirect impacts	Aspectual crossover Social aspect
Long-term impacts	Long reaction times of post-social aspects

Philosophically, the reason why unexpected repercussions occur is because the laws of aspects still pertain, transcending us, even when overlooked (see §3.1.4.1). If, as subjectivism holds, there are no transcending laws, it is difficult to account adequately for unexpected impacts.

This suggests that identifying which aspects have been overlooked during design or use can help us identify the broad areas in which the most serious unintended impacts are likely to occur, and that, if we then take account of (our best knowledge of) the laws of these aspects, we can often gain an indication of what the unintended impacts are likely to be, and over what timescale. This is important in IS development (Chapter VI). The author has found in practice that this is an intuitive, rather than rational, process, probably because the kernel meanings of aspects are grasped by intuition rather than by reason (§3.1.4.10), but it is possible some structured method could be devised for it.

4.3.3.4 Overview

This view of HLC as multi-aspectual human functioning in which the computer system is an object provides a basis on which we can begin to analyse and discuss the various components of the complexity of HLC. Table 4-8 summarises how each component may be addressed by a specific portion of Dooyeweerd's thought.

4.4 The Normativity of Computer Use

In discussing the Socrate case earlier, we referred to "dysfunction". But on what basis may we call this "dysfunction"? Sometimes, for example, it might be good, in the longer term, to get passengers used to a different way of buying tickets. We are faced with the philosophical issue of normativity, of good and evil: what types there might be, how they mix in the complexity of everyday experience, on what basis to differentiate one from the other, and what hope may we have that norms can be realized? Here we examine briefly the normativity of each of HCI, ERC, and HLC, and how this should guide research and practice.

The approach explored here is that the aspects, especially the later ones, are normative and define different kinds of good and evil because they are spheres of law (see §3.1.3). And, because they are irreducible to each other (§3.1.4.2), every aspect must be considered in its own right.

4.4.1 Normativity of HCI: Usability

Usability, or ease of use, was, for long, assumed to be the overriding norm for computer use as a whole and still is the norm for HCI. But it is now acknowledged to cover many factors, which can be understood multi-aspectually. For example, in their *A Practical Guide to Usability Testing*, Dumas and Redish (1999, p. 4) define usability as:

Usability means that the people who use the product can do so quickly and easily to accomplish their own tasks. This definition rests on four points:

1.   *Usability means focusing on users*
2.   *People use products to be productive*
3.   *Users are busy people trying to accomplish tasks*
4.   *Users decide when a product is easy to use*

At first sight, there is a laudable focus on the human being, but closer examination reveals a heavy emphasis on the economic aspect: "product", "quickly", "tasks", "use products to be productive", "busy", "accomplish tasks". Many elements of it would, thus, be irrelevant for many types of IS, such as games, unless one distorts the meaning of words like "productive". On the other hand, many Web sites today are designed mainly to look stunning, often at the expense of usability; are the Web designers wrong, right, or what?

The proposal here is that neither Dumas and Redish's emphasis on the economic nor Web designers' emphasis on the aesthetic are right or wrong in themselves because these are just two among many aspects. The shalom principle (§3.4.3) applies to HCI, and absolutization of any aspect will jeopardise usability.

Table 4-9 lists several normative factors under each aspect.

Table 4-9. Aspects of usability

Aspect	Norm	Usability issues (examples)
Biotic	Health	No undue muscle strain, RSI
Psychic	Sensing	Minimise cognitive load: Few keystrokes, clicks, movements; Easy recognition of shapes, sounds.
Analytic	Clarity	Ease with which the user can distinguish or find the UI things that are important to them (e.g. toolbars present the main tools immediately visible).
Formative	Shaping	Good structure of information (e.g. hyperlinks). Learning how to use UI to achieve various computer activity.
Lingual	Expression	Understandability, accuracy, veracity of what is presented (links to ERC). Obviousness of how to achieve their application goals via the UI.
Social	Respect	Ensure that cultural meaning is clarified and its presence as an assumption is clearly indicated to the user. Use standardized UI style.
Economic	Frugality	Freedom from clutter; good use of screen area. Good response times. Efficient access to relevant info.
Aesthetic	Harmony	Consistent visual, aural style. Balance, elegance, harmony.
Juridical	Due	Ensure that the style of display is appropriate to the type of information being displayed, c.f. Gibson's affordances as adapted to UI by Greeno [1994].

Sometimes there might seem to be conflict between such aspectual norms. For example, the juridical norm of appropriateness can make it difficult to standardise the style of UI (Basden et al., 1996). One way to resolve this is to take into account the qualifying aspect of HCI. If, as suggested earlier, this is the lingual aspect, then its norms of conveying information, understandability, and truth-telling should always be honoured. The aesthetic aspect of graphic design should serve that—which is a major theme of that authority on graphic design, Edward Tufte (1990). However, the lingual norms should not themselves be absolutized because HCI only gains its meaning by referring beyond itself to ERC and HLC. The norms for ERC and HLC are even more important.

The application of the shalom principle to the HCI norm of usability has yet to be properly explored. A practical device is offered later.

4.4.2 Normativity in ERC: Justice to (Virtual) World

The aspects of the represented content are those of the domain meaning which are signified by the symbols that constitute the program or database of the IS. The qualifying aspect of the represented content could be any, depending on the purpose of the IS, and so the normativity of ERC could be treated in a manner similar to that of HCI: ensure all other aspects represented in the content support and serve it.

But, as in the case of *Elsie*, the purpose changed during use, and so such a strategy could be dangerous to the flexibility of the IS if followed too rigidly.

A better strategy, which offers greater future-proofing flexibility, is that the represented content should do justice to the full meaning of the domain of application. This is a juridical norm.

The responsibility for ensuring this falls, of course, on the IS developers, as discussed in Chapter VI, but a few notes may be made here regarding actual user engagement with the represented content when they use the IS. It is impossible to represent fully all relevant meaning, not just because of deficiencies in the knowledge acquisition process discussed in Chapter VI, but because the lingual aspect of representation is non-absolute (§3.1.4.4). But justice may still be done if certain norms are noted (which arise from reflecting on the juridical norm of "due").

- **The norm of diversity:** In general, every aspect of the lifeworld of the domain should be represented (§3.4.5). This is especially important in software for ill-structured domains.

- **The norm of fitting the user:** But aspects may be omitted if, as discussed in Chapter VI, there is a reasonable expectation that those aspects will be active in the users as knowledge of the domain meaning, including their tacit and other lifeworld knowledge. In a calculator (to take the extreme example) the represented content is almost exclusively of the quantitative aspect, and its use relies almost completely on the user knowing how it is appropriate to handle quantitative information in the domain of application.

- **The norm of richness:** It is not enough to represent each aspect thinly; it must be represented in the richness with which we encounter it in the real world. For example, under the analytic aspect of distinction, we classify and identify objects, and in everyday life we have several degrees to which we can do this. So when the player of *ZAngband* (see §4.3.2) finds an object they must be able to detect its general type immediately (two-handed sword, soft leather shoes, gold ring, etc.), but they must also be able to identify the object's quality—such as how effective the sword is in being wielded and doing damage—and in *ZAngband* this requires the player to zap a Rod of Perception or read a Scroll of Identify. There is also a third, special or secret degree of identification. These degrees of identification make the game much more interesting.

- **The norm of multi-aspectual trustworthiness:** The user should not be lulled into a false sense of reality. In the mid-1990s film *Gladiator*, the (virtual) Colosseum looked very realistic (visual, psychic aspect). But the crowds, seen from above, looked unreal because the people were almost equally spaced, rather than huddling in groups as real crowds do (social aspect). Beware of maximizing realism in one aspect and ignoring others. It is often preferable to deliberately degrade visual faithfulness in order to avoid false impressions of reality (as in *ZAngband*'s diagrammatic, 2D view of the territory).

- **Norms for virtual worlds:** What is the ERC norm for virtual worlds, such as in games? What kind of virtual reality "rings true"? Bartle (2004, p. 66) discusses why one Internet game succeeded where another failed, including: "If what you see looks and behaves like reality, you feel you're 'there' more than if it looks and behaves like a gridwork of platonic solids. Recognising that people have expectations of degrees of reality within the virtual world. Fulfilling these expectations leads to increased immersion and denying it leads to decreased immersion." While fictitious *entities* (hobbits, magic weapons, etc.) are acceptable, they must behave as though subject to *laws* we know. The author does not know of any attempt to explain this. But Dooyeweerd's fundamental distinction between law and subject sides (§2.4.4) might provide a starting point.

- **The norm of flexibility:** Ensure user is aware of opportunities to use it in ways not anticipated, and stimulate them to take advantage of these. Build up their courage and hope where necessary.

These are not intended to be a complete list, but rather to suggest directions for future research and development of Dooyeweerd's ideas for ERC.

4.4.3 Normativity in HLC: Shalom

No longer do we judge the "success" or "failure" of IS solely in technical terms. Landauer (1996) did a lot to widen our view to HLC. But he focuses on productivity as the norm of HLC, as that which may be expected to be fulfilled and yet, paradoxically, is not. But such a treatment is too narrow for all IS in that it elevates a single aspect (in this case, the economic). Three more general norms have been suggested:

- The objectivist norm of meeting objectives
- The subjectivist norm of user satisfaction
- The criticalist norm of emancipation

The first is problematic because over 50% of IS fail to meet objectives, because objectives change, and because, like *Elsie*, a system may be "good" even when it does not meet its objectives. The second is problematic because "satisfaction" to one person may involve "dissatisfaction"—or harm—to others, and in any case it is not clear what "satisfaction" is. The third is problematic for a similar reason: what is

Table 4-10. Possible IS benefits and detriments

Aspect	Benefits to stakeholders	Detrimental impacts
Biotic	Improves health, vitality	Threatens, reduces health, vitality
Psychic	Makes people feel better, happier	Frustrates or saddens people
Analytic	Greater clarity in understanding	Confusion
Formative	Better planning, goal achievement	Destruction
Lingual	Better recording, communication, Higher information satisfaction	Reduced access to information, poorer quality information
Social	Leads to more respect, friendliness Better fulfilment of roles	Enmity, hatred, suspicion, Jeopardises role fulfilment
Economic	Better frugality, economy Better management	Waste, squandering, superfluity Mismanagement
Aesthetic	Harmony and integration Beauty, fun, interest	Fragmentation,Misfitting Boredom, Ugliness
Juridical	Justice to all, emancipation Assumption of responsibilities	Injustice, oppression Shirking responsibility
Ethical	Encourages self-giving, generosity	Selfishness, self-centredness
Pistic	Vision, commitment Orientation to true Absolute	Disloyalty, lack of morale Orientation to false Absolute

"emancipation"? A common definition is "freedom from unwarranted constraints", but this presupposes clear meaning of "freedom", "unwarranted" and "constraint", all of which are in fact far from clear. And all three suffer from the problem of realizability because (with the possible exception of Marxist versions, which exhibit other problems) modern thought has divorced "Is" from "Ought" (see §2.4.5).

But objectives, satisfaction, and emancipation may, with benefit, be understood by reference to the intrinsic normativity of aspects (§3.1.4.9) and the shalom principle (§3.4.3). The aspects, because they are spheres of law, can guide the setting of objectives that are appropriate, they can define "satisfaction" in terms of mixed aspectual diversity, and then can define the "evils" from which we seek emancipation and the "good" we seek to be emancipated into.

The shalom principle urges us to seek positive ("good") functioning in every aspect, and promises that if we do then "good" repercussions are likely to ensue. Here it means that use of the IS should enhance the user's HLC functioning in each aspect and degrade it in none, compared with not using the IS. That is, it should bring benefit and not detriment in each aspect. Table 4-10 lists examples of benefit and detriment in each aspect that an IS might bring; such a table can stimulate analysis during evaluation of IS in use, and also in IS development (Chapter VI).

This may be taken further, beyond a vanilla-flavoured aspectual democracy, to something that rectifies distortions in the status quo. Often, the root of distortions is undue elevation of an aspect, and an IS can be used to ameliorate them by reactivating aspects of human living that are currently ignored. Sadly, usually the opposite happens because it is easier to make a case to implement an IS that assists the currently-elevated aspect. For example, virtual reality is used in architecture to allow quicker assessment of the visual aspect of designs—but some believe that architects already give undue attention to that aspect.

Realizability of norms is addressed philosophically by Dooyeweerd in two ways. First, be reconnected Is with Ought, because both are founded in the same aspects so that what "Is" contains within it the potential to fulfil its normative cosmic destiny (see §3.4.6).

Second, the aspects are in harmony (§3.1.4.3). Dooyeweerd believed there is no inherent disharmony between aspectual norms, in contrast to those who assume conflict between the demands of different aspects (e.g., economic and ethical). The harmony rather than competition between the economic and other aspects is supported by a wide range of current thinking. While we might not maximise profits by being generous, experience shows (Collins & Porras, 1998) that generosity and justice bring sustainable profits. This is likely to be the same with IS. The late Enid Mumford, pioneer of the socio-technical approach to IS, was one who believed in this. Stahl, in his tribute to her (Avison, et al., 2006, p. 373) says, "Most importantly, she had shown theoretically as well as practically that the assumption of reflective responsibility is not only possible, it is even economically viable in a market environment."

But to achieve shalom—benefit in every aspect—is nevertheless a challenge, for two reasons. A positive reason is that the nature of the harmony between the aspects requires creative exploration. A negative reason is the orientation of the human heart towards a substitute for the true Absolute (see §3.3.6), such as when those involved in IS have hidden agendas. As discussed later, the ultimate solution for this is religious, but some of its effects can be partly ameliorated by the ethical aspect of self-giving.

4.4.3.1 The Ethical Aspect of Self-Giving

If the active stakeholders are functioning well in ethical aspect of self-giving, as expressed, for instance, in the well-known norm of "Love your enemies, do good to those who hate you, bless those who curse you" (Luke 6:27-28, Good News Bible) then it is possible that negative functioning can have positive repercussions. There are at least three ways in which this aspect is important.

It sets up an attitude in a community or organisation which allows all to become less defensive or competitive and, in their turn, more self-giving. This multiplies.

It blocks the normal negative chain of repercussions ("dog eat dog"). Conflicts can be defused, even those based on clashing perspectives, because we are no longer seeking our own advantage, pushing our own views, or protecting our own interests, but seeking blessing for others, listening to and understanding their views, and protecting their interests.

Third, The very success of information systems depends on this aspect. In Western cultures, we take the entering of information for granted. But Walsham points out (2001, p. 55) that in some cultures people are unwilling to "give away" information by entering it into an IS, because they fear others might take advantage of them. "Giving away" information in this way involves the ethical aspect of self-giving.

Thus Dooyeweerd's insight that the ethical aspect of self-giving is to be kept distinct from the juridical aspect of what is due is remarkably apposite.

4.5 Practical Devices

Dooyeweerd's philosophy very readily leads to practical devices that can assist research and/or practice in an area. Most of these are based round his suite of aspects and assist analysis. Those presented here are ones the author has found useful.

4.5.1 Aspectual Analysis

Aspectual analysis involves noting the ways in which each aspect expresses itself in the situation being analysed, often seeking balance (as in §3.4.3.1). Several types have already been demonstrated, each resulting in a list or table:

- Analysis of texts (§4.2.1, Table 4-1)
- Brainstorming multi-aspectual functioning (§4.2.1, Table 4-2)
- Analysis of distinct yet interwoven multi-aspectual functionings (§4.2.2, Table 4-3)
- Analysis of software types (§4.2.3.2, Table 4-5)
- Analysis of subject and object functioning (§4.3.1, Table 4-6)
- Analysis of represented content (§4.3.2, Table 4-7)
- Analysis of usability (§4.4.1, Table 4-9)
- Analysis of benefits (§4.4.3, Table 4-10)

These can be used as examples to follow, though they only give a few example items in each aspect and in practical analysis one would expect more.

Table 4-11. Aspects of Web design guidelines

Guideline	Aspect Q SKPBPAFLSEAJEP	Guideline	Aspect Q SKPBPAFLSEAJEP
Design Process:		**Links:**	
Set and State Goals	FL	Position Important Links Higher	F
Set Performance and/or		Show Links Clearly	PA
Preference Goals	F	Indicate Internal vs. External	
Share Design Ideas	L E	Links	F S
Create and Evaluate		Use Descriptive Link Labels	L
Prototypes	AF	Use Text Links	P
Design Considerations:		Avoid Mouseovers	P
Establish Level of Impt'nce	Q F	Repeat Text Links	F E
Reduce Users' Workload	E	Present Tabs Effectively	A
Be Consistent	A A	Show Used Links	F
Provide Feedback to Users	L	**Graphics:**	
Include Logos	PA	Use Graphics Wisely	P
Limit Maximum Page Size	S E	Avoid Using Graphics As Links	A
Limit Use of Frames	S E	Avoid Graphics On	
Content and Content Organization:		Search Pages	P E
Establish Level of Impt'nce	Q F S	**Search:**	
Provide Useful Content	F J	Consider Importance of	
Put Important Information		Search Engine	FL
at top of Hierarchy	AF	Indicate Search Scope	L
Use Short Sentence/		Enhance Scanning	FL
Paragraph Lengths	L E	**Navigation:**	
Provide Printing Options	P	Keep Navigation Aids	
Titles and Headings:		Consistent	A A
Provide Page Titles	F	Use Text-Based	
Use Well-Designed Headings	F	Navigation Aids	P F
Page Length:		Group Navigation Elements	F
Determine Page Length	Q S	Place Navigation On Right	S
Determine Scrolling vs.		**Software/Hardware:**	
Paging Needs	SK	Determine Connection Speed	P E
Page Layout:		Reduce Download Time	E
Align Page Elements	S	Consider Monitor Size	S
Establish Level of Importance	S F	Consider Users'	
Be Consistent	S A	Screen Resolution	S P
Reduce Unused Space	S EA	Design for Full or	
Put Important Information at		Partial Screen Viewing	P
top of Page	S A	**Accessibility:**	
Format for Efficient Viewing	S E	Use Color Wisely	P
Font/Text Size:		Design for Device	
Use Readable Font Sizes	S P	Independence	P
Use Familiar Fonts	S P	Provide Alternative Formats	P F
Reading and Scanning:		Provide Redundant	
Use Reading Performance or		Text Links	FL E
User Preference	L	Provide User-Controlled	
Enhance Scanning	FL	Content	LS
Determine Scrolling vs.			
Paging Needs	S		

4.5.2 Aspects as Checklist: Guidelines for UI

All aspects have normativity (even the deterministic ones). This normativity offers a basis for establishing sound practical guidelines for developing or evaluating UIs or whole computer systems. The shalom principle of simultaneous realization of norms discussed in §3.4.3, emphasises the importance of attending to each aspect. While it is appropriate on occasion to focus attention on one aspect (usually the qualifying) we should always do so in a way that gives all the other aspects their due. If we over-emphasise an aspect, and in the extreme absolutize it, we begin to ignore other aspects, and the result is that the success or fruitfulness of our activity is jeopardised. Thus, for example, a Web page that has superb graphics but is otherwise devoid of useful content it will fall into disuse.

Web pages are user interfaces, and we can see the normativity of many of the aspects recognised in the more mature published Web design guidelines. Table 4-11 shows the "Research-Based Web Design and Usability Guidelines" of the National Cancer Institute (2005) and the main aspects of each guideline (aspects indicated by the first letter of their name, from Q = Quantitative to P = Pistic). Many have two aspects, sometimes because they cover two things [e.g., "set goals" (formative) and "state goals" (lingual)] and sometimes because the main idea is of two aspects (e.g., sharing is both lingual and ethical). We do not differentiate between qualifying and founding aspects here, but could do if a more precise analysis were needed.

We can use aspectual analysis as a basis for critique. The first thing that strikes us is how many aspects are represented here. This is, of course, what one would expect from a good, mature set of guidelines such as the NCI guidelines are. Second, we might look for imbalance among the aspects. The spatial and formative aspects appear more often than most other aspects; we can ask ourselves whether this is appropriate. Perhaps more significant are some gaps, at least in this 2005 version, some of which are quite surprising:

- The pistic aspect of vision of who we are is completely absent, yet one might expect some mention of the designers' vision for the Web site. (It is possible that "Set goals" implies some pistic vision for the site.)

- The ethical aspect of self-giving is present only in sharing design ideas. Guidelines on how to give the reader more than is actually due to them, and thus create a site that feels generous, would be useful.

- The juridical aspect is almost absent, only represented tangentially in the concept of providing "useful" content. The juridical aspect would be relevant in terms of giving both the topic and the readers their due.

- Perhaps most surprising is the almost complete absence of the social aspect—the two inclusions are rather tangential. Since Web sites are read by people from any and every cultural group, with varying background knowledge, expectations and world views, we might expect a whole set of guidelines on appropriate use of cultural connotations, humour, idiom, and on respecting cultural sensitivities.

- The kinematic aspect is almost entirely absent. Animation can be used to show movement, but have the designers of these guidelines overlooked this, treating animation as a mere sensitive or aesthetic decoration?

This aspectual analysis of these guidelines is not meant primarily as a criticism of the guidelines, which are excellent when compared with many others that are available, but rather to show how aspectual analysis can be useful as an evaluation tool, and how it might be used to suggest future improvements.

Figure 4-3. Fir tree of aspectual repercussions

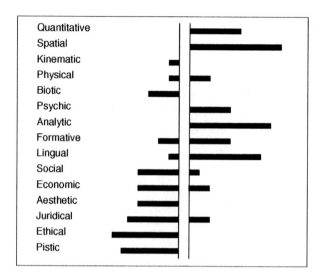

4.5.3 The Aspect Tree: Evaluating Computer Use

Reference to Dooyeweerd's aspects may be made the basis of various methods for practically evaluating a situation of HLC computer use. To provide an overview of aspectual normativity (e.g., in HLC) the simple visual device of a "fir tree" may be used. The aspectual fir tree is a double-sided bar chart showing each aspect, with bars to the left indicating negative functioning or repercussions, and bars to the right indicating positive, as illustrated in Figure 4-3. The length of the bars can be used to indicate the amount (number and/or strength) of positive and negative in each. This devise shows at a glance where the main benefits or problems might lie. Notice that there can be both positive and negative in any aspect.

It can be useful in a number of ways, especially if we take an intuitive grasp of aspectual normativity and analyse the complex lifeworld of use rather than some pre-structured account or prediction thereof. For example, Mitev's description of the Socrate failure (see §4.1.1) has a lifeworld feel, and we can make a simple count of the number of times each aspect is referred to (as expressed by the phrases in the quoted text), either as a functioning or as a repercussion. We obtain the chart illstrated in Figure 4-4.

Even though only a single paragraph has been analysed—and so this picture will be grossly misleading—several things become clear: that this failure involved negative functioning in many aspects, not just the technical (formative), that the aspects in which the significant negative repercussions occur might not be those in

Figure 4-4. Aspectual view of Socrate failure

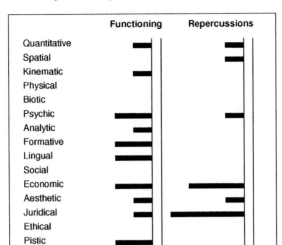

which the negative functioning occurs, and that the most serious repercussions (that Mitev was interested in) were in the juridical aspect. That Mitev was not aware of Dooyeweerd's aspects, but merely adopted a lifeworld stance, is testimony to the power of this approach.

The aspect tree device may be used for both retrospective analysis, as here, or prospective analysis, during design or prediction. It may be a single tree, showing either functioning or repercussions or both together, as in Figure 4-3, or one in which functioning and repercussions are separated, as in Figure 4-4.

But, as a pseudo-numerical device, the aspect tree can give an over-simplified picture if we are not careful. The following guidelines should be noted:

• Never take the length of a bar as some kind of absolute value of aspectual functioning, and never compare two bars of similar length to conclude that functioning in one aspect is "better" than in another.

• But look at the overall patterns and groups. Often, as in Figure 4-3, the main positive functioning or benefit lies in the earlier aspects, while the main negative functioning lies in the later ones. This should give cause for concern because the impact in the earlier aspects is likely to be more visible and to accrue in the short-term while the detrimental impact in the later aspects might become manifest only over the long term. This gives a false impression of the "success" of a computer project if evaluation is undertaken too soon.

- Then look at the longer bars. Do they truly indicate major repercussions or functionings, or do they indicate undue attention to these aspects during analysis? Make a specific study of these aspects, to determine which it is.

- Look likewise at the short or zero bars. Were these aspects overlooked during analysis? Make a specific study to check, then redraw the tree.

- Remember that everybody's understanding of the aspects is variable—by interviewees, authors and analysts—even though we all begin from an intuitive grasp of their meaning (see §3.3.5). So, any further analysis undertaken should include appropriate checks.

In this way, the aspect tree device is not an end in itself so much as a stimulant to focus further analysis, which could be carried out using Winfield's MAKE method, described in Chapter VI.

4.5.4 Questioning Assumptions and Presuppositions

Finally, the author has found Dooyeweerd very useful for questioning assumptions and presuppositions, even in the immediacy of discussion. Frequently the entire discussion revolves around one pole of the Nature-Freedom ground-motive, and drawing attention to everyday factors aligned to the other pole can get discussion out of ruts. So can drawing attention to aspects which are overlooked (see §3.4.3.1); an example is the questioning of the power-relationships "lens" next.

4.6 Relating to Extant Frameworks

Dooyeweerd enables us to construct a framework by which human use of computers may be understood. But we have made at best only passing reference to extant discourse in the area. Here we discuss how extant issues might be addressed from within a Dooyeweerdian framework and how that framework might relate to extant frameworks for understanding use.

4.6.1 Approaches Centring on Limited Aspects

A psychological approach has been a dominant way of understanding HCI since the 1970s. It has spawned much research and many theories, models, and methodologies, from Miller's (1956) famous finding of the "Magic Number 7, plus or minus 2" and the widely-known GOMS model for analysing user interactions (Card, Moran, &

Newell, 1983) to what are verging on comprehensive FFUs like Green and Petre's (1996) cognitive dimensions. By and large, however, this approach is limited to HCI and to considering the psychic, analytic, and perhaps formative aspects thereof. This approach has very little to say about ERC or HLC.

There is a business-oriented approach to HUC, which likewise focuses on a single aspect—the economic. This focuses on HLC and has little to say about HCI or ERC, and is rather limited even for HLC. For example, Landauer (1996) focuses almost solely on productivity, and ignores matters like justice.

Dooyeweerd would see such approaches as scientific rather than philosophical because science focuses on single aspects (see §3.3.8), and thus is unable to furnish us with a comprehensive framework for understanding which is sensitive to the rich diversity and coherence of everyday experience. As such, these approaches could be "plugged in" to our multi-aspectual approaches to HCI and HLC discussed earlier, contributing their specialist expertise, but should never be used in isolation.

4.6.2 "Making a World of Difference"

In his book *Making a World of Difference: IT in a Global Context*, Walsham (2001) provides an insightful discussion of HLC, ranging from individual use, through groupware, organisational use, inter-organisational and up to the level of society (the latter being more of interest to Chapter VIII). He spends two early chapters reviewing a number of approaches, first on the macro level of contemporary society, modernisation, globalisation, and the "information age", then on the micro level of computers at work, covering issues of the relationship between data, information and knowledge, the role of IT, improvisation versus routine work and the appropriation of IT, and finally power and politics. Almost all the approaches are theoretical in nature.

In Part 2, these approaches are assembled into a basket of conceptual and analytical tools, which he employs to analyse a dozen case studies at the four levels from individual to inter-organisational, with a core interest at each level. At the individual level his core interest is in identity of those who use IT (vision of themselves: the kernel functioning of the pistic aspect). At the groupware level, it is teamwork. At the organisational level it is enterprise reorganisation. At the inter-organisational level it is trust. In Part 3, he addresses cultural and cross-cultural issues such as cultural values, with further studies. In Part 4 he discusses the study, suggesting directions and topics for future research. He concludes by saying, as mentioned in Chapter I of this book, that his professional interest could be encapsulated in the question "Are we making a better world with IT?"

The basket of analytical and conceptual tools that Walsham used is impressive for its variety and its ability to cover both macro and micro issues. The macro-level tools include views on modernity by Giddens, on the "risk society" by Beck, on globalisation by Robertson, and on the "information age" by Castells. The micro-level tools include views on data, information, and knowledge by Checkland and Holwell and various thinkers in knowledge management, the lens of social construction of technology, Latour's actor-network theory, views on improvisation by Suchman, Weick, and Ciborra, views on cross-cultural appropriation of IT by various authors, various views on surveillance, and the lens of Foucault's notion of power-relationships. Most of these ideas are not discussed here, only the way Walsham uses some of them; for references see Walsham (2001).

These may be assumed (because of the way they are used) to largely constitute Walsham's framework for understanding these areas of IT use and the societal issues: what he considers important and meaningful, what he considers good or problematic, what cultural baggage he comes with, the scientific paradigms he will entertain and conceptual frameworks he employs.

What interests us is to what extent Walsham's framework for understanding might suit our requirements here, and to what extent Dooyeweerd might engage with it. A general critique will be made first. Then two of Walsham's cases will be outlined to allow discussion of how his basket of tools is used.

4.6.2.1 General Critique

The general critique identifies the limitations in scope of Walsham's work.

First, Walsham's study must first be commended for covering both IT use and what we have called technological ecology. He brings these together into a single volume and shows some points of contact between the areas. Though he refers to it, he does not discuss ISD in any depth, and his work casts no light on the nature of computers nor on the shape of technological resources available to IS developers. But in the two areas he does cover, Walsham shows a commendable interest in and appreciation of non-Western cultures. He is one of few to do so. But, as will become plain later, it is undermined by one of his conceptual tools.

Second, throughout his work runs an interest in the issue of identity: who people believe themselves to be in relation to their work. This is commendable because identity is of the pistic aspect of IS use, which is too seldom considered. But a possible over-emphasis on identity means that many important aspects of IT use were not reported fully, if at all, such as social impact on families, to what extent customers and others received their due or change of lifestyle (such as increased or decreased use of road transport). Therefore, Walsham's work should not be used as examples of the general picture of IS use.

Third, the focus of IS use is on work, especially professional work. This means that his framework for understanding is not attuned to other types of use, including home use, computer games, computer art, and the like. Most of his cases, even of individual working, involve organisational systems, so it is not clear to what extent his framework is suited to, for example, studying personal use of word processors or photographic software.

However, these criticisms should not necessarily be seen as weaknesses, only as indicating limited scope. To demonstrate weaknesses, two of his cases will be examined.

4.6.2.2 Two Cases

One case is of individual use of IT, while the other, drawn from the opposite end of the spectrum, is of inter-organisational activity.

ComCo produced plotters, had a team of engineers to maintain them, but then outsourced the plotter business. The engineers became agents. The company had an IT system, Traveller, to enable them to communicate, such as log their visits. Walsham cites stories from three engineers, called Gary, Keith, and Neil. Neil loved fixing machines, did not see his job as including answering customers' other questions, and subverted and bypassed the Traveller system. Keith used the system, and ComCo itself, to further his own goals (setting up his own company). Gary wanted to give good service to customers, was loyal (in spirit as well as letter) to ComCo, used Traveller, but also went beyond the call of duty, such as keeping his own stock of spares.

The EPS (electronic placement system, the system mentioned earlier in the chapter) was intended to assist insurance brokers in placing risks with underwriters in the London Market. Currently brokers queue outside underwriters' offices with slips of paper on which all the details of the risk are recorded, along with information from subsequent negotiations. The EPS would replace face-to-face meetings with electronic transmission of risk information and make such queuing unnecessary. Also a swathe of underwriters could be targeted with a risk in one go, and those who accepted it most quickly would get the business. The EPS was not widely used however, even several years after its launch. Walsham discussed some of the reasons for this, including diverse views of the merits or demerits of electronic negotiation, and whether the inherent complexity of negotiation is "seriously impaired". But the main reason, according to Walsham, is that the system did not support the principle of Utmost Good Faith, which requires, and assumes, that (p.155), "a broker must display all known relevant information about the client and the insurance risk to underwriters upon presentation of the risk to them. ... A key issue concerned the difficulty of the broker inputting all relevant information ..." A more fundamental problem was that face-to-face interaction facilitates the devising of innovative insurance products, for which the London Market is renowned. Walsham sums up the case with (p. 157), "Electronic trading offers some opportunities for speed, efficiency and the bridging of time and place. At the same time, complex insurance risks need delicate and sophisticated negotiation." This boils down to a balance between economic and aesthetic aspects.

4.6.2.3 Critique Arising from Cases

Walsham's description of both cases exposes a diversity of lifeworld issues, and may be commended for displaying the humanity of all the participants. Support for this view can be found in aspectual analysis of the text (similar to that undertaken for Mitev's case), which shows almost all human aspects represented.

But while the descriptions of cases exhibit a lifeworld diversity and meaning, the interpretation of them does not. This might be because his basket of tools seems more like a collection than a coherent whole, in that Walsham simply throws them all in together, without showing the relationships between them. He admits (p. 62):

The conceptual tools ... will be used to support analysis of these topics, but not in a mechanistic way. In other words, I will not use all of them explicitly in all the chapters, but rather use the conceptual tools selectively to illustrate specific analytical points.

There is therefore every possibility (except perhaps for Walsham's own integrity) that he has selected the tools he prefers. Even when there is no undue selectivity, the analysis will be partial or distorted because some of the tools assume a theoretical rather than lifeworld attitude (§3.4.5).

4.6.2.4 Queuing Time

It seems from the number of places it is mentioned in the text, that one of the main benefits sought for the EPS was reduction in time spent in queuing. This is reasonable only on the assumption that queuing is solely an economic issue (in Dooyeweerd's sense). But presumably (and this is not recorded) brokers had, over the years, devised informal ways of ensuring queuing time was not simply wasted time, such as by discourse and social activity in the queue, or by using it as thinking time. These are lifeworld issues, to which a framework for understanding should be sensitive.

But it is not clear how Walsham's basket of tools encourages the researcher or analyst to consider these. It might not prevent these issues emerging, but it provides no positive method for surfacing them, except methods that depend on participants remembering issues and being able and willing to make them explicit. As a result, it seems, Walsham was too easily satisfied with the view of queuing as it is seen by the formal system: as nothing but wasted time, and there is no indication that he probed more deeply about this particular issue.

But Dooyeweerd, with his ready-to-hand suite of aspects, might analyse queuing thus, by considering each aspect:

The main visible problem with queuing is time-waste (economic aspect), but probably the main motivator of dislike is fear of boredom (sensitive and aesthetic aspects). However, opportunity exists to use the time imaginatively (formative aspect), for social or communicative (social, lingual) purposes or for thinking (analytic, formative). Moreover, it is also likely that a certain attitude has developed about queuing, either grumbling or accepting (ethical aspect) perhaps bolstered by a (pistic) view that "I am too important to have to queue" or alternatively "This is what we have always done."

Such a scenario immediately recommends itself to our lifeworld attitude as reasonable, whatever the details.

In practical analysis, once the aspects have been learned, their kernel meanings are so intuitive that the previously mentioned issues can occur to the analyst within the space of a few seconds. Then non-leading questions might be posed to find how the queuer responds to queuing in every aspect. Such aspectual analysis does not need to take long. As has already been found (Basden & Wood-Harper, 2006), Dooyeweerd's suite of aspects can act as an excellent surfacing tool, which actually stimulates people to think of things normally overlooked. Moreover, Kane (2006) has found that the aspects used in interviewing, rather than constraining interviewees, can actually free them to mention things they might consider either irrelevant or embarrassing.

4.6.2.5 The Lens of Power-Relationships

However, there is something much more problematic, because less explicit in Walsham's work. The framework for understanding that Walsham used to comment on both cases, while it exposed some useful factors, led him into distorting his understanding of what was going on rather than being fully sensitive the lifeworld of the people concerned.

The Foucauldian conceptual framework of power-relationships, which Walsham brings as a lens to the interpretation of both cases mentioned previously, distorts it. Among the description of the EPS case is (Walsham, 2001, p. 156):

Shifting power relations between underwriters and brokers were also a subject of concern to both parties. On the one hand, the EPS system could facilitate simultaneous risk transmission, giving the broker the opportunity to "flood" the market by sending the risk to a large number of underwriters at the same time, rather than queuing outside an individual underwriter's office. The underwriters feared a drastic change in trading conditions, from a leisurely but measured discussion of the terms of the risk, to a situation where the risk was eventually placed with only those participants who responded quickly and at the lowest price. // However it is by no means obvious that the underwriter would be the loser in the longer term ...

While there might be some competition between brokers and underwriters, to see the relationship between underwriters and brokers as purely or mainly one of (competitive) power distorts the case. The key is the word "measured", a juridical word in this context: the underwriters were concerned that each risk received its due. This juridical concern, while it can involve and be mixed up with issues of power, goes beyond power. The power-relationships lens is fundamentally unable to recognise it.

The distortion is even more evident in the ComCo case. While his Foucauldian power-relationships lens might be appropriate to Keith and to some extent Neil, it was not appropriate to Gary's story. Walsham's interpretation (p. 91) is that "Gary was able to draw on his deep knowledge of plotters to leverage power over ComCo who placed high value on his expertise." But there is no mention in the text itself of the story of Gary seeking in any way to "leverage power over ComCo", not even a hint of it. On the contrary, Gary seems to have exhibited the very opposite, his whole attitude being characterised instead by loyalty to both Comco and customers and by self-giving generosity. And Walsham even admits (p. 69), "Gary ... was content ..."

4.6.2.6 Enriching Walsham's Basket of Tools

The tendency to see things in terms of power-relationships is by no means restricted to Walsham. This is a growing tendency among some IS academics, and it needs to be challenged, but that is for another occasion. What interests us is the effect of the power-relationships lens. There is often two casualties, both illustrated in Walsham's cases. One is a concern to ensure what is due, the other is self-giving or generosity, motives which those who adopt a power-relationships lens tend to find difficult to see. Self-giving in particular is almost the very antithesis of power-relationships. It is ironic that in many non-Western cultures, in which Walsham shows a particular interest, generosity is part of the way of life, in sharp contrast to Western competitiveness.

This is not to diminish the importance of Foucault's insight into power and its pervasiveness, but rather to object to making this insight the main key to understanding situations. At the risk of over-simplifying, it seems the notion of power has been useful in the IS community to move away from a "managerial" approach; this may, in Dooyeweerdian terms, be seen as a dialectical swing from the Nature to the Freedom pole of the Nature-Freedom ground-motive (§2.3.1.4). It can also be seen as an absolutization of an aspect. Foucault's actual notion of power is obviously rich and complex (in Dooyeweerdian terms multi-aspectual) but the way it is referred to in IS has a strong formative element: the attempt to shape things, the aspect of will. What tends to happen is that this aspect is absolutized, making other aspects invisible. This is not surprising, when one considers Foucault's debt to Nietzsche. Dooyeweerd's suite of aspects might provide a lens-system that is more sensitive to the lifeworld, because of his sophisticated understanding of the lifeworld (§3.4.5), especially in non-Western cultures like Sub-Saharan Africa, where generosity is more of a way of life. Aspectually-centred perspectives (§3.3.4) act as lenses that bring into focus a particular type of meaning. The formative aspect brings certain issues of power into focus. The pistic aspect brings issues of identity into focus (Walsham's particular interest). But with Dooyeweerd's suite of aspects we have not one but 15 lenses with which we can bring many different things into focus.

This, coupled with the suggestion mentioned earlier that Dooyeweerd's suite of aspects could be used to get under the system-visible surface to the lifeworld diversity of meaning underneath, means that this part of Dooyeweerd's philosophy, his suite of aspects, might enrich Walsham's basket of tools in a number of ways. Exploration of this is needed.

4.6.3 Winograd and Flores—and Beyond

Soon after it was published, Winograd and Flores' seminal work *Understanding Computers and Cognition* (1986), made a profound impression on this author. It did not so much create a new way of looking at computers in him, as undergird and express what he had already felt and believed for over 10 years. Moreover, though Polanyi's "tacit dimension" (1967) was his mainstay at the time, Winograd and Flores also helped him understand the difference between distal and proximal user interfaces (discussed earlier) before he discovered Dooyeweerd.

Winograd and Flores (W+F) questioned the prevailing "rationalistic" approach to computers especially found in AI and suggested an approach based on phenomenology, hermeneutics, and language theory. The first challenge was to the way computers were understood in terms of the Cartesian subject-object relationship, as objects distal from, and operated upon by, humans. In place of this W+F offered the ideas of "thrownness" and "breakdowns" based on Heidegger's notion of being-in-the-world. Their second challenge was to the assumption that cognition is the manipulation of knowledge of an objective world, and that we can hope to construct machines that exhibit intelligent behaviour (as AI hoped to do). Instead, using Maturana's notion of autopoiesis, they argued that cognition is an emergent property of biological evolution and that interpretation arises from cognition, and that computers themselves can never be made truly intelligent. Their third challenge was to assumptions that language is constituted in symbols with literal meanings, that such symbols can be assembled into a knowledge base, and that they are used within organisations as a means of transmitting information. Instead, in accord with Searle's speech act theory, the listener actively generates meaning especially as a result of social interaction, and language is action, responsible for creating social structures, not just being used within them. It is impossible, they argued, for computers to use language in the way humans do (even though they might process natural language).

Using this approach, they suggested "A new foundation for design" of computer systems. The aim of AI and KBS should be redirected, away from an attempt to make computers "intelligent" or to support "rationalistic problem-solving", toward building useful systems that are "aids in coping with the complex conversational structures generated within an organization" (p. 12). They continue, "The challenge posed here for design is not simply to create tools that accurately reflect existing domains, but to provide for the creation of new domains." This, they hope, will open the way to social progress and "an openness to new ways of being" (p. 13). They outline the design of a Coordinator system to support cooperative work.

W+F's work is still avidly discussed, and even inspirational, 20 years later (Weigand, 2006). It deserves to be because it provides a framework for understanding three of the areas of research and practice (HUC, nature of computers, and ISD) and touches on that of technological ecology. It is seen as a flagship of the language-action perspective, which focuses on computer use in organisations and especially the use of language in changing them.

4.6.3.1 Problems in Winograd and Flores

Though Winograd and Flores have inspired many in academia, it is not clear what actual impact on the design of computer systems and user interfaces as a whole their views have had. Much that has happened could be traced instead to Shneiderman's notion of direct manipulation, the much-quoted example of the Apple Macintosh, or the everyday, practical creativity of, for example, computer games designers.

One of the problems with W+F is their focus on organisational IS and the work context. That computers are seen as bundles of conversations, and useful in coordinating networks of speech acts, raises questions about the extent to which their framework can throw light on other types of computer use, such as games, virtual reality, computer art, or devices to make scientific calculations or simulations, which are difficult to see in terms of speech acts, except perhaps metaphorically or by theorizing them.

Their whole section devoted to "Design" contains guidance only at a rather high level ["Speech act origination ... Monitoring completion ... Keeping temporal relations ... Examination of the network ... Automated application of recurrence ... Recurrence of propositional content" (pp. 159-161)]. One might expect readiness-to-hand would be important in computer games, but their guidance is all in terms of "conversations" between people in the application area of decision support in management.

But those are comments on the scope of W+F's proposal rather than a criticism of it as such.

A more substantial criticism is made by Spaul (1997), who questions W+F's claim to have opened the way to social progress. Weigand (2006) also mentions a similar concern by Suchman. Winograd and Flores offer no way of differentiating social "progress" from its opposite—nor even the more limited notions of benefits or detrimental impact—because neither Heidegger nor Maturana nor Searle offer a solid philosophical normativity. There is little benefit in making computers more "ready-to-hand" if what we use them for is harmful. The use of W+F's tool would be "unreflective".

Spaul argues that critique of, and change in, social conditions and structures fundamentally cannot be based on Heidegger because critique and change need a distancing of oneself from the social milieu. (A flaw in Spaul's argument must be exposed which, as far as we know, has not yet been discussed. Spaul conflates HLC with HCI, and proximal engagement with the computer is of the latter, while critique of social conditions is of the former. There is no reason to believe that critical distance in HLC, which is what Spaul seeks, is incompatible with a proximal relationship in HCI, which is what W+F seek. Indeed, in the case of *Elsie*, it was its very ease of use that led to change of the social structures of the relationships between surveyor and client.) Nevertheless, Spaul's highlighting of the fundamental problem with Heidegger is valid, and has been mentioned by others also.

In the end, Spaul suggests that we need to combine Heidegger with the Cartesian distancing of thinking subject from thought-about object and briefly suggests how a Habermasian notion of the difference between lifeworld and systemic life might achieve this. Spaul's suggestion would seem doubtful because of the supposed opposition of Heidegger and Descartes when viewed from the Nature-Freedom ground-motive, were it not for Dooyeweerd's claim that he can resolve this antinomy.

It seems that Heidegger must be combined with other ideas. Winograd and Flores felt constrained to appeal to evolutionist philosophy to try to account for how cognition is not in Cartesian separation from the world but is part of that very (biological) world, and, when they attempted to work their idea out in their Coordinator system, they had to supplement Heidegger with Searle's speech-act theory. Spaul suggests Descartes and Habermas. They all assumed, rather than justified or argued, the validity of such combinations. Combining the ideas of two different philosophers might be considered creative and fruitful, but it raises serious questions, first about the choice of philosopher to accompany Heidegger, and second about the sufficiency of Heidegger. Ironically, Spaul points out the radical incommensurability between Searle and Heidegger.

That Heidegger needs to be combined with other philosophers, whether Searle or Descartes, suggests a radical insufficiency as a basis for formulating frameworks for understanding. Why is it that Heidegger, or perhaps other existential phenomenologists, is not sufficient to offer a basis for practical IS development, for normativity, nor for critical distance?

4.6.3.2 Shift to Another Ground-Motive

Dooyeweerd made a comprehensive study of Heidegger and, though many of their ideas overlap, and both thinkers exposed Kant's failure to make the synthesis between understanding and sensibility into a critical problem, Dooyeweerd criticised Heidegger too for being "unable to pose it in a truly critical way" because "he clung to the immanence standpoint [see §2.3.3] even more tightly than Kant had done" (Dooyeweerd, 1984, II, p. 536). As to many before him, so to Heidegger, the transcendental, autonomous ego was still sovereign (Dooyeweerd, 1984, I, p. 112) and he presupposed a theoretical synthesis (Dooyeweerd, 1984, II, p. 536). This might help apprehension and amelioration of the earlier mentioned problems, in the following manner (which is a summary of the full argument).

If we presuppose an autonomous ego we rule out a normativity that transcends us. But a transcending normativity is required for a critique of the social milieu. That problem is inherent, Dooyeweerd argued, in the NFGM. Moreover, presupposing a theoretical synthesis makes it difficult to engage fully with the diverse coherence of everyday life, including designing a system like the Coordinator. That problem is inherent to the immanence standpoint (§2.3.3).

There are two ways forward. One is to allow Dooyeweerd to be the companion of Heidegger, instead of Searle, Maturana, Descartes, or Habermas, supplying both a genuine normativity and a fully lifeworld attitude, while at the same time being in deep sympathy with Heidegger's notions of *Dasein*, and so forth. That avenue should be explored, in its potential to design systems and also critique social structures.

The other is to found Winograd and Flores' work directly in Dooyeweerd rather than in Heidegger. The framework developed in this chapter can do this. It largely satisfies W+F's aims, and also addresses Spaul's criticism, as follows:

- Use of computers is multi-aspectual human functioning in which the user is fully engaged by virtue of responding as subject to all aspectual law. When considering HCI, in contrast to the Cartesian subject-object relationship, which implies distal separation, Dooyeweerd's law-subject-object relationship implies proximal engagement.

- Our engagement with the world no longer has to invoke evolutionist philosophy because it is fundamental to Dooyeweerd's notion of law-subject-object.

- The design of a system like the Coordinator, the purpose of which is qualified by the lingual aspect, must of course make use of humanity's best knowledge of that aspect, which will probably include Searle's speech-act theory in the role of a scientific conceptual framework rather than a philosophical framework. But Searle is not as useful for other types of software that are qualified by other aspects. The design of a computer game would be qualified by the aesthetic aspect. This concerns ERC.

- The HLC use of IS to critique social structures involves the user's functioning in post-social aspects, especially the juridical. This in no way requires Cartesian separation. Viewing *Elsie* though a Dooyeweerdian lens, it was not primarily critical distance that led to the change in social structures, but rather a functioning in the ethical aspect of (being willing to be) self-giving.

- While, strictly, Dooyeweerd does not force us to differentiate HCI, ERC, and HLC, his prompting us to focus on types of multi-aspectual human functioning made it easier to do so and provided philosophical grounding for doing so.

Though this is only a summary of a fuller argument, it demonstrates the utility of Dooyeweerd in first exposing what might be the root of the problems that attend a framework for understanding like that proposed by Winograd and Flores, then suggesting on the one hand how Dooyeweerd could overcome some of those problems while retaining the original philosophical basis, and on the other how the framework could be grounded in Dooyeweerd directly.

4.7 Conclusion

Winograd (2006, p. 73) says, "The field of interaction design is in its infancy and we are still struggling with finding the appropriate foundational questions and concerns for new kinds of interactions." In his short article, it is clear that he is looking beyond HCI. The reinterpretation of human use of computers (HUC) using Dooyeweerd's philosophy that has been presented here not only goes beyond HCI, and even HLC, but has raised a number of foundational questions and concerns, and suggested some new directions for research. Here the process by which the framework was developed is reviewed and the framework itself is summarised.

The starting question was with what attitude we should approach HUC. The decision was made to seek to understand everyday use of computers, rather than limited aspects thereof. So a few cases were examined, including one of which the analyst (this author) had intimate everyday knowledge. This led to:

- Use of computers is multi-aspectual human functioning, which may be examined by aspectual analyses of various kinds. The social aspects are particularly important. Example: Socrate.

This analysis revealed several different types of things in each aspect, which led to distinguishing several multi-aspectual functionings. We then sought philosophical grounds for the difference, by examining their qualifying aspects:

- Overall use is constituted of several multi-aspectual functionings, enkaptically bound with each other:
    - o   HCI (qualified by the lingual aspect; example: *Elsie* UI)
    - o   ERC (qualified by any aspect, but which is related to general purpose of IS; example: *ZAngband* computer game)
    - o   HLC (qualified by any aspect related to the type of use to which the user puts it; example: *Elsie* in use, Walsham's EPS)

This put us in a position to seek deeper understanding of each. First the structure of each was examined, especially the relationship between human beings and the IS, and then the normativity that governs each. This yielded:

- **Structure of HCI:** May be understood in terms of law-subject-object and *Gegenstand* relationships (and thus differentiate distal from proximal).
- **Structure of ERC:** May be understood as the lingual aspect reaching out to all spheres of meaning of the domain (especially for knowledge bases and virtual worlds).
- **Structure of HLC:** May be understood as aspectual repercussion, with unexpected impacts analysable by reference to transcendental cosmic law, and aspectual crossover leading to indirect impacts (example: the cases cited in §4.1 earlier).

- **Normativity of HCI:** Usability as multi-aspectual normativity (example: Web site design).

- **Normativity of ERC:** What is represented should do justice to the domain meaning, whether virtual or modelled (example: virtual reality).
- **Normativity of HLC:** Bring shalom to human living: the IS should enhance HLC in various aspects and harm it in none (example: *Elsie*).

Three practical devices were briefly described: aspectual analysis, aspectual checklists, and aspectual tree.

The benefits of this framework is that it not only enables but encourages us to address a wide range of issues sensitively and in new ways. Yet it also provides the means to examine each in depth, fully aware of all the other issues. So this framework will help form more useful guidelines to get the best out of IS, and will reduce the detrimental impact of unexpected repercussions of IS use. It is an ethical framework because it provides a broader understanding of what is to be sought or avoided. It is holistic because it provides a means to see the points of contact and conflict between areas. Some links were made to other areas, ensuring coherence with them. This framework does not necessarily replace others such as those by Walsham (2001) and Winograd and Flores (1986), but can critique, underpin, and enrich them.

But, as Winograd and Flores indicated, the human use of computers will eventually be linked with another issue: what is the nature of computers and information?

References

Avison, D., Bjørn-Andersen, N., Coakes, E., Davis, G. B., Earl, M. J., Elbanna, A., et al. (2006). Enid Mumford: A tribute. *Information Systems Journal, 16*, 343-382.

Bartle, R. A. (2004). *Designing virtual worlds*. Indianapolis, IN: New Riders.

Basden, A., Brown, A. J., Tetlow, S. D. A., & Hibberd, P. R. (1996). Design of a user interface for a knowledge refinement tool. *International Journal of Human-Computer Studies, 45*, 157-183.

Basden, A., & Wood-Harper, A. T. (2006). A philosophical discussion of the root definition in soft systems thinking: An enrichment of CATWOE. *Systems Research and Behavioral Science, 23*, 61-87.

Brandon, P. S., Basden, A., Hamilton, I., & Stockley, J. (1988). *Expert systems: Strategic planning of construction projects*. London: The Royal Institution of Chartered Surveyors.

Card, S. K., Moran, T. P., & Newell, A. (1983). *The psychology of human-computer interaction.* Hillsdale, NJ: Erlbaum.

Castell, A. C., Basden, A., Erdos, G., Barrows, P., & Brandon, P. S. (1992). Knowledge based systems in use: A case study. In *British Computer Society Specialist Group for Knowledge Based Systems, Proceedings from Expert Systems 92 (Applications Stream)*. Swindon, UK: British Computer Society.

Collins, J. C., & Porras, J. I. (1998). *Built to last: Successful habits of visionary companies.* London: Century.

Davis, F. D. (1986). *A technology acceptance model for empirically testing new end-user information systems: Theory and results.* Unpublished doctoral dissertation. MIT Sloan School of Management, Cambridge, MA.

Dooyeweerd, H. (1984). *A new critique of theoretical thought* (Vols. 1-4). Jordan Station, Ontario, Canada: Paideia Press. (Original work published 1953-1958)

Downton, A. (1991). *Engineering the human-computer interface.* Maidenhead, UK: McGraw-Hill.

Dumas, J. S., & Redish, J. C. (1999). *A practical guide to usability testing* (2nd ed.). Exeter, UK: Intellect.

Eason, K. (1988). *Information technology and organisational theory.* London: Taylor and Francis.

Eriksson, D. M. (2006). Normative sources of systems thinking: An inquiry into religious ground-motives of systems thinking paradigms. In S. Strijbos & A. Basden (Eds.), *In search of an integrative vision for technology: Interdisciplinary studies in information systems* (pp. 217-232). New York: Springer.

Ferneley, E., & Light, B. (2006). Secondary user relations in emerging mobile computer environments. *European Journal of Information Systems, 15*(3), 301-306.

Friedman, A. L., & Cornford, D. S. (1989). *Computer systems development: History, organization and implementation.* Chichester, UK: John Wiley & Sons.

Green, T. R. G., & Petre, M. (1996). Usability analysis of visual programming environments: A cognitive dimensions framework. *Journal of Visual Languages and Computing, 7,* 131-174.

Greeno, J. G. (1994). Gibson's affordances. *Psychological Review, 101,* 336-342.

Jones, G. O., & Basden, A. (2004, April 19-24). Using Dooyeweerd's philosophy to guide the process of stakeholder engagement in ISD. In M. J. de Vries, B. Bergvall-Kåreborn, & S. Strijbos (Eds.), *Interdisciplinarity and the integration of knowledge: Proceedings of the 10th Annual Working Conference of CPTS* (pp. 1-19). Amersfoort, Netherlands: Centre for Philosophy, Technology and Social Systems.

Kane, S. C. (2006). *Multi-aspectual interview technique.* PhD Thesis, University of Salford, UK.

Landauer, T. K. (1996). *The trouble with computers: Usefulness, usability and productivity.* Cambridge, MA: MIT Press/Bradford Books.

Laurel, B. K. (1986). Interface as mimesis. In D. A. Norman & S. W. Draper (Eds.), *User centred system design: New perspectives on human computer interaction* (pp. 67-85). Hillsdale, NJ: Erlbaum.

Miller, G. A. (1956). The magic number seven, plus or minus two: Some limits on our capacity for information processing. *Psychological Review, 63*(2), 81-97.

Mitev, N. N. (1996). More than a failure? The computerized reservation systems at French Railways. *Information Technology and People, 9*(4), 8-19.

Mitev, N. N. (2001). The social construction of IS failure: symmetry, the sociology of translation and politics. In A. Adam, D. Howcroft, H. Richardson, & B. Robinson (Eds.), *(Re-)defining critical research in information systems* (pp. 17-34). Salford, UK: University of Salford.

National Cancer Institute. (2005). *Evidence-based guidelines on web design and usability issues.* Retrieved October 19, 2005, from *http://usability.gov/guidelines/*

Norman, D. (1990). Why interfaces don't work. In B. K. Laurel (Ed.), *The art of human-computer interface design.* Reading, MA: Addison-Wesley.

Polanyi, M. (1967). *The tacit dimension.* London: Routledge and Kegan Paul.

Shneiderman, B. (1982). The future of interactive systems and the emergence of direct manipulation. *Behaviour and Information Technology, 1*(3), 237-256.

Spaul, M. W. J. (1997). The tool perspective on information systems design: What Heidegger's philosophy can't do. In R. L. Winder, S. K. Probert, & I. A. Beeson (Eds.), *Philosophical aspects of information systems* (pp. 35-49). London: Taylor and Francis.

Tufte, E. (1990). *Envisioning information.* Cheshire, CT: The Graphics Press.

Ulrich, W. (1994). *Critical heuristics of social planning: A new approach to practical philosophy.* Chichester, UK: Wiley.

Walsham, G. (2001). *Making a world of difference: IT in a global context.* Chichester, UK: Wiley.

Weigand, H. (2006). Two decades of the language-action perspective. *Communications of the ACM, 49*(5), 45-46.

Winograd, T. (2006). Designing a new foundation for design. *Communications of the ACM, 49*(5), 71-73.

Winograd, T., & Flores, F. (1986). *Understanding computers and cognition: A new foundation for design.* Reading, MA: Addison-Wesley.

Chapter V

A Framework for Understanding the Nature of Computers and Information

"What is a computer?" and "What is information?" are questions that the reflective user will sometimes ponder. These questions do not refer to what the computer means to us, subjectively (or intersubjectively) nor as enabling particular applications, whether controlling or emancipatory in nature; those issues were addressed in Chapter IV. What is at issue is the nature of the computer regardless of application and of (inter-)subjective meaning. What is it that differentiates computer from, for example, mechanical machine or electronic gadget on one hand, or from other information technology such as writing, printing, film, or video on the other? Pretheoretical experience continues to assert that there is a difference.

But perhaps the comparison that has been most discussed is that between computer and human—the artificial intelligence (AI) question: can computers think or understand? A lot follows from any answer offered.

More recent is the question about the nature of cyberspace. John Perry Barlow's (1996) *Declaration of the Independence of Cyberspace* is a polemic that claims cyberspace is a different type of reality, a reality of mind, information, thought, in which body and matter are either not needed or irrelevant. If this is so, he claims, then we should have different social arrangements, different ethics, different views of what is considered criminal or legal, different legislative frameworks and different freedoms. The legal systems of the old, matter-based reality no longer apply.

In direct opposition to Barlow is the feminist notion of embodied knowledge. Not only does knowledge need a body, but propositional, conceptual knowledge is not true knowledge at all, or is at most only one kind of knowledge. There is a considerable amount of knowledge in our bodies as opposed to minds. The elevation of mind and logic over body and feeling is a conspiracy of masculinity. Haraway's *Cyborg Manifesto* (1991) is perhaps the best-known version of this, a polemic like Barlow's, but much longer and dressed up in academic observations and questions, but a polemic nonetheless.

How do we respond to such claims? How might we get behind the polemic, and engage critically with them? Is it possible that both claims could contain useful insight?

In attempting to understand the nature of computers and information, we cannot accept any of these views uncritically—nor do we accept their claim to set the agenda for debate. Nor do we accept naïve realism. Rather, we seek a framework that enables us to understand computers and information primarily as they present themselves to us in the lifeworld, as Dooyeweerd recommended (§2.4.2).

This chapter explores how Dooyeweerd's philosophy might provide a basis for understanding the nature of computers, information, and programs. It does not attempt to arrive at a single "best" definition, but rather to provide a foundation for fruitful discourse. It begins by critically examining what is meant by the question "What is?," then presents a Dooyeweerdian answer, and ends with discussing one extant framework and some of the earlier questions.

5.1 What is Meant by "What is?"?

The question, "What is a computer?" is often answered by reference to a definition. Webster's Dictionary (1971) defines computer as, "a calculator esp. designed for the solution of complex mathematical problems; *specif.* an automatic electronic machine for performing simple and complex calculations." But, for three reasons, such definitions do not help much. First, definitions tend to theorise the type of thing they define and privilege specialist views over everyday experience. This definition is far too narrow to understand the nature of computers today. Second, definitions change. In the 1993 edition of Webster's Dictionary the second part has been replaced with "a programmable electronic device that can store, retrieve, and process data." Third, definitions presuppose the meaning of such words as "programmable", "electronic", "device", "store", "retrieve", "process", and "data", which all have to be understood if the definition is to help us understand the nature of computers. Because of these problems, the nature of computers becomes a matter for philosophical analysis rather than definition. Philosophical ontology becomes important.

5.1.1 Philosophical Understanding of the Nature of Things

The question of the nature of computers, information, program, and so forth, is part of the wider question of the nature of Being or Existence: what does it mean to be a particular type of thing (a computer). On what basis may we differentiate one type of existence from another?

It may be no coincidence that those in this area often refer back to Greek thought, or early modern thought which was influenced by Greek thought, because it was the Greeks who took seriously the ontological questions about the nature of things and they did so from within the Form-Matter ground-motive (FMGM; see §2.3.1.1). Under the Nature-Grace ground-motive (NGGM, §2.3.1.3) only the sacred was worthy of our philosophical attention. Under the Nature pole of the Nature-Free-dom ground-motive (NFGM, §2.3.1.4), physics and mathematics seemed to give the complete answer to the nature of things, so no further exploration was required (the Romantics were a dissenting voice). But this obliterated human freedom, so a swing to the Freedom pole occurred. Under the Freedom pole, post-Kant, it is presupposed that the "thing in itself" cannot be known, so ontology is replaced by, or made subservient to, epistemology.

As a result, it is fashionable in some areas of IS, especially those of ISD or societal matters, to eschew "essentialism", and thus seemingly to forbid any attempt to discuss the nature of computers as such. Extreme versions of this view would, at a stroke, wipe out this whole area of research and practice as completely misguided. But that approach fails to do justice to our everyday experience, that there is such a type of thing as "computer", "information", and "program".

Less extreme versions might allow discussion of the nature of computers, but would always derive this from our (inter-)subjective views of what a computer is, which are diverse and open-ended. But such views do not allow the thinker to take seriously that the computer presents itself to us as something with an integrity beyond our (inter-)subjective beliefs about it, or our application of it. Moreover, they provide no grounds for discussing the real potential and limitations of what is computer, such as expressed by the AI question of whether computers can truly understand, nor for either hope or despair regarding those, nor of the ethical and legal implications of any supposed "new reality".

So those who wish nevertheless to explore the nature of computers revert to FMGM thinking because it seems to be the only one available so far. Questions about the nature of computers are couched in terms of brain versus mind (though other versions will be discussed later). Systems theory's notion of emergence, such as in Wilber (2000), may be seen as a multi-level version of FMGM.

But Dooyeweerd, in his complete break with immanence-philosophy, which underlies FMGM, NGGM, and NFGM, offers a different way of understanding Being, under the Creation-Fall-Redemption ground-motive (CFR, §2.3.1.2), and a different approach to ontology that does not need to be subsumed into epistemology, yet answers the challenges that fired Kant and all since him (see §2.4.5). It reinterprets this whole area, to throw up new issues and shed new light on existing ones.

5.1.2 Some Issues

The nature of computers is closely bound up with human experience. It has links with use, discussed in Chapter IV, but mainly with HCI rather than either ERC or HLC. This provides a useful link between these two areas of research and practice: how the user experience computers in everyday living should at least be commensurate with how we understand the nature of computers, information, and program as such, even though not identical with it. Therefore, at least, HCI may be a good place to begin in order to discern a number of issues.

The first issue is **diversity of our experience of computers**. The computer presents itself to us in many ways—as hardware, as pixels and sound, as symbols, as useful content, and so on. A certain science might focus on one of these, but it cannot account for them all. Newell (1982) recognised this, and suggested that computer systems are multi-levelled beings, understandable in a number of distinct ways. A similar approach may be seen in discussions of the relationship between data, information, and knowledge.

The second issue is **potential of computers**; on what basis may we explore and discuss the potential of computers? What will they be able to do, and not do, once current limitations are overcome? What is the status of a statement like "Computers can think"? Will we end up with cyberspace as a kind of reality that is completely independent of matter and flesh? Or are there fundamental barriers to either of these? If so, or if not, why?

The third issue is the **innards of the computer**. Our everyday understanding of the computer tells us there are files "in" the computer. There are bits and bytes, numbers, data structures, and so forth. This chapter on the nature of computers is (hopefully) "in" my computer. But how can I say they are really "in" the computer, if, when I open the case, all I see is the motherboard, cables, fan, and so forth? There are three practical reasons why it is valid to believe these are "in" the computer:

- One is the rather obvious reasoning that since it is we who placed the files on disk, and can retrieve them, so they must be there, and since it is we who input various data, so that data, or some result of processing it, must be there.

- The second is by analogy with human functioning. For example, human semantic memory is often likened to a data structure inside the computer.

- The third is that since we see something on the screen (in each aspect) then it is reasonable to believe that something "inside" the computer "caused" this to be displayed. For example, I see shapes formed of pixels in various colours, suggesting there is a bitmap somewhere "in" the computer, which was "caused" by various bit manipulations, or I see the number 3984, suggesting there is a numeric variable in the program with that as its value, or, knowing that 3984 is the height of Ben Lawers (a mountain in Scotland), I can say that the height of Ben Lawers is held inside the computer.

But these presuppose rather than address the nature of innards. They are not fundamental and may still be argued against philosophically. So can we find any philosophically sounder understanding of the innards? I have not encountered so far any attempt to do this using conventional streams of philosophy, possibly because approaches based on sensory experience make innards irrelevant while those based on a substance-idea (see the following) presuppose innards and thus cannot make it into a philosophical question to be asked. It is merely assumed to be the case that such things are "in" the computer (though there is debate about what these are), and the very posing of the question is taken as evidence of the inferiority of the pre-theoretical approach. So the question still remains open: on what basis can we validly say that such things as bits, numbers or content are "in" the computer?

The fourth issue is **persistence and change**. Though continual replacement of old technology has been the norm over the past 30 years, and "legacy systems" is usually a term of disparagement, there is no fundamental reason why this should be so. Increasingly, people are coping with the technology they have, merely upgrading it, so there is a need to understand persistence and change. "My computer", an Amiga 1200 on which this book was written, began life in 1993, but since then it has undergone change: added memory, added SCSI, added faster processor, replaced hard disks, removed SCSI, new motherboard, new keyboard, new mice, new version of the operating system, added software—but it is still "my computer". How may we account for this experience of identity despite change?

Perhaps more important is software and data. This book, as a computer file, has been rewritten five times since it began as a project in 2003, has completely new structure, argument and even purpose, might spawn another book, and probably there is none of the original text remaining—yet it is still the same book. Such questions are important not only with distinct projects but also in legacy systems.

Acknowledging the continuing identity of a thing is important, not just to settle arguments, but for legal reasons and to instil vision that inspires.

The fifth issue is **norms for computers**: What is a "good", "true", or "real" computer? How can we make it more "true"? In everyday life, such questions are frequently asked of things.

These issues all concern the structure of the thingness of the computer (and information and program): what is necessary for a thing to be a computer?

5.1.3 The Need for Dooyeweerd's Approach

As quoted earlier (Dooyeweerd, 1984, III, p. 53) immanence-philosophy "has never analysed the structure of a thing *as given in naïve experience.*" It is not that nobody has tried to analyse the structure of things, but that nobody has done so as things present themselves to us in everyday experience. Dooyeweerd (1984, III, pp. 3-53) examined a number of answers that have been offered to the fundamental question of what it means for a thing to be that type of thing.

- **Fiction:** That there is no "thingness" as such, only a fictitious union of sensory impressions; that view is obviously unsatisfactory.

- **Reduction to sensory function (pp. 28-36, 102):** That we can understand the nature of things by, and only by, sensory experience of them; that prevents any understanding of innards, and of norms.

- **Reduction to mathematical-logical relations, as in Russell (p. 25):** This provides no basis for understanding the diversity things present to us, without importing other meaning *a priori*; it cannot even differentiate between hardware and software.

- **Aristotelian substance concept (pp. 17-18):** There is a "computer-substance" of which all computers are made [for example silicon electronics plus programming, or a certain type of causality (Searle, 1990)]; this makes it difficult to account for the variety of our experience and consideration of potential becomes open-ended speculation that degenerates into dogmatic positions (such as on whether computers will ever be able to understand; see later). This view also contains an antinomy (pp. 11, 17).

- The "**modern Humanistic concept of substance, following Descartes**," which Dooyeweerd dismissed with (p .26), "Whereas the Aristotelian idea ... was at least intended to account for the structures of individuality as they are realized in the concrete things of human experience, the modern concept of substance was meant to eliminate them."

- Kant's identification of "the 'things' of naïve experience ... with **the *Gegen-stände* of natural scientific thought**" (p. 28); Dooyeweerd continued, "This procedure immediately resulted in the elimination of the datum of naïve experience." As a result, we can never know the "thing in itself" even in principle, and so ontology is absorbed into epistemology.

- **Heidegger's account of being-in-the-world**; though perhaps an attempt to provide an everyday account, cannot account for innards nor for normativity.

- **Process philosophy**, a variant of the above which emphasises the dynamic over the static in substance; Dooyeweerd saw it as a "meaningless alternative" (p. 18).

Dooyeweerd, contrary to this, believed we can discuss the nature of things as they present themselves to us in everyday experience, but in a different way, based on the CFR ground-motive.

5.2 A Dooyeweerdian Approach to the Nature of Computers

As explained in Chapter II, Dooyeweerd believed our starting point for understanding the nature of anything must be that everything functions and exists within a cosmic framework of Meaning and Law-Promise (see §2.4.3, §2.4.4). So we do not ask, first, "What *is* computer?" but "What *means* computer?," not "How *does* computer *behave*?" but "What *law* enables computer?" Therefore, instead of seeking to identify a self-dependent essence or substance (nor even process or causality) that is "computerness" or "information", by reference to which all discussion within this area can occur, such things as being and behaviour are derived from Meaning/Law (see §3.1.5). We may delineate a number of distinct ways of being and functioning by reference to aspects. What differentiates the thing that is computer from other things (such as lump of silicon, electronic device, or digital device) is the internal structural principle, which involves all aspects led by the qualifying aspect (see §3.2.5).

To illustrate the general approach that will be adopted, consider the book you are holding.

- **It is a lingual thing:** A discussion of philosophical frameworks for understanding information systems.

- **It is a formative thing:** A structure of chapters, paragraphs, etc.

- **It is a physical thing:** Half a kilogram of paper.
- **It is a juridical thing:** Someone's property.
- **It is an economic thing:** A product with a cost.
- **It is a pistic thing:** The author's vision for how we should understand information systems.

The book has several distinct modes of being (§3.1.5.3). The being of a thing is an interlacement of several "aspectual beings", each of which is a reification of its meaning in that aspect. The nature of computers can be understood in a similar way, but a closer analysis of this is needed.

5.2.1 In Relation to Human Beings

"In veritable naïve experience," Dooyeweerd believed (Dooyeweerd, 1984, III, p. 54), "things are not experienced as completely separate entities." To understand computerness, therefore, we must understand it in relationship.

In Chapter IV three ways in which the computer can relate to its user were distinguished: as something with which the user interacts (HCI), as represented content (ERC), and as an artefact used as part of human living (HLC). To understand the nature of computers regardless of application, only HCI provides a useful starting point because with both ERC and HLC the main (qualifying) aspect varies with the application. HCI is qualified by the lingual aspect. Perhaps this is what Winograd and Flores meant when they said (1986, p. 78), "Computers do not exist, in the sense of things possessing objective features and functions, outside of language."

Chapter IV gave two main facts about the structure of HCI as understood from a Dooyeweerdian viewpoint:

- It is multi-aspectual human functioning concerned with the human being's direct experience of the computer (specifically via its user interface (UI)). This multi-aspectual functioning is led by the lingual aspect.
- In all post-physical aspects the computer functions as object as part of human functioning, while in the physical aspect it may function as subject.

The view here is echoed in Milewski's (1997) "delegation" proposal, that we should understand agents not in terms of their innate characteristics but in terms of the relationship they have with users. But our view extends to any human, including developers too, and it also allows for subject-functioning in the physical aspect.

That the latest subject-functioning aspect of the computer is the physical is what makes computers so useful. The later aspects, especially from the analytic, are non-determinative and so all subject-functioning in them will be non-determined. But, barring quantum effects, the laws of the physical aspect are determinative, and as a result the behaviour of the computer is determined and predictable. That means that, in computer technology, humanity has at its disposal the possibility of meaningful functioning in the later aspects that is reliable and predictable. Other implications of the difference between subject- and object-functioning will be considered later (§5.5.6).

Notice that Dooyeweerd does not make relatedness as such the key to understanding the nature of things, as Heidegger might, but rather the cosmic meaning of that relatedness. The diverse cosmic meaning which is the computer, whether as subject or object—which will be called meaningful-functioning—can reveal the nature of computers.

5.2.2 Human Experience of the Computer as a Whole

Table 5-1 shows some examples of how the computer functions in each aspect in relation to the human being, regardless of its application. These are ways in which we encounter the computer in our everyday life as users and/or developers.

The left-hand column refers to what we experience directly via the UI, the middle column refers to what we have learned to believe is "inside" the computer, and the right-hand column is sundry other functionings, which have little to do with it as a working piece of information technology and more as any other material artefact that is owned and is part of our lives. The left and middle columns may be compared with Table 4-6. Differentiating in such a way accords with our intuition, but on what philosophical grounds may we do this?

Our experience of or in the left-hand revolves around the qualifying aspect of HCI, which is usually the lingual. For example, its screen content is understood (lingual functioning), the text and graphics on the screen are structured (formative), they are taken note of as types of data (analytic), the text characters are seen and recognised by virtue of being known fonts (psychic), and the screen emits light (physical). In the anticipatory direction, the understanding is facilitated by cultural connotations (social), there is a limit to the amount of text on screen (economic), and so on.

But the right-hand column shows the computer functioning in ways that are not always necessary to serve the leading aspect. For example, being visible to us (e.g., from the back) is psychic functioning, but not relevant to its lingual functioning above. (The computer might serve a contingent function, such as making money for its manufacturer or acting as paper weight. The success of Apple computers might in part lie in the harmony between the two aesthetic functions, which itself is meta-aesthetic.)

Table 5-1. Aspects of computer

Aspect	Functionings of Computer Meaningful in Each Aspect		
Quantit'ive	A lot of stuff on screen		One computer
Spatial	Screen layout, size		Space taken up on desk
Kinematic	Animation on screen		Fan air flow
Physical	Light emitted from screen, Vibrations from speakers Pressure I exert on mouse	Electromagnetic fields	Force exerted on desk
Biotic / Organic	Activates nerves in ear, eye My hand pushes mouse	Voltages	Repetitive strain injury
Psychic / Sensitive	Colours, shapes on screen, Sounds from speakers Key hits, Mouse moves	Memory bits, signals	Case is beige colour, Fan noise
Analytic	Icons, numbers Mouse gestures	Pieces of data	
Formative	Tables, lists, paragraphs Syntax of my command	Data structures algorithms	How the screen, wires, keyboard are connected, arranged
Lingual	Represented Content		Manufacturer's, logo Labels on sockets
Social	Cultural implications		Sound from computer annoys others in office
Economic	Limited screen area Max keyboard rate	Limited memory size	Cost of purchase
Aesthetic	Style of UI		Style of case as decor in room
Juridical	Appropriate expression of info		Ownership of computer
Ethical	?		
Pistic			

5.2.3 The Innards

The middle column refers to things "inside" the computer, and requires more elaborate treatment.

Dooyeweerd offers us a basis for considering innards. He addressed the question of things "hidden" from our naïve experience. First, he argued that even though it is through sensory functioning that most of our experience comes, "Naïve experience ... is by no means restricted to the sensory aspect of its experiential world." (Dooyeweerd, 1984, III, p. 102)[1]. So the fact that we cannot directly see or hear bits, numbers, content, and so forth, in a computer does not rule it out from being part of our naïve experience. It does not mean such things are in any way a mere theoretical abstraction from the sensory or physical.

But, if they are hidden how can we experience them? Our sensory function can be "opened" by means of techniques and technological apparatus. Dooyeweerd gave examples of microscopes, telescopes, and using developed physical theory (for experiencing cells, galaxies, and atoms, respectively; there are various reasons why things may be "hidden"). Our experience of such things may be indirect, but it is still everyday and not theoretical; even though such techniques and technological apparatus are the product of the theoretical attitude, their concrete actualisation in life brings them into the sphere of our naïve experience. Specifically, we have apparatus by which we may experience the innards of the computer indirectly in each aspect. For example:

- Psychic aspect: Memory dump software.

- To interpret this memory dump at the analytic aspect, a lot of memory-dump software also shows the bytes interpreted as ISO characters, or via some other bit-to-data coding.

- To interpret structure and purpose of this data (formative aspect) the user needs to know the program structures, or a tool that knows this (for example the amazingly useful program called Structure Browser, which allows its user to move around the structures of the Amiga operating system).

- In the lingual aspect, the program has meaning only insofar as the user knows the parameter interface and what the program is supposed to do in terms of its application.

Thus we have a soundly philosophical way to underpin our intuitive belief that the computer "contains" various things inside it, and these are meaningful at various different aspects.

5.2.4 Excursus: Reinterpreting the Biotic-Organic Aspect

To Dooyeweerd, the kernel meaning of the biotic-organic aspect is life functions, vitality. But we have reinterpreted this somewhat from the customary life functions, to that of hardware components. What justification do we have for this? (This excursus is mainly of interest to Dooyeweerdian scholarship in that it shows how the Dooyeweerdian view might be taken further, and may be skipped.)

We could argue that our framework for understanding the nature of computers needs to differentiate hardware from both (physical) materials and (psychic) bits and signals, and that between the aspects of these two lies, very conveniently, an empty slot which is the biotic aspect. But philosophical convenience is no good reason.

It is obvious that the computer, not being alive, does not function as subject in the biotic aspect. But it is also difficult to see how the computer functions as object in this aspect, since it is not a means of life for any living thing, nor is the content of the program it is running necessarily about a biotic topic (those would in any case be HLC and ERC). We seek to identify the biotic meaning that is germane to the computer being a computer, if such exists. Dooyeweerd posed similar questions in his discussion of Praxiteles' sculpture of Hermes and Dionysus (Dooyeweerd, 1984, III, p. 112ff.).

We have three reasons for treating the hardware of the computer as its biotic aspect. The first line of reasoning goes as follows (remembering that the computer functions as object rather than subject in the biotic aspect).

- We start by considering the biotic aspect of our interaction with the computer (HCI).
- At the biotic level of the human being, we speak about organs like stomach, fingers, eyes, ears, in contrast to the chemical-physical material of which these on one hand and feelings, sensations and motor impulses (psychic-sensory aspect) on the other.
- With what, of the computer, do organs engage? Hands grasp a mouse, rather than moveable plastic, and eyes see the screen, rather than light. Our organs engage, not with physical material, but with manufactured components.
- Therefore perhaps it is valid and useful to say that the discrete, manufactured hardware components of the computer function as biotic objects to the user's biotic subject-functioning.

A secondary reason to support this view, which might arouse controversy in Dooyeweerdian circles, is that one important feature of living things is that they maintain a distinctness from their environment, an active equilibrium state different from the environment, even as they interchange physical material and energy with it—unlike many physically qualified things like inter-stellar dust or river currents. And organisms repair themselves and thus maintain their integrity as organisms. Computers too maintain a distinct active equilibrium and maintain their integrity (e.g., by checksums built into memory cells).

Naïve experience offers a third reason: for several decades it has seemed meaningful to compare and contrast machine with human body, which suggests they lie within the same sphere of meaning (i.e., aspect).

For these reasons, therefore, the hardware that is the computer will be seen here as its biotic object-aspect, but to differentiate between biotic aspect involving living things, and this hardware aspect, we will use the word "organic". However, this is contentious and requires further debate within the Dooyeweerdian community.

5.2.5 Aspectual Beings that Constitute the Computer

Sometimes it is convenient to talk solely in terms of what is meaningful about the computer within each aspect, but usually it is more convenient to talk about "things" related to the computer, that is, nouns rather than adjectives or verbs. If Being is complex and founded on aspectual meaning, then the computer exists in many different aspects—it is many "aspectual beings". This accords with naïve experience but is foreign to most theoretical ways of understanding Being; see the aspectual beings of a book, above. Dooyeweerd held (1984, II, pp. 418-419) that "On the immanence standpoint it is impossible to recognise the modal all-sidedness of individuality." (One possible exception to this is Newell's multi-level view of computers, discussed later.)

Table 5-2 shows a host of aspectual beings of the computer: what we deem "things" or activity that are meaningful about the computer in the physical to lingual aspects. The aggregations from left to right, as well as the vertical relationships across aspectual boundaries are mentioned later. Now is a good time to peruse the table.

This illustrates quite clearly Dooyeweerd's claim that aspects are modes of being. The computer as such exists as materials, as hardware components, as memory, and so forth, as raw pieces of data of various types, as structured data, as applications content. It exists as all of these at the same time, not one after the other. Aspectual beings are merely ways in which the computer is meaningful reified into things.

Six aspectual beings have been identified for the computer. In fact, Dooyeweerd contended, all things are meaningful in all aspects, though sometimes only latently. There is nothing which has an aspect missing. If there were, then that thing would never be able to be an object in that aspect. (This seems to be something of a dogma to Dooyeweerd, and some might question it. But it is a reasonable dogma, because it reflects our everyday experience, and we will adhere to it here.) For example the aesthetic aspectual being of the computer is what its lingual aspectual being anticipates and makes possible, and yet also what serves to either serve or undermine that lingual function. This does not refer to any aesthetic use to which the computer is put, which was differentiated in Chapter IV as HLC, but the aesthetic aspect of HCI, which includes the style of the user interface.

It might be asked why the digital bit is not of the analytic aspect of distinction, and has been linked with the psychic aspect. There are two reasons. The bit need not be digital but could be continuous as in analog computers discussed next. And the bit equates, in human functioning, to activation of neurones (psychic) rather than to mental concepts (analytic).

Table 5-2. Aspectual beings of computer

Aspect	Atomic	Aggregations		
Physical	Fields, Quanta, Waves, Particles / Atoms, Crystals, Energy troughs	Doped Silicon, Conductance, Light-emitting phosphor	Behaviour of P-N junctions Light emissions	Active whole computer / Electric connect to grid
Organic (Biotic)	Electronic component Conductor Voltage, current Mechanical component	Circuit board Speakers Mech'l assembly	Larger assemblies (disk, mouse)	Computer as hardware
Psychic	Bit Signal Coloured pixel Sound Key press, etc.	Byte, word Register Shape, pattern, texture Sound FX Gestures: click, d-click drag	Alloc'd memory machine instr'n CPU Bitstream Shapes, bkgnd, animation Complex sounds	Machine code program TCP/IP Window, screen Bkgnd music Sequence of gestures
Analytic	Datum: integer, letter, etc. Data change Visual or aural datum Enter a letter Select	Set of data DB transaction Data input		All the data and its manipulation
Formative	Relationship between data Transformation Relshp bet things displayed A whole user interaction (qn, ans, help, etc.)	Record, data structure Algorithm, inference engine Diagram Rational sequence of user interactions		Database Program Document, website User session
Lingual	Symbol with its signification / Piece of knowledge	Knowledge gained or refined Virtual reality		Application, multimedia title game, etc.

5.2.6 Analog Computers

The discussion thus far has assumed digital computers. But this Dooyeweerdian approach allows for analog computers too. In analog computers, continuously variable voltages and currents signify numeric quantities in what seems a more direct way than by digital coding (e.g., 1-5 volts might map to a level of activation). This assigning of lingual meaning (semantics) to voltages involves the intervening aspects, but in a rather simple one-to-one way, with the result that it can be difficult to separate out the different aspects.

- **"Organic" aspect:** A voltage or current (e.g., 4.2 v).
- **Psychic aspect:** The voltage-level as a level of activation (e.g., as proportion of voltage-range 1-5 v).
- **Analytic aspect:** The sensed thing as a quantity (e.g., 80% of range).

- **Formative aspect:** The distinct variable with a purpose, and among other voltages.
- **Lingual aspect:** The semantic meaning of the purposeful, related variable, such as what angle to raise the gun.

Though most of the discussion in this work will be in terms of digital computers and information, it should always be borne in mind that most can be extended to analog technology.

5.2.7 Meaningful Wholes

These aspectual beings do not have any existence apart from the whole that is the computer. The term "meaningful whole", or just "whole", will be used to refer to the entire thing in its unity as a thing, as it appears to us in everyday life; see §3.2.3. A thing like a computer system is a meaningful whole. It is the meaningful whole that presents itself to us first to our everyday experience. The coherence of such wholes is possible, philosophically, because we presuppose the harmony of the aspects (see §3.1.4.3).

[An early version of the proposal developed here may be found in Basden and Burke (2004), which addresses the related question, "What is a document?" It finds a similar dynamic, multi-aspectual meaningful whole, and also discusses aspects of responsibility. The reader will find a different slant on these issues therein.]

"Aspectual being" is, however, not a phrase that Dooyeweerd used. He seemed not to refer to the actual being as such, but the structural laws or principles that make such beings possible. He spoke of "individuality structure", but this usually referred, not to the concrete being but to a general type of such beings, to the law-side "internal structural principle" (§3.2.5) by which such beings might exist. We want to reify aspects of the thing, so that we can use nouns or noun-concepts when thinking about it, rather than having to restrict ourselves to other parts of speech. Thinking of a thing's aspectual beings is largely equivalent to thinking about its aspects, so the two may usually be used interchangeably. But they have different characteristics and sometimes we will find one more helpful, sometimes the other. Thinking about aspectual beings enables us to consider relationships among things in a computer.

5.2.8 Relationships Among Things in a Computer

Aspectual beings cannot be seen as parts of the whole, not in the way pages or chapters are parts of the book. So how do these things relate? Table 5-2 listed many aspectual beings of the computer, with two types of relationship between them, horizontal and vertical.

5.2.8.1 Relationships Between Beings within an Aspect

Within each of the aspects except the physical we have several lists of things, beginning with basic or atomic things that cannot be subdivided in this aspect, followed by various degrees of aggregation of these things. This is the part-whole relationship, which Dooyeweerd characterised as being between things qualified by the same aspect; for example a bit is part of a byte, which is part of a chunk of allocated memory, or an integer is part of a record which is part of a table in a database. This enables us to understand aggregations of things as in Table 5-2.

But it cannot elucidate the relationships between things in different aspects. There is something wrong in saying that a bit pattern is part of an integer or text string: a category error.

5.2.8.2 Relationships Between Beings of Different Aspects

The relationship between aspectual beings and their whole is foundational enkapsis (§3.2.6.2). Dooyeweerd illustrated by reference to Praxiteles' sculpture *Hermes and Dionysus*: the relationship between the sculpture and the block of marble from which it is made. Whereas Praxiteles' sculpture involved two main aspects (physical marble, aesthetic work of art), in a computer, at least six aspects are involved, from physical to lingual—these are the aspectual beings we identified above: material, hardware ("organic"), bits (psychic), pieces of data, structures and processing, and content.

In foundational enkapsis, foundational inter-aspect dependency (§3.1.4.6) plays an important part. Each aspectual being depends on those of earlier aspects in order that it may be "implemented": profit level (§4.3.1.1) is "implemented" in numbers, which are "implemented" in binary-coded bit patterns, which are "implemented" in voltages, whose components are "implemented" in silicon. A similar account, in both directions, may be made of the user interface screen, beginning with phosphor and glass to make a cathode ray tube.

5.2.9 Implementation

When we use one aspect to implement the next, we "add" the meaning of each aspect to what we already have, and it is a different kind of meaning with each move. Thus, starting with the:

- Physical aspect of materials (like silicon, phosphor, glass);
- To implement the "organic" aspect of hardware components like IC chips, collect the materials together into distinct 'organs' or hardware components;
- To implement the psychic aspect, interpret certain voltages, etc., as digital states (e.g., 5v = 1, 0v = 0), and the components that hold those voltages as memory cells, registers, etc.;
- To implement the analytic aspect, add a coding system like ASCII or binary (e.g., bit pattern 01100001 is the number 97 under binary coding);
- To implement the formative aspect, add structuring and processing (for example, the number 97 might be subtracted from 114 to yield a new number, 17);
- To implement the lingual aspect, add semantic meaning (e.g., these three numbers might represent expenditure, income and profit); and
- To implement the social aspect, add the cultural connotations of these numbers (e.g., "Profits only 17M? Hmmm: a bad risk").

Because of the fundamental irreducibility in meaning between aspects (§3.1.4.2), how an aspectual being in one aspect may be implemented in an earlier aspect is not determined. This provides a philosophical account for several things that are usually taken for granted:

1. It gives freedom of implementation, at every level:

- Some computer components are made of Gallium Arsenide instead of Silicon.
- Different voltage levels carry the bits 1, 0.
- The bit pattern 01100001 is the letter "a" under the ASCII code.
- The number 97 might mean, not expenditure, but the number of students in my class.
- Profits of 17 might indicate a healthy rather than weak performance.

As a result, the same software can run on hardware from different manufacturers, why it is possible to make advances in hardware without necessarily upsetting the working of the software.

2. It enables virtual data (as it is called in database circles). What is one being in one aspect might be many beings and even many activities in the earlier

aspect without an actual "static" thing. For example, the information "profits last year" might not be stored in the database or computer as a single datum (analytic aspect) but, whenever it is called for, a quick calculation of profit is made on the basis of two other figures, income and expenditure. Such "virtual data", though a single lingual aspectual being, is stored as multiple analytic aspectual beings together with the formative aspectual functioning that is the subtraction process. Another example: whereas in most computers a memory bit is implemented as a static electric charge, in one digital system the author once worked on in the mining industry, where there is much electric interference, the single bit was implemented as a phase change in alternating current waveforms.

3. But it makes it impossible to interpret something in one aspect unambiguously in the next aspect if we do not import meaning from that aspect. For example, given the bit pattern 01100001 we cannot tell whether it implements the quantity 97, the letter "a" or anything else, unless we already take into account its meaning within the analytic aspect. Likewise, we can have no idea what a piece of program code (analytic and formative aspects) without meaningful names or comments stands for (lingual aspect)—the curse of programs written without comments!

4. It makes randomizing and file compression possible. Randomisation involves an operation on the bit pattern of a number that makes sense in the psychic aspect (for example involving exclusive-or) but makes no sense at the analytic aspect. File compression involves bit-level (psychic aspect) operations that alter the coding without altering the analytic data: for example compressing a file encoded as ASCII characters as a ZIP file.

5. It also enables us to understand errors of various types. What is an error at one level is explicable at the next lower level. For example, if a program's memory cell is overwritten by another program (such as a virus) then, from the point of view of the psychic aspect, all that has occurred is that a bit pattern has changed, and in principle we could know which program did this. But at the analytic aspect of data, the value in the variable has suddenly changed, and the change is completely inexplicable even in principle.

Thus this understanding of the nature of computers, as multi-aspectual meaningful wholes for which many of the things we often refer to are aspectual beings enkaptically bound into the whole, is widely applicable and provides a framework within which we can address many issues.

5.3 Information and Program

The nature of information is treated in a similar way, but the nature of program goes further.

5.3.1 Data, Information, and Knowledge

Surprisingly, perhaps, there is still debate about the relationship between data, information, and knowledge. It has been made a current issue by knowledge management in organisations, where one speaks of the collection of databases or a data warehouse as "the company's knowledge base". Knowledge and information are somehow "in" the data warehouse.

Alavi and Leidner (2001, p. 109) suggest, "data is raw numbers and facts, information is processed data, and knowledge is authenticated information." Checkland and Holwell (1998) review a number of views to suggest an extra link: capta. Then "the attribution of meaning in context converts capta into something different, for which another word is appropriate: the word 'information'" (p. 90). Information then contributes to "larger-scale, slower-moving knowledge." ["Capta" was actually used by Langefors (1966) to denote something different: what is "captured" from perceptions, and which then becomes information.]

Checkland and Holwell's account presupposes processing and a temporal sequence from data through to knowledge. This might be adequate for the example they give of a manager accessing a database of sales figures, selecting them (capta), applying context (information) to contribute to knowledge of the market. But there are four problems. One is: is there always a temporal separation? That assumes there can be such a thing as "data" that is not yet information or knowledge. Though the data might not yet contribute to that manager's knowledge without processing, it itself is not "just data". It is a sales figure, which was at one time information and knowledge for someone else. Second, they presuppose a distal relationship (analytical) between the user and the data. But here we need to account for the proximal relationship too, the immediate perception of what is on the computer screen, which is, simultaneously to us, shapes, digits, numbers, information, and what it signifies. Third, Tuomi (1999) put forward an "iconoclastic argument" that "the often-assumed hierarchy from data to knowledge is actually inverse: knowledge must exist before information can be formulated and before data can be measured to form information." Fourth, their account starts with data as pre-given, but how does it relate to the the medium, whether bits in computer memory or marks on the page?

5.3.1.1 A Dooyeweerdian Understanding

A Dooyeweerdian view of information, similar to that developed for computers above, can overcome these problems. Information, data, knowledge, and so forth, are the same thing, just seen from different aspects:

- **Psychic aspect:** "Bits" and "states" refer to Shannon's and Bar-Hilell's views of what they call "information", to differentiate it from its biotic or physical medium. They concern pre-conceptualised bits, with signals and signal paths. On screen this is pixels, on paper this is visible deliberate marks, and heard, this is sound.

- **Analytic aspect:** "Data" may usefully refer to what we have called raw pieces of data.

- **Formative aspect:** "Information" may usefully refer to data as part of something which has been processed or structured.

- **Lingual aspect:** "Knowledge" may usefully refer to what the information is about (though, we use "knowledge" here in a different way from Dooyeweerd in Chapter III).

- "Wisdom" may be added; refer to our taking all aspects into account when we consider what the information is about (De Raadt, 1991).

This links these to the medium, it does not presuppose a temporal sequence or processing, nor a distal relationship, but rather can allow for a Gestalt immediacy in which the bit-perception *is* the data *is* the information *is* the knowledge. The sales figures are bits, data, information, and knowledge simultaneously. There is nothing in a database (or a book) that is "only" data without being at the same time all the others. This view is thus comfortable with Tuomi's iconoclastic argument, in that our knowledge is involved in this immediate experience. Bits anticipate data, which anticipates (processed and structured) information, which anticipates signification (knowledge). (See §3.1.4.5.)

The manager's processing of the sales data is not a matter of making it into information that it was not before, but rather of creating something new by analytical and other functioning. Indications follow of how Dooyeweerd can be applied to three topics. They are very brief; fuller exploration of these has yet to be carried out.

5.3.1.2 Knowledge Management

The picture is further confused by knowledge management issues that have arisen over the past decade. Walsham (2001), after discussing Checkland and Holwell's view, moves to Blackler's "types of knowledge" as embrained, embodied, encultured, embedded, encoded, Tsoukas' "processes of knowing" deriving from expectations, dispositions, interactions, situations, Nonaka's "knowledge conversion" (socialization, externalization, internalization, combination) and Lam's "sharing knowledge across cultures". But Walsham does not make the link explicit.

Most of these relate to ways of knowing in various aspects (§3.3.3) rather than to what emerges from information. But the concept that data, information, and knowledge are "in" a data warehouse makes it easy to assume that we may treat data, information, and knowledge as kinds of substance that are transformed into each other. Therein lies the problem.

In rejecting any substance-concept as an account of the nature of things, Dooyeweerd can clarify the issues. After data (analytic aspect) and information (formative aspect) should come not "knowledge" but "signification" (lingual aspect). It is signification that may be said to be "in" the data warehouse, that is shared as organisations "share knowledge", "create knowledge", and so forth. Unfortunately, the term "knowledge" has now stuck as the word referring to this, including in Newell's "knowledge level" discussed later.

5.3.1.3 Long-Term Digital Preservation

This multi-aspectual understanding of information is of more than speculative interest. It becomes important, for example, in long-term preservation of information on computers—digital archives. What happens, for example, in 500 years' time, when the coding between bits and symbols is lost and cannot be guessed at? What is the nature of such archives?

Dollar (2000, p. 58) distinguishes "logical and physical structure, intellectual content, and context that were apparent at the time of creation or receipt," clearly indicating different aspects (in this case, despite the inexact use of "physical", they are, respectively, the analytic, psychic, lingual, and social). But the digital preservation community has not yet agreed a set of such aspects, and this Dooyeweerdian view could help bring agreement and also separate out issues in research.

Another important question is when a record stays the same, and when it becomes a different record (e.g., when moved to a different medium or "cleaned up"). This was discussed in the case of documents by Basden and Burke (2004), and makes use of Dooyeweerd's deliberations on becoming and change (§3.2.4): a thing may change in one aspect yet stay the same in another.

That post-physical innards require specific technology or techniques in order to make them accessible as part of our experience, and that inter-aspect implementation is innately indeterminate, as discussed earlier, should at least warn us of the major problems inherent in long-term digital preservation.

5.3.1.4 *Virtual Beings*

What is the ontic status of beings we encounter in virtual world such as games like *ZAngband* (Chapter IV), MUDs (multi-user dungeon games), or MMORPGs (massively-multiplayer online role playing games)? How is it possible that MMORPG players buy and sell pieces of virtual equipment in the real world (on eBay)?

The answer that these are "only information" (or a cyberspace reality of "pure mind", discussed later), while it might be satisfactory as a theory, does not accord with our everyday experience of them. Is this a case where everyday experience must bow to theory, or is there a more satisfactory way to understand these phenomena that is philosophically sound?

Slightly better is Dooyeweerd's discussion of the creative imagination of the artist (1984, III, pp. 113-116). That which the artist imagines, such as the aesthetic idea of a beautiful human body sculpted in marble, is an intentional object of fantasy. He expands on this in his work (1984, II, p. 387ff.). Depicted in Figure 5-1, the person who is functioning as subject in the analytic-logical aspect has a thought (intentional logical concept) and this might or might not be a logical object (thought-about denizen of the subject side). Virtual being or characters, according to this picture, have no logical object.

But this complex subject-object relationship is unsatisfactory as an expression of everyday experience of virtual worlds, in which the virtual beings are not mere passive things like the intentional idea of the human body, but are highly active, in some cases, exhibiting many aspects of human living. *Pacé* Dooyeweerd, this author finds it more useful to conceive of virtual characters and items as actually existing in the subject-side cosmos and functioning in a range of aspects from the psychic onwards, as shown in Table 5-3. But their existence is purely as object-functioning and never subject-functioning. It is programmed in the computer. Their existence in pre-psychic aspects is ignored, but if one must consider them, one can revert to the aspects of the hardware of the computer.

This view can be fitted into Dooyeweerd's, but is much more fruitful than his for understanding the important issues in virtual worlds. It is certainly more useful than the non-Dooyeweerdian belief that they are "only information" or are living beings

Figure 5-1. Dooyeweerd's understanding of thoughts

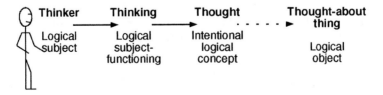

Table 5-3. Aspectual beings of virtual character

Aspect	The virtual beings ...
Psychic	appear on screen and make sounds
Analytic	are distinct from surroundings (or not)
Formative	plan and act intelligently (or not)
Lingual	communicate
Social	threaten and challenge (or not)
Economic	haggle over prices, are frugal and efficient (or not, often humourously)
Aesthetic	(deliberately) do fun or funny things
Juridical	do virtual justice or injustice
Ethical	are generous or mean
Pistic	are committed to a deity

in a reality of pure mind. For example, such a table like could be useful in design of virtual characters, to ensure they are fully-rounded.

5.3.2 Program and Software

What is a program? It might be true, but it is not sufficient, to say that it is a statement in computer language of what we want the computer to do. Consider the following:

I have an idea of a program I want to write and ponder it. I put those ideas down on paper. I write the program to implement those ideas and get it working properly. I store it on disk, and print in on paper. I run it, supplying whatever input is asks for. I run it again with different input. I give a copy of it to a friend and they run it.

How do we understand this? What is the difference between each and how do they relate to each other? Such questions, seldom considered, are important in, for example, legal cases when delineating various rights. Since Dooyeweerd himself did not have much experience of computers and so did not discuss such issues directly, we must apply his ideas to understand the nature of computer program, and will do so in two ways.

5.3.2.1 Program as Law Side

A program may be seen as a law side for a virtual world, which enables that virtual world to be and occur. The virtual subject side is the program actually running. This parallel is especially clear when the virtual world is that of a computer game or virtual reality, but it is also valid when the "virtual world" is merely a few bits of data that model something.

- Both law side and programs enable being and occurrence.
- Both subject side and running program are what exists and occurs.
- Law side and programs are universalia while both subject side and running-program are specific instantiations of this.
- Both law side and programs define what is meaningful.
- Both law side and programs comprise distinct aspects of meaning-law.

Self-modifying programs (LISP programs or neural nets might be cited as instances) make this parallel difficult, since our cosmic law side does not get modified by the subject side, until we realise that program self-modification is always itself enabled by the program. But this is an open question that is still to be addressed.

The parallel between program and law side could be used to test the shape of Dooyeweerd's notions of law and subject sides. It would not test its validity, which is a pre-theoretical stance, but it could help test and refine our understanding of its shape, in the vein of Dennett (1998) when he suggests that AI can provide new ways to test philosophy in general. We might construct special programs to undertake such testing, but this might not be necessary because a survey, examination, and exploration of a wide variety of existing programs would yield many insights. If it is objected that programs are built of propositional statements, which can never fully express meaning, we must remember that program statements do not have to be propositions, but often include non-propositional elements like neural nets and spatial or fuzzy constructs (see the later discussion). However, it should always be remembered that programs are always to be understood in terms of human subject-functioning, which is a response to the "real" law side, and so the philosophical relationship between this and the pseudo law side that is program should be kept in mind.

5.3.2.2 Program as Performance Art

However a rather more penetrating analysis of the manifestations of program might follow Dooyeweerd's analysis of art. In discussing Praxiteles' sculpture of

Hermes and Dionysus, he said (Dooyeweerd, 1984, III, pp. 116-117) (remember that Dooyeweerd's use of "objective" and "subjective" has nothing to do with whether something is fact or opinion):

First, observe that the vital (organic) function of Hermes and the boy Dionysis was objectively intended *in the artistic conception of these figures. ... the artist indeed had a productive vision of two living deified human bodies. The organic vital function of Hermes and the boy Dionysus was thus undoubtedly implicitly intended in his productive fantasy. This aesthetic* intention *is realized in the objective structure of the statue, as a thing. ... in its aesthetic structure, the intentional vital function has been objectively represented or depicted.* And this objective representation belongs to the reality of the marble ...

In this we see the relationship between what creator intended and its representation in some medium, and its reality in that medium. A computer program is not fine art, but the programmer's idea is likewise represented in some program and gains a reality in that medium. But a program is more like performance art, of which Dooyeweerd said (1984, III, p. 110):

It would be incorrect to assume that all works of fine art display the structure of objective things. This will be obvious if we compare plastic types (i.e., painting, sculpture, wood carvings, etc.) with music, poetry and drama.

Works of art belonging to the last category lack the constant actual existence proper to things in the narrower sense. They can only become constantly objectified in the structure of scores, books, etc. ... such things as scores and books, are, as such, symbolically qualified. They can only signify the aesthetic structure of a work of art in an objective way and cannot actualize it.

This is why artistic works of these types are always in need of a subjective actualization lacking the objective constancy essential to works of plastic art. Because of this state of affairs they give rise to a separate kind of art, viz. that of performance, in which aesthetic objectification and actualization, though bound to the spirit and style of the work, remain in direct contact with the re-creating individual conception of the performance artist. The latter's conception, as such, cannot actualize itself in a constant form, though modern technical skill has succeeded in reproducing musical sound-waves by means of a phonograph.

Considering these two quotations together, we may draw the following parallels:

- There is an intentional object, which is the thought-up idea which the creator wants to express, and does so in the medium.

- The program (written down: written-program) itself is like a music score, both of which may be printed on paper or stored on disk.

- Both are symbolically (lingually) qualified.

- Both written-program and music score are symbolic significations of what the creator intended, in a chosen language of notation, which differs from natural language. The language might use text as a medium, such as C or Java, or might use graphics, like music scores or visual programming languages.

- The playing of the music is like the running of the program in the computer (running-program); this results in a actualisation of both music and program.

- Just as the score is not the music, so the written program is not the running-program.

- Just as the score persists while the performance is transient, so the written-program persists while the execution of running-program is transient.

- It is the human performance artist who plays the music and the human user who runs the running-program, and supplies input.

- One performance of a piece of music differs from the next; one run of a program differs from the next, usually in that different input is supplied but even when the same input is supplied there are other differences, such as location in memory.

- The score and program both express a general ability to perform in a range of different situations, in one the ability to perform this music, in the other the ability to execute the program.

- This allows someone else to play the music, run the program.

- The recording of a performance equates with a recording of the running of a program (by means, for example, of a keystroke recorder or a video of the screen). What the I.S. developer creates is like what the composer creates: a symbolic signification of what s/he intends.

This gives us a foundation for understanding the different manifestations of program exemplified previously.

It also indicates a point of contact with program as virtual law side. Both program and music occur in relation to law as written down. It is clearer if we consider a game played rather than a piece of music (both qualified by aesthetic aspect). A game has rules that make the concrete playing of it possible and directs it without determining its outcome. The playing of the game, the performing of the music and the running of the program positivises their laws, just as the actual being and occurring of the cosmos positivises the cosmic law side.

Note that the process of programming usually involves running a program development system, which might comprise, for example, a text editor and a compiler. The programmer is seen as its user in the manner of Chapter IV, but not user of the program s/he is writing. This fascinating process of programming, in its wider context of knowledge elicitation and system development, will be the topic we seek to understand in Chapter VI.

5.4 Computer System Levels

The history of how computers have been understood looks very like a progress along the aspects. Initially, computers were hardware (analog or digital) then, in the 1960s, they were primarily seen as bit-manipulation machines with a memory, stored machine-code programs and a central processor unit. From the 1970s, they were seen as data storage and manipulation devices, and "high level" programming languages were the norm. Then the structure of such data and programs became important. In the 1980s the content (what is signified) became important, and in the 1990s, organisational meaning became important.

In each of these phases, most people saw computers in only one way; they could be said to have been using frameworks for understanding based on a single aspect. But in 1981 Allen Newell, an eminent figure in early AI, gave the first Presidential address to the American Association for Artificial Intelligence, which was published as "The Knowledge Level" (1982). He put forward a multi-level view of the nature of computers—a multi-level framework for understanding which set the direction of research for many in AI, KBS, and HCI for the next 20 years. This classic paper is still referred to. As mentioned in Vignette 4, the author found Newell's FFU very attractive, highly intuitive, and of immense help in separating out ideas in his research and in structuring his practice and teaching.

What is even more interesting from a philosophical point of view is that Newell seems to have done what Dooyeweerd suggested had never been done from within immanence-philosophy: he analysed the structure of a thing, the computer, from the point of view of naïve experience (see §3.2.1). What Newell arrived with is very similar to the multi-aspectual nature of computers developed earlier; indeed the author owes a large debt to Newell and might not have developed the proposal in this chapter without him.

5.4.1 Newell's Levels

Newell's main concern was to explore the intuitive distinction between knowledge and the symbols that hold it and to understand "What is knowledge?," "How is it related to representation?," and "What is it that a system has when it has knowl-edge?," and he presupposed that the answers to these questions can be the same for both human beings and computers. His paper addressed two main issues: the levelled nature of the computer, and why it is that at one level behaviour of an agent is deterministic while at another level it is not.

To Newell, any computer system can be described at several distinct levels:

- **Device level, whose medium is electrons and magnetic domains in physical materials:** Looking at hardware as physical materials ("device" refers to such as semiconductor P-N junctions).

- **Circuit level, whose medium is voltages and currents in electronic components:** The view of hardware we discussed earlier.

- **Logic level (bit level), whose medium is bits in computer memory and registers:** Bit-manipulation machine ("logic" refers to digital bits, not to reasoning).

- **Symbol level, whose medium is symbols in data structures:** Computer as symbolic program.

- **Knowledge level, whose medium is knowledge:** What the symbols are about, "aboutness": computer as agent.

Each level is a way of seeing the computer, and "Neither of these ... definitions of a level is the more fundamental. It is essential that they both exist and agree" (ibid, p. 95). Different levels describe the same system, not different parts thereof, and do so in equally valid ways, so a description at a level is complete, in the sense of not leaving gaps that must be filled in by reference to descriptions from other levels.

Newell worked out his notion of levels to some detail. Each level provides a set of concepts and vocabulary for describing a system that includes (Newell, 1982, p. 95), "a *medium* that is to be processed, *components* that provide primitive process-ing, *laws of composition* that permit components to be assembled into systems, and *laws of behavior* that determine how system behavior depends on the component behavior and the structure of the system."

"It is noteworthy how radically the levels differ. The medium changes from electrons and magnetic domains ... to current and voltage ... to bits ... to symbolic expres-sions" ... to knowledge. On the other hand, "some intricate relations exist between and within levels."

Each level defines a distinct technology. If a system has a description at one level then it will always be possible to describe it at the next lower level. Because of this, lower level technologies are used to implement higher ones, and it will always be possible to realize any level's description as a physical system.

But the reverse is not always the case. "Computer systems levels," said Newell (1982, p. 97), "are not simply levels of abstraction. That a computer has a description at a given level does not necessarily imply it has a description at higher levels." "Within each level," stated Newell (1982, p. 95), "hierarchies (aggregations) are possible," but merely aggregating things at one level does not necessarily take us up to the next level—though Newell never clarified exactly what it is that takes us up to the next level.

While much of this has been discussed, most have overlooked a curious claim Newell repeated for his suite of levels:

They (levels) are not just a point of view that exists solely in the eye of the beholder. This reality comes from computer system levels being genuine specializations, rather than being just abstractions that can be applied uniformly. (ibid, p. 98)

Nature has a say in whether a technology (and therefore a level) can exist. (ibid, p. 97)

To repeat the final remark of the prior section: computer system levels really exist, as much as anything exists. They are not just a point of view. Thus, to claim that the knowledge level exists is to make a scientific claim, which can range from dead wrong to slightly askew, in the manner of all scientific claims. (ibid, p. 99)

This is a strong ontological claim. But, to Newell's regret (1993, p. 33), "no one has taken seriously—or even been intrigued with—the proposition that the knowledge level was not invented," and to my knowledge this is still true.

Following this, Newell addressed the problem of non-determinate knowledge level behaviour, which we discuss later.

Newell did not analyse his notion of levels philosophically. Toward the end of Newell (1982), he suggested the knowledge level resembles Dennett's (1978) intentional stance, but there are significant differences from Dennett (such as in the number of levels/stances, and in Newell's ontological claim), and Newell called for closer analysis (1982, p. 123).

A philosophical analysis is certainly needed because ontic irreducibility between levels, stances, or technologies is a philosophical rather than a scientific issue. This might explain the rather mixed reception the notion of the knowledge level received that Newell reported (1993, pp. 34-36) by six communities. One embraced the notion because it gave them "a way of talking about what knowledge a system must have without regard to how it is to be represented," but did so uncritically. One adopted it as a framework to stimulate a new way of looking at learning. Three communities, to which it is potentially relevant, largely ignored it, not because they disagreed with it but because they were not interested in philosophical matters like defining intelligence or they were happy with current philosophical underpinnings. The only community that made it central was the SOAR community, in which Newell himself was closely involved.

It is difficult, however, to find a philosophy that critiques Newell's proposal immanently, doing justice to it and respecting rather than dismissing his ontological claim. Much philosophy of the past century, including Dennett's, ignores ontology, and that which does not is usually reductionist in flavour and thus fundamentally incapable of allowing for a plurality of levels. Dooyeweerd's philosophy, however, might allow us to undertake a philosophical analysis of Newell's proposal.

5.4.2 A Philosophical Analysis of Newell's Proposal for Levels

We can immediately detect a similarity between some of Dooyeweerd's aspects and Newell's levels, shown in Table 5-4 with reasons which are elaborated later.

The similarity is strengthened when we compare the two notions of levels and aspects as such:

- Both provide different ways of describing the same thing.
- The levels and aspects occur in the same sequence.
- Both levels and aspects exhibit irreducibility of meaning.

Table 5-4. Comparing Newell's levels and Dooyeweerd's aspects

Aspect	Level	Reason
Quantitative Spatial Kinematic		
Physical	Device	Energy, electrons
Organic-Biotic	Circuit	Distinction from environment
Psychic-Sensitive	Logic 'Bit'	Signals, states
Analytic	Symbol 1	Decision, distinction
Formative	Symbol 2 (probspace)	Processing, structure
Lingual	Knowledge	'Aboutness'
Social Economic Aesthetic Juridical Ethical Pistic		

- Both levels and aspects exhibit inter-aspect/-level dependency.
- What Newell calls system might be Dooyeweerd's enkaptic meaningful whole.
- Levels, like aspects, involve laws.
- Just as each level defines a distinct technology, so each aspect defines a distinct area of science.
- Newell stated (1982, p. 95), "Within each level hierarchies are possible" just as within each aspect aggregation occurs.
- Newell's exploration of the knowledge level might be seen as his exploration of the lingual aspect as it relates to computers.
- That knowledge is generative reflects its formative aspect (on which the lingual depends).

Further, four similarities may be found in the two approaches:

- Newell (1982, p. 123) spoke of the relationship between symbol and what it stands for as "aboutness", but "the knowledge level does not itself explain the notion of *aboutness*; rather, it assumes it." This is reminiscent of the kernel meaning of the lingual aspect (signification) being graspable not by theoretical thought but only by intuition (§3.1.4.10). We may see Newell's detailed exploration of the relationship between knowledge level and symbol level as an exploration of the lingual aspect as it relates to computers, which might contribute to Dooyeweerdian scholarship.
- Newell claimed that his levels are not derived from a priori theory but derived primarily from years of practice in artificial intelligence (ibid, p. 92); Dooyeweerd's aspects are derived from years of reflection on everyday life (§3.1.6).
- Newell made a strong ontological claim for his suite of levels (though he recognised that this claim could "range from dead wrong to slightly askew, in the manner of all scientific claims" (ibid, p. 99). Likewise, Dooyeweerd made a similar, though subtly different, claim, that the aspects are not just a point of view, though his suite of aspects should be subject to criticism and refinement (Dooyeweerd, 1984, II, p. 556). The subtle difference is that the aspects cannot be said to "exist" so much as "pertain", since they are the very framework that makes existence possible (§3.1.3).

There are however some differences between them, such as that Newell did not recognise what Dooyeweerd called anticipatory dependency. Nevertheless, we have good reason to propose that Newell's levels are remarkably similar to Dooyeweerd's aspects, and we can propose Dooyeweerd's philosophy as one to provide a philosophical grounding for Newell's levels.

5.4.3 Level-Aspect Correspondences

The correspondence between the individual levels and aspects shown in Table 5-4 may now be examined more closely. The device level, concerned with physical materials, is obviously the physical aspect of the computer. The knowledge level's concern with "aboutness" closely matches the lingual aspect's "signification". But the other level-aspect correspondences are less obvious. We must avoid being misled by the labels used for levels or aspects and focus on what each thinker referred to when speaking of it.

That Newell's bit level corresponds with the psychic aspect becomes clearer when we consider alternatives. Briefly, the argument is as follows. The physical, biotic, formative or lingual aspects are ruled out because the bit, as a digital state of being on or off or of switching between them (a "signal") has no meaning within them, which leaves the psychic or analytic. The psychic aspect seems more appropriate when we treat such states as object-functioning as part of human (or even animal) experience: the colours on screen are aggregations of pixels each of which holds a state. Finally, this is supported by a strong similarity between things at a level of a computer system and aspects of a human (or animal) subject. The psychic, as opposed to analytic, operation of human memory and recognition [Dooyeweerd drew attention to animals' distinctions of, e.g., mates (Dooyeweerd, 1984, I, p. 39)], involves the activation states of neurones without logical functioning.

What Newell (1982) identified as a single symbol level corresponds with two aspects, the analytic aspect of distinction and conceptualisation, in which basic types of data—integer, boolean, text, and so forth—are meaningful, and the formative aspect of deliberate shaping, in which data structures and algorithms are meaningful. However Newell later said (1993, pp. 36-37, our emphasis):

Gradually it has become apparent that between the knowledge-level description of a Soar system and the symbol-level description (the one that talks about recognition memory, working memory, the decide process, impasses, and chunking*) there is an organization in terms of problem spaces, which in many ways is like another computational model. One need only talk about* operators, states, desired states, selection knowledge for operators, *etc. This must be a symbol-level organization (we know of no way to have a genuine system level between the symbol level and the knowledge level), but different from ... the recognition memory, etc.*

From the text we have emphasised, it is clear that Newell himself had grasped intuitively the distinction in kernel meaning of Dooyeweerd's analytic and formative aspects. But while Newell felt constrained by his ontological commitment to a single symbol level made a decade earlier, Dooyeweerd does offer a "way to have a genuine system level between the symbol level and the knowledge level."

Finally, it seems strange to align the biotic aspect, usually seen as having a kernel meaning of life functions, with an electronic circuit level. There are several reasons for this, which have been discussed earlier, the main one being that one important way in which physical meaning is inappropriate to both a living organism and a manufactured circuit or piece of hardware is that both are distinct from their environment whereas in the physical aspect, fields and forces pervade all.

5.4.4 Enriching Newell's Notion of Levels

If Dooyeweerd can provide a philosophical foundation for levels, then we have a basis for not only affirming Newell's notion of levels, as we have demonstrated, but also critically enriching it. For example:

- We have already noted the possibility of enriching the symbol level by reference to two aspects.

- While Newell held that each level provides a distinct description and has distinct types of laws of composition and behaviour, Dooyeweerd's aspects involve inherent normativity; so could this provide Newell's levels with the ability to give normative guidance to those working at each level?

- Because Dooyeweerd's suite contains more aspects than Newell's set of levels, this suggests that there might be other levels above the knowledge level—and indeed, Jennings (2000) has argued for a social level just above the knowledge level.

- Though Newell, working within a positivist tradition, did not mention meaning, it seems, to this reader at least, that he was reaching for meaning, in Dooyeweerd's sense of "referring beyond".

Dennett (1998, p. 284) criticised Newell's lack of clarity about "aboutness". To Newell, the role of a symbol is to give "distal access" to knowledge, and he depicted this (1982) as an arrow from the symbol to what it "accesses". This might be further knowledge in the agent, but, as Dennett pointed out: "Those (distal access) arrows ... lead one *either* to more data structures or eventually to something in the external world—but he is close to silent about this final anchoring step. ... Newell sweeps (this issue) under the rug right here." However, Dooyeweerd might come to Newell's rescue in explaining why Newell swept "aboutness" under the rug: it concerns the kernel meaning of the lingual aspect, symbolic signification, which can be grasped only by intuition, not by theoretical thought, (§3.1.4.10) and hence theoretical discussion of this "final anchoring step" will always feel vague.

In short, Dooyeweerd seems to be what Newell was reaching for in his theory of levels. By reflection on what AI researchers and developers had come up with, Newell wanted to see both computers and human beings through a lens of multiple levels that are ontically, irreducibly distinct ways of describing this, each of which has a distinct set of concepts, medium, laws, technology, and so forth, and in which both deterministic and non-deterministic behaviour can be incorporated. (We might note also that systems thinkers, such as Wilber, Bunge, and Hartmann, are also reaching for level plurality, though they do not work it out in the way Newell did; see later.) This is precisely what Dooyeweerd offers.

5.4.5 Some Practical Implications of Aspectual Levels

It might appear from the discussion that Newell's levels were selected as a framework to analyse using Dooyeweerd simply because it seemed promising as a recipient of Dooyeweerdian attention. In fact, the author first discovered Newell's levels in the early 1980s, before returning to academic life, and long before he discovered Dooyeweerd. See Vignette 4 in the Preface. Immediately they appealed to him because they accounted for what he was experiencing in information systems at the time. When he returned to academic life, he used the levels to structure his undergraduate and postgraduate teaching in a number of modules in order to ensure that he covered a wide range of relevant issues in a way that does not confuse them, and because it imparted the "wisdom" that integrates the human and ethical with the technical. He still does. Table 5-5 shows how he structured various courses according to the levels. (Such tables could be used more generally in course design, covering all aspects.) Note that a level was added above the knowledge level, so far called the tacit level, the main theme of which is cultural connotations—and thus is social in nature. Gradually, his use of these is taking on a more Dooyeweerdian flavour, with separation of the symbol level into analytic and formative aspects.

5.5 Computers and Human Beings

An issue that has taxed us since computers arrived is the similarity and difference between computers and human beings—can computer think, understand, and so forth?—which is the central question of artificial intelligence. Boden (1990) has made a collection of key papers on this issue, but it shows that while the debate

Table 5-5. Structure of courses

Level (aspect)	Database	Multimedia (+ web sites)	User Interface	Psychology in HCI
Device/ Materials (physical)	Magnetic v optical tgy	-	-	Chemistry of nerves
Circuit (organic)	Disk electronics, mechanics	MM displays, sound systems	UI devices	Nervous system, ears eyes muscles
Logic/bit (psychic)	Disk tracks, sectors, checksums Data security	As UI; Anti-aliasing, Rendering Animation speeds Sound, Lip Sync	Graphics, sound samples, fonts, Windows, Gestures	Memory, Pattern detection, recognition; Stim+Resp; Behaviorist psy
Symbol 1 (analytic)	Basic data types, fields, Indexing, Transaction processing	Basic types of info to show	Types of info to show	Concepts
Symbol 2 (formative)	Records, Data models Keys	Structure: Page, links, Animation paths, 3D models	Structure: complex info	Semantic, procedural memory Attention, Cognitive psy
Knowledge (lingual)	Normalisation, Content, Knowledge management	The story; Accuracy, etc.	The content	Human meaning and behaviour; Psycho-analysis
Tacit (Social)	Cultural connotation	The feel	Nuances	Social psychology

continues, it does not seem any nearer a resolution than at the start. To address the AI question we must first establish the basis for debate and nature of such a comparison. Should it be focused on the material-mental (brain-mind) dichotomy (McCulloch and Pitts), or on the contrast between the deterministic machine and the non-determinacy of human free action (Newell), or on whether machines could ever truly understand (Searle), or on Dennett's broader notions of intentionality, or on the Turing test based on how questions are answered? It is not the intention to attempt to make a substantive contribution to the debate here, let alone a comprehensive resolution, but rather indicate what might be a new, fruitful way forward toward such a contribution. We will examine two of the bases for debate, Newell's attempt to account for non-determinacy of knowledge level behaviour while assuming that computers and humans are essentially the same, and Searle's thought-experiment of the Chinese Room, by which he hoped to demonstrate the opposite.

5.5.1 Determined and Non-Determined Behaviour

In the second half of his (1982) paper, which has already been discussed, Newell noted that knowledge level behaviour is unpredictable and undetermined (for example, he cited Frank Stockton's (1895) story *The Lady or the Tiger*) but symbol level behaviour is predictable and determined (because it consists of mechanical processing of symbols). Why is determinacy suddenly lost between the symbol and knowledge levels?

Newell tried to account for this by defining knowledge as the logical closure of, that is, what could ever be deduced from, all that is represented in the agent's mind. In brief, his argument proceeds: since this is infinite, in any concrete decision-situation, the agent must make use of only part of this "knowledge", so we can only predict what part this will be by looking at exactly what symbolic representations the agent has—which means that to understand the behaviour of the agent at the knowledge level, a knowledge level description of the agent is "radically incomplete" and must be augmented with some symbol level description.

But this argument proved "rather hard to understand" (Newell, 1993, p. 33), and his definition of knowledge as logical closure is so completely at variance with everyday experience of what knowledge is that its veracity must be questioned. Also, why should the knowledge level be unlike other levels of description in being "incomplete", other than that to believe so is required by Newell's argument?

Furthermore, Newell restricted his attention to the ability of an observer to predict the agent's behaviour, and completely ignored the issue of whether behaviour can be non-determined as such. So it does not really address the AI question. We might also note that his initial assumption that knowledge level behaviour is non-determined while symbol level behaviour is determined might itself be questioned, because the only direct experience of non-determined knowledge level behaviour is in human beings and our only experience of determined symbol level behaviour is of computers (since we have no unambiguous access to what we might think of as symbol structures in the human mind but do have access to these in computers because we programmed them).

Thus Newell's attempt to discuss the AI question in terms of determinacy leaves many questions unanswered.

The root of the problem lies in the presuppositions of the community of thought in which he was working, which are brought into sharp relief by reference to Dooyeweerd. (Dooyeweerdian answers come later.)

- The very formulation of the problem, that determinacy and non-determinacy require explanation and are fundamentally incompatible presupposes the Nature-Freedom ground-motive.
- That Newell felt compelled to explain why non-deterministic behaviour emerges from determined behaviour at the level below arises from the presupposition of self-dependent substance, on which all else depends. In this case, the "sub-

stance" is deterministic in nature (whether it be symbols or, as suggested by the first part of the paper, physical, does not matter here).

- That Newell believed knowledge level behaviour to be non-determined and symbol level to be determined arises from the AI presupposition that computers are, in principle, completely equivalent to human beings at the symbol and knowledge levels. It is by the presupposition that he assumed that symbol level behaviour is determined and predictable while knowledge level behaviour is not.

Newell's problems, it seems, arise from his deep-seated immanence-standpoint, as Dooyeweerd predicted would be the case (§2.3.3). Before trying to transplant his ideas to a transcendence-standpoint, another immanence-standpoint stand-off will be examined.

5.5.2 The Chinese Room

Searle (1990) crystallised the debate by proposing a thought experiment to demonstrate that the claims of what he called "strong AI", that appropriately programmed computers can genuinely understand (possess intentionality), and that programs thereby explain the human understanding, are baseless. To summarise: Suppose I do not understand Chinese, and cannot even recognise Chinese writing from any other shapes. I am in a room with a batch of such Chinese writing. From time to time more pieces arrive through a hole in the wall. I also have a rule book in English (which I understand well) that tells me how to reply (by drawing shapes) to each received pattern on the basis of formal properties like its shape and which take into account all previous patterns received and sent. Also, occasionally, I receive questions in English, and reply to those. "Where," asks Searle rhetorically, "in this room is the understanding of Chinese? And how does it differ from my understanding of English?" He argued that computers running a program are like following the rule book, and cannot understand in the way human beings do, and that programs are not even necessarily part of our understanding and hence do not constitute a useful explanation:

... the programmed computer does not do "information processing". Rather, what it does is manipulate formal symbols. ... The computer ... has a syntax but no semantics. (p. 85)

He argues that biological causality is necessary for understanding, and that physical causality can never achieve this; humans operate by one while computers operate by the other.

5.5.3 The Debate

Various counter-arguments have been attempted by AI supporters, including:

- The systems reply, that while I do not understand Chinese, the system of room, rules, me, etc., as a whole does, as an emergent property.
- The robot reply, that understanding involves action in the world so the Chinese Room would understand if only the symbols I draw in reply are sent to robotic arms, etc.
- The brain stimulator reply, that all we need is for the program to simulate the operation of brain cells rather than rules directly.
- The combination reply, that putting all of these together is enough for genuine understanding.
- The other minds reply, that we cannot know what is in another mind except by the behaviour we see, so if the Chinese Room behaves aright we may say it understands Chinese.
- The many mansions reply, that eventually we will build computers with the right type of causality and these will truly understand.

Searle has countered all of them successfully (1990, p. 72ff).

Boden (1990) gives more substantial arguments against Searle's view. First, she suggests that Searle's argument involves a category mistake, in that it is not the brain (me in the Room) that understands, but the person—and in this Dooyeweerd would agree because it is the meaningful whole, not the part, that functions; see §3.2.6.2).

But her next two arguments are weaker. Second, following the rule book does involve understanding—an understanding of English in which the rules are written. She seems to miss the obvious rejoinder, that understanding the English in which the rules are written still does not enable understanding of Chinese, which is what Searle was concerned with.

Third, she argues against Searle's statement that computer has syntax but no semantics by arguing that for the internal rule-following program to run there must be some procedural element, and that this constitutes semantics. But her attempt to locate semantics in the Room in the procedural following of rules is the wrong semantics: Searle is concerned with the semantics of Chinese, not of rule-following.

So Searle's question, "Where is the understanding of Chinese?" seems to remain unanswered. We have two ideologically-motivated camps, each of which simply rejects the arguments of the other, and between which there is no real communication. Nothing has been resolved, and the debate has not even thrown much light on the issues.

5.5.4 A Critique of the Debate

Before an answer to Searle's question is attempted, Dooyeweerd would urge us to examine the debate itself, especially its presuppositions that both sides adopt without question. His different starting point for philosophy can be used to expose some of these.

First, we might detect a version of the Nature-Grace ground-motive (§2.3.1.3), in which human beings are "sacred" and computers, "secular". When a debate occurs between supporters of two opposing poles of a dualistic ground-motive no logical resolution is possible because it is the nature of ground-motives that they are religious in nature (in the sense defined by Dooyeweerd) and this involves tenaciously held commitments, in defence of which reason is harnessed. Resolution involves shifting to a different ground-motive, which is attempted next.

Second, and commensurate with the NGGM, we see in the debate about what types of causality is necessary for true understanding and intentionality, a presupposing of a substance-idea. That is, both proponents and opponents presuppose that there is some "substance" (in the Aristotelian sense of fundamental principle on which all else depends, and in this case the substance is not a static "stuff" but a type of causality), and their contention is about what this substance is that is necessary for understanding. The debate centres on what a computer is and can do "in and of itself".

One problem with substance-presuppositions is that we end up with dogmas for which little justification can be offered. Two of these can be seen in Searle.

- He claims that, "the programmed computer does not do 'information processing'. Rather, what it does is manipulate formal symbols. ... The computer ... has a syntax but no semantics" (p. 85). But on what basis may we reject information processing or semantics while accepting symbol-manipulation and syntax? Indeed, on what basis may we reject or accept either of these on their own? As has already been noted, if we open up the case of a computer we find no symbols there being manipulated!

- He claims that biological, not physical, causality is necessary for "information processing" and understanding, and that the former cannot be reduced to the latter. Yet Searle does not justify this belief, nor does he explain why it is that biotic causality can support understanding yet physical causality cannot. He holds this as a dogma.

His opponents hold equivalent, reverse dogmas.

5.5.5 Toward a Resolution

Both Searle and Newell work within the substance presupposition. Dooyeweerd would urge us to critically examine all these presuppositions, and might offer the following counter presuppositions:

- The CFR ground-motive accepts both determinacy and non-determinacy, though in different aspects of the same thing. It refuses to allow a divorce between sacred and secular.

- That existence, and the structure of things, may be derived from meaning, provides a way of allowing for multi-aspectual behaviours without having to see one as emerging from another.

- The distinction between meaningful- and subject-functioning, which we introduced earlier, allows us to see computers and humans as alike in one way but not in another. Both symbol level and knowledge level behaviour are human subject-functioning, but both computers and humans can function meaningfully in both. Note that to Dooyeweerd, symbol level functioning is also non-determinate because it is analytic and formative functioning that interprets the physical operation of the computer.

If we do not wish to adopt all these presuppositions, we might note that they help us in different ways. The third leads us to question certain assumptions. The second removes the need to explain by emergence. The first dissolves the problem.

However, there is a secondary level to the debate, which is not so bound by the NGGM, namely the recognition that it is possible there are two distinct types of causality. This could, of course, result in dualism. But it might alternatively be seen as a subset (two in size) of the full set of aspects, in which each aspect defines a distinct type of causality or repercussion (as was elaborated in §3.1.5.5). If we define understanding (or information processing) in terms of post-biotic functioning, then, owing to inter-aspect dependency it will necessarily involve biotic functioning, affirming Searle's view, but not in the way he would expect. It also affirms his opponents' views that physical causality is also necessary (though both he and they conflate the physical with the analytic aspects).

In the same way, as has already been explained, syntax and symbol manipulation may be seen as of the formative aspect while semantics and what Searle calls information processing may be seen as functioning in the lingual aspect.

What Searle holds as dogma—the distinction between biotic and physical, and that between semantics and syntax—is revealed to be an outcome of, and groundable in, Dooyeweerd's wider theory of aspects. But the way both humans and computers function in these aspects becomes clearer when we apply Dooyeweerd's non-Cartesian notion of subject and object.

5.5.6 Subject- and Object-Functioning

As explained in §2.4.5. §3.1.5.4 and §5.2.1, an entity can function in an aspect as either object or subject (agent). If it functions as object, then it does so as part of some other agent's subject-functioning. Everything can function as object in any aspect. But humans can function as subject in all aspects, animals can function as subject only as far as the sensitive aspect (though possibly higher primates might extend into the next couple of aspects), plants, as far as the biotic, and non-living things only as far as the physical.

Since the computer is not living, the latest aspect in which computer functions as subject is the physical. That is, it can function "on its own" and without human functioning only as far as the physical aspect. As explained earlier, the computer's object-functioning in later aspects is meaningful only by virtue of our ascription of those aspects' meaning to its physical subject-functioning.

If we think only in terms of subject-functioning then we cannot validly say such things as "The computer thinks." But if we think in terms of its object-functioning then we can do so. And doing so is neither metaphor nor anthropomorphism.

5.5.7 A Fresh Look into the Chinese Room

Dooyeweerd can now let us approach Searle's question, of where is the understanding of Chinese, in a different way than either Searle or his opponents do. We argued that it is valid to say "The computer knows X," but only if we speak, not in terms of subject-functioning, but by what we might call meaningful-functioning, aspectual functioning as either subject or object. In terms of meaningful-functioning, we can say that genuine knowledge of Chinese is located in the rule book (after all, somebody wrote it, so it is an object of human understanding-functioning)—a rather obvious answer which, interestingly, nobody in the debate seems to have seriously considered, though Boden does get near it sometimes.

What Searle was at pains to argue against is subject-functioning, and we would agree with him. But what some of his opponents might include is the meaningful-functioning. With this differentiation, we are able to welcome contributions from both camps as insights into the whole issue. Their apparent incommensurability dissolves when we move to the Creation-Fall-Redemption ground-motive which Dooyeweerd presupposes, because it allows for a cohering diversity in which Meaning is central.

Searle did in fact allude to what we have called meaningful-functioning (1990, p. 72):

We often attribute "understanding" and other cognitive predicates by metaphor and analogy to cars, adding machines, and other artefacts, but nothing is proved by such attributions. We say "The door knows *when to open because of its photo-electric cell," "The adding machine* knows how (understands how, is able) *to do addition and subtraction but not division,' and 'The thermostat* perceives *changes in the temperature."*

He quickly dismissed this approach with, "but I take it no philosophical ice is cut by such examples. ... the issue would not be worth discussing" and thereby failed to explore it and discover that it is by no means "metaphorical" and is rich in terms of multi-aspectuality. His quick dismissal led him to overlook what might be the only way to resolve the question in a fruitful manner.

Searle's claim that the computer "has a syntax but no semantics" may now be examined. We agree that semantics is "in the minds of those who program them and those who use them" on the grounds that it is our (lingual) subject-functioning, which ascribes lingual meaning to the computer. But is this not true of syntax too? Syntax is formative meaning ascribed to the computer. The only thing the computer "has" of itself without reference to us is its physical subject-functioning. All else is its object-functioning ascribed by us within the cosmic meaning-framework.

Table 5-6 summarises this, comparing strong AI, Searle and Dooyeweerd in aspectual terms. In these terms, strong AI believes that computers can operate "on their own" and without reference to humans (that is, function as subject) in all aspects from physical to lingual (and beyond). Searle believes that they can do so only as far as the formative aspect of syntax. But Dooyeweerd has two answers: in terms of subject-functioning (S-F) the computer can only function as subject in the physical aspect, but in terms of subject- and object-functioning taken together (meaningful-functioning, M-F) it functions in all aspects.

To the meaningful-functioning question, the answer is "Yes!" So statements like "The Prospector program found a molybdenum deposit" are as meaningful and as valid as "Jim Smith found a molybdenum deposit using Prospector"—so long as we see the first as M-F and the second as S-F. To do so is neither anthropomorphism nor a metaphor. In terms of meaning, computer and human are alike because both function within the same meaning-framework, but they function in different ways, one as object, one as subject. Likewise, it is valid—under lingual object-function-

Table 5-6. Views of functioning of computer

Aspect	Dooyeweerd S-F	Dooyeweerd M-F	Searle	A. I.
Lingual (semantics)	No	Yes	No	Yes
Formative (structure)	No	Yes	Yes	Yes
Analytic (typed data)	No	Yes	Yes	Yes
Psychic (digital)	No	Yes	Yes	Yes
Physical (materials)	Yes	Yes	Yes	Yes

ing—to say that the computer "knows", "understands", "has an intention towards", and the like.

The benefit of this approach is that the debate moves away from conflict based on dogma to seeing both sides as part of a wider picture because it exposes their presuppositions to scrutiny and re-grounding in a different ground-motive.

5.5.8 Cyberspace and Bodiless Reality

Barlow's (1996) claim, mentioned at the start, that cyberspace is a mind without any body and that this is a new reality which demands a new type of ethics and legality presupposes that there can be information without physical substratum. We might like or dislike his polemic, but it demands philosophical analysis. A Dooyeweerdian approach can help us undertake this:

- Barlow's view is at variance with not only the content but also the structure of everyday experience and can only be maintained under a dogmatic theorizing attitude.

- Body and mind serve in Barlow's argument as aspects. But he confuses irreducibility with dependency (see §3.1.4.2, §3.1.4.6). The Dooyeweerdian notion of irreducibility can support Barlow's attempt to question assumptions about the relationship currently assumed between mind and matter, but to propose that mind needs no matter presupposes there is no inter-aspect dependency. Our everyday experience asserts that there is.

- Barlow's suggestion that a completely new type of ethics and legality is needed shows that these are two more of the aspects he assumes. But on what basis has he the right to speak of these things as meaningful issues if his presuppositions about aspects are flawed?

- The opposing of mind and matter is inspired by the Form-Matter ground-motive. This is not a "truth" but a presupposition, and a deeply problematic one, as Dooyeweerd showed. It can only be held as a dogma and prevents a full pre-theoretical view.

In place of this, cyberspace may be understood with the frameworks worked out here in the following way, which is a summary of a fuller argument. Cyberspace refers to our multi-aspectual functionings with computers in use (discussed in Chapter IV as HCI, ERC, and HLC), but one in which humanity as a whole is involved (see Chapter VIII). Its primary aspects are the lingual and social. The functioning of these depends foundationally on that of the physical to formative aspects as explained previously. The physical aspect (matter) is important because this is how subject-subject interaction is enabled, and without it there would be nothing to which to attribute later-aspect meaning (which we do tacitly). This includes mind. It also includes ethical and juridical meaning. The HLC of cyberspace is governed by the same ethical and juridical aspectual norms as govern the rest of our lives because the aspects pertain. But the full potential of these aspects has yet to be opened up (§3.4.6), so we can neither defend ourselves against Barlow nor support him by reference to our current views of them, but only by reference to the kernel norms of these aspects (due and self-giving) and the shalom principle which involves all aspectual norms. The virtual law sides that are the programs that are running on the computers can only operate within the real ones via the human imagination. But this can perhaps help in the opening process to which humanity is mandated. And this, ultimately, is where Barlow probably wants to point, even if he might point in a partly wrong direction.

A full analysis of cyberspace, and of Barlow's claims, must await another occasion, and should take into account extant discourse. But it should be clear that a Dooyeweerdian approach does not take sides "for" or "against" Barlow's *Declaration of the Independence of Cyberspace* but rather provides a basis for critical exploration.

5.5.9 On Comparing Computer to Human Being

Comparisons between computers and human beings can take many forms. One compares the functioning in various aspects. As we have seen, the computer and human can be similar in this way—and this way is replete with diverse possibilities. A second compares the roles each play in this functioning, as subject and object. In this comparison the computer is unlike the human being in all post-physical aspects.

A third is to theorise the aspectual functioning into definitions. Though this seems attractive, in promising crisp comparisons, it ultimately fails and such comparisons are of limited value. The Turing test, for example, defines "intelligence" in terms of behaviour, but it defines this only as surface behaviour, and even this is restricted to lingual ability to communicate.

A fourth is to attempt to compare the human self with the computer's "self". Not only can we never experience what it is like to be a computer, but Dooyeweerd held that the human self is ultimately beyond the grasp of theoretical thought, because it is trans-aspectual, and even beyond the grasp of intuitive understanding, because it is supra-temporal (see §3.3.6). Therefore, even if we find a way of defining the computer, we can never furnish a similar definition of the human self. So this comparison is impossible in principle. Any AI challenge that demands comparison between them is meaningless.

This is not, however, dodging the AI question. It reinterprets it. Instead of trying to find a common theoretical definition of both human and computer, we can compare how human being and computer respond to the aspectual framework of law-and-meaning, and this provides new insight into the challenge.

5.5.10 Ideology in AI

Colburn (2000, pp. 80-81) sums up the debate about whether computers can understand with:

If the idea that mental processing can be explained by computational models of symbol manipulation is an AI ideology, there seems to be a countervailing ideology, embodied by Searle and many others, that no matter what the behavior of a purely symbol manipulating system, it can never *be said to understand, to have intentionality.*

Dooyeweerd sought a philosophical method of dialogue that avoided clashes of ideology, not by denying ideologies but by understanding them using immanent critique and setting them within the same framework, so that we no longer, in Colburn's words "talk past each other".

The basis of ideology is that humankind is inescapably religious (§2.4.1). Dooyeweerd's notion of ground-motives as spiritual driving forces (§2.3.1) can throw light on the diversity of ways in which the AI question of whether computers are like human beings is addressed; see Table 5-7. Humans exhibit behaviour or property X and computers exhibit Y, and then the question is to what extent and in what ways X = Y. Under the FMGM, X is mind or information, and Y is physical matter of which the computer is made. Under this dualistic ground-motive, the only way to harmonise X and Y is by giving absolute priority to one, and if necessary reduce the other to it. Materialists give priority to matter while holders of the cyberspace

Table 5-7. Accounting for extant views of human and computer

View	Human	Computer	To harmonise
Form-Matter GM	Mind	Matter	Materialist: reduce mind to matter Cyberspace: deny matter
Nature-Grace GM	Sacred	Profane	Dogma: Must not
Nature-Freedom GM	Non-determined	Determined	Systems: mystical emergence Quantum: physics not determined Newell: Incomplete KL, log. closure
Searle	Biological causality	Physical causality	Dogma: Cannot
Dooyeweerd	Subject-functioning	Meaningful-functioning	Cosmonomic notion of law-subject-object relshp

perspective give priority to mind. Under the NGGM, X is sacred "divine spark" and Y is profanity. The sacred-profane divide implies a normative and not just ontic divide, so those operating under this motive hold as a dogma that they must not attempt to see computers as similar to humans. Under the NFGM, X is non-determinacy and Y is determinacy. Various ways have been attempted to harmonise these. Some merely hold as a dogma that all freedom is illusory, others suggest that even physical behaviour is non-determined, and yet others resort to philosophical idealism. Systems theory resorts to "emergence", which is ultimately mystical because it is presupposed rather than critically probed. [The unsatisfactoriness of Newell's account of non-determinacy might reflect an attempt to "think together" the two poles of the NFGM, which Dooyeweerd pointed out is always doomed to failure (Dooyeweerd, 1984, I, p. 65).]

Searle's answer, that X is biological causality and Y is physical, which are fundamentally different, however, must be understood in a different way, in terms of aspects rather than ground-motives. As we have seen earlier, he seems to offer no grounds for this difference, holding it as a dogma, and offers no explanation of why it is that biological causality can "process information" while physical causality can only "manipulate symbols".

However, Dooyeweerd's ground-motive, CFR, and his notion of aspects help solve these problems. The CFR sees both humans and computers as part of a created, integral reality. Each aspect enables a different type of "causality" (repercussion), which accounts for the fundamental difference between biotic and physical causalities. Though Searle holds the dogma that human and computer are fundamentally different, in Dooyeweerd, as we have seen, while the difference is maintained and accounted for, in terms of subject-functioning (see Table 5.6). (Searle was perhaps reaching for what Dooyeweerd offers in his notion of irreducible law spheres). The similarity is also maintained and accounted for, in terms of meaningful-functioning (see Table 5-6). And, under Dooyeweerd, Searle is simply wrong to hold that computers can, of themselves, "manipulate symbols" while holding they cannot "process information" because both these are object-functionings.

We can go deeper into the religious root. Most who have addressed the AI question may be seen to have presupposed that we can answer it by seeking some substance, process, or type of causality that, in itself and on its own, can explain the difference or similarity—that is, in terms of immanence-philosophy. We can see this in Boden's (1990, p. 103) suggestion that the main question we must all address is "What things does a machine (whether biological or not) need to be able to do in order to be able to understand?" The strong AI position suggests intentionality may be rooted in physical or logical causal processes, Searle claims it must be rooted in biological causal processes, and Boden herself suggests it is rooted in symbolic causal processes in which "the brain is the medium in which the symbols are floating and in which they trigger each other" (Boden, p. 99).

Both supporters and opponents of strong AI seek some kind of "thing-reality" as Dooyeweerd (1984, III, p. 109) called it and this is what led to problems. As already quoted (§3.2.4), Dooyeweerd believed that "To seek a fixed point in [thing-reality] is to seek it in a 'fata morgana', a mirage." This is why he gave primacy to Meaning.

5.6 Conclusion

One might expect that trying to understand the nature of computers, information, and programs would be a theoretical exercise, especially if three decades of debate in artificial intelligence are taken as a model. So what is everyday experience in this area of research and practice? Is everyday experience even possible? The starting point in trying to understand the nature of computers is to take them as they present themselves to us in our everyday lives as users, developers, and so forth, and to seek an understanding of them that pertains regardless of application.

But "as computers present themselves to us" is so closely tied up with their application, that we have to be careful in how we approach the issue of their nature. So we had first to settle the issue of what we mean by the nature of a thing. After noting a number of problems with conventional assumptions about Being as such, we turned to Dooyeweerd's approach, which founds Being in cosmic meaning. The first principle of the framework developed here arose from Dooyeweerd's contention that in everyday experience things are not experienced as completely separate entities:

- The nature of computers is to be understood by reference to human beings. It functions as object, not subject, in all but the physical aspect.

Computers exist *qua* computer by virtue of human subject-functioning in various aspects.

This principle, seeing the computer as functioning as aspectual object, proved useful, later in the chapter, in throwing fresh light on the AI question of whether computers can understand, by reference to Searle's thought-experiment, the Chinese Room. That Dooyeweerd's non-Cartesian subject-object relationship is grounded in cosmic law and meaning rather than in self-dependent entities, as both Cartesian and anti-Cartesian notions are, allows us to meaningfully state that computers can understand, as long as this is recognised to be object-functioning rather than subject-functioning. Moreover, Dooyeweerd's delineation of the ground-motives could account for three dualistic approaches to the AI question. By these means, the AI question can be approached in a completely new way, with a hope of resolution.

Referring back to Chapter IV, it was determined that the subject-functioning that is relevant here was HCI, which is qualified by the lingual aspect regardless of application. This leads to the second, and central principle:

- The being of a computer is multi-aspectual, multi-levelled; a computer is a meaningful whole constituted of a number of aspectual beings, in which the lingual, the represented content, is key.

It was noted later that this accords very closely to Newell's (1982) notion of computer system levels, which likewise arose from his reflection on the everyday life of AI researchers and practitioners. Newell's proposal was discussed and enriched.

This principle also offers a very practical guide to teaching many computer-related topics: separate the issues out into each aspect.

That the computer is a multi-aspectual being immediately raises the question of the relationship between the beings, and two were found:

- While a part-whole relationship may be found among beings within each aspect, the aspectual beings of different aspects are bound together in the whole by foundational enkapsis, in which inter-aspect dependency plays an important part.

This enabled us to address the nature of implementation of the various levels (aspects) of the computer, each in terms of earlier aspects. On this basis, a number of issues could be considered, including freedom of implementation, the possibility of virtual data, the impossibility of interpreting the computer seen from one aspect mechanically from a description at another, randomizing and file compression, long-term digital preservation, and an understanding of errors.

Finally, the nature of information and program were considered. Information was likewise understood in multi-aspectual terms, while programs could be seen as a virtual law side, and also in terms of Dooyeweerd's discussion of the nature of performance art.

Some benefits of this framework are that it can throw fresh light on issues where debate has deteriorated, or is likely to deteriorate, into dogmatic positions, so that the opposing positions may begin to understand and accept something of the validity of each other without negating their own positions. It is a fuller framework, which expresses something of the richness of our everyday experience of computers.

With this understanding of the nature of computers, information, and program, combined with an understanding of use (Chapter IV), it is now possible to make the consideration of IS development (Chapter VI), the creation of basic technologies (Chapter VII) and societal views of IT (Chapter VIII), somewhat richer than it might otherwise have been.

References

Alavi, M., & Leidner, D. E. (2001). Review: knowledge management and knowledge management systems: Conceptual foundations and research issues. *MIS Quarterly, 25*(1), 107-136.

Barlow, J. P. (1996). *A declaration of the independence of cyberspace.* Retrieved March 16, 2005, from *http://homes.eff.org/~barlow/Declaration-Final.html*

Basden, A., & Burke, M. (2004). Towards a philosophical understanding of documentation: a Dooyeweerdian framework. *Journal of Documentation, 60*(4), 352-370.

Boden, M. A. (1990). Escaping from the Chinese room. In M. A. Boden (Ed.), *The philosophy of artificial intelligence* (pp. 89-104). Oxford, UK: Oxford University Press.

Checkland, P., & Holwell, S. (1998). *Information, systems and information systems – Making sense of the field.* Chichester, UK: Wiley.

Colburn, T. R. (2000). *Philosophy and computer science.* Armonk, NY: M. E. Sharpe.

De Raadt, J. D. R. (1991). *Information and managerial wisdom.* Pocatello, ID: Paradigm Publications.

Dennett, D. C. (1978). *Brainstorms: Philosophical essays on mind and psychology.* Montgomery, VT: Bradford Books.

Dennett, D. C. (1998). *The intentional stance.* Cambridge, MA: Bradford Books/MIT Press.

Dollar, C. M. (2000). *Authentic electronic records: strategies for long-term access.* Chicago: Cohasset.

Dooyeweerd, H. (1984). *A new critique of theoretical thought* (Vols. 1-4). Jordan Station, Ontario, Canada: Paideia Press. (Original work published 1953-1958)

Haraway, D. (1991). *Simians, cyborgs and women.* New York: Routledge.

Jennings, N. R. (2000). On agent-based software engineering. *Artificial Intelligence, 117*, 277-296.

Langefors, B. (1966). *Theoretical analysis of information systems*. Lund, Sweden: Studentlitteratur.

Milewski, A. E. (1997). Delegating to software agents. *International Journal of Human-Computer Studies, 46*, 485-500.

Newell, A. (1982). The knowledge level. *Artificial Intelligence, 18,* 87-127.

Newell, A. (1993). Reflections on the knowledge level. *Artificial Intelligence, 59*, 31-38.

Searle, J. R. (1990). Minds, brains and programs. In M. A. Boden (Ed.), *The philosophy of artificial intelligence* (pp. 67-88). Oxford, UK: Oxford University Press. (Original work published 1980 in Behavioral and Brain Sciences, 3, 417-424)

Stockton, F.R. (1895). *A chosen few: Short stories*. New York: Charles Scribner's Sons.

Tuomi, I. (1999). Data is more than knowledge: Implications of the reversed hierarchy for knowledge management and organizational memory. In *Hawaii International Conference on Systems Sciences, Proceedings from the Thirty-Second HICSS* (pp. 147-155). Los Alamitos, CA: IEEE Computer Society Press.

Walsham, G. (2001). *Making a world of difference: IT in a global context*. Chichester, UK: Wiley.

Webster's Dictionary. (1971). *Webster's third new international dictionary of the English language unabridged.* Springfield, MA: Merriam.

Wilber, K. (2000). *A theory of everything: An integral vision for business, politics, science and spirituality*. Boston: Shambhala.

Winograd, T., & Flores, F. (1986). *Understanding computers and cognition: A new foundation for design*. Reading, MA: Addison-Wesley.

Chapter VI

A Framework for Understanding Information Systems Development

Lemmings migrate every three or four years, following each other in crowds along paths until—as myth has it—they fall off cliffs into the sea. Some readers might remember the computer game *Lemmings*, which was built on this myth. Its object was to prevent such self-destruction by guiding a line of lemmings to their proper destination by various means. According to an interview the author once read, the idea for the game came one day as its creator was doodling with the "repeat brush" facility of the Amiga's standard paint package.

Most IS do not come into being because of such creative sparks (a good thing too?) but all must be developed. Information systems development (ISD) usually involves a much more structured process of analysis of the users' requirements (discussed in Chapter IV), deciding what technological artefact or system should be constructed, designing it, implementing it (programming), testing it, preparing it for the user (e.g., documentation, training), delivering it, and subsequent maintenance. During any of these activities mistakes can be made and the ISD project can fail. Large ISD projects are especially prone to such failure.

ISD is human activity. The central practical question is: What should guide ISD? The main theme of most research in ISD is methodology to guide that activity. The central philosophical question addressed in this chapter is: What is the nature of ISD, including its norms? These are the ways this chapter tries to formulate a framework for understanding ISD.

This chapter explores how Dooyeweerd's philosophy might help us understand the challenges and issues in ISD as they are seen from an everyday perspective (see §2.4.2). The information system that is developed includes both the technical artefact or system and the human context of its use, which is often organisational. The communities of practice and research in this area include those involved in programming, system design, systems analysis, organisational analysis, knowledge elicitation, modelling, and many more. First this chapter reviews the history of ISD and paradigms, and shows briefly why a new paradigmatic approach might be useful. Then it applies Dooyeweerd's notion of multi-aspectual functioning to understand what goes on in ISD, and derives a tentative framework for understanding it.

6.1 Approaches to ISD

Until recently, reflection in this area of research and practice has been directed at the development, not of games, but of "serious" IS used in professional settings. So it is no surprise to find the discussion of frameworks for understanding, including methodologies for guiding, ISD initially theorised practice quite narrowly, and then gradually re-admitted aspects of the lifeworld of both users and IS developers. A brief history is included of such theorisation and a discussion of several philosophically-motivated paradigms; the purpose of this is to situate the main proposal of this chapter in extant philosophical thinking in this area. The reader may, on a first reading, skip directly to the outline of everyday issues which are re-introduced by expanding on the author's own experience outlined in Vignette 3 in the Preface.

6.1.1 Brief History of ISD

In the early days ISD was programming, a technical creative activity, which could be quite unstructured, though techniques found to work in one project would be carried over to others. But, as Hirschheim, Klein, and Lyytinen (1995, p. 29) put it, "projects failed due to the lack of methodical guidelines and theoretical conceptions of IS." ISD methodology became an important topic for research. Hirschheim et al. give a brief historical overview of the field, as seven generations of ISD methodology which they discern to have arisen since the mid 1960s, each in response to a particular problem that was perceived:

- **Life-cycle methods:** Concerned to control the whole life of an ISD project (from user requirements analysis, through the actual programming and testing stages, to delivery), usually by means of standardisation.

- **Structured approaches:** Concerned to increase productivity of the development team and ensure that the IS developed are more maintainable.

- **Prototyping and evolutionary approaches:** Concerned about this rigidity, and that it is more important to get the right system rather than get the system right, by exposing users frequently to versions of the system and responding to their criticisms.

- **Socio-technical, participatory approaches:** Concerned to ensure participation of users so that they, rather than the development team, are in final control of ISD.

- **Sense-making and problem-formulation approaches:** Concerned to ensure that multiple perspectives (not just those of users or developers) have influence in ISD.

- **Trades union led approaches:** Concerned that workers' rights and industrial democracy should prevail in ISD and/or in the social situation in which the IS is used.

- **Emancipatory approaches:** Concerned about barriers to effective communication due to power and social differentiation, by encouraging the questioning of dominant forms of thinking and access to information.

(Note, however, that unstructured programming has continued to this day, especially for small programs or in amateur situations, such as early games development.)

This picture, however, should perhaps be extended with two other "generations". One is approaches to the development of knowledge based systems (KBS) such as *Elsie* in Chapter IV. The other is so-called agile system development methods (SDMs), many of which have developed since 1995, such as Beck's (2000) Extreme Programming. These seek to achieve those benefits but without a heavy overhead of questioning discourse that often attends sense-making and emancipatory approaches. Agile methods, more than most, aspire to an everyday stance rather than one driven by a particular theory or method.

6.1.2 ISD Paradigms

At an even more abstract level are ISD paradigms, which drive the development of ISD approaches: the world-views that researchers and practitioners in the area hold as part of their culture. Burrell and Morgan's (1979) classic model of sociological paradigms has been used to understand ISD paradigms. Four paradigms are differentiated by the interaction of two dimensions: objectivism-subjectivism and order-conflict. Objectivism assumes that the world is "given" and we can apply models and methods derived from the natural sciences to study it, whereas subjectivism denies both of these and instead probes the subjective experience and beliefs of individuals. "Order" assumes the social world is, or should be, stable, consensual and integrated, whereas "conflict" assumes change which is conflictual and coercive in nature. The four paradigms are:

- Functionalism (objectivism-order)
- Social relativism (subjectivism-order)
- Radical structuralism (objectivism-conflict)
- Neohumanism (subjectivism-conflict)

The model has been used in two ways, indirect and direct.

6.1.2.1 Indirect Use: Systems Approaches

Jackson (1991) uses the Burrell-Morgan model to inform a study of systems approaches, which themselves affect ISD methodologies. All presuppose that some situation (system state) needs to be changed. The organisation-as-system approach treats the whole organisation as a system, which has goals and the means to achieve them by the operation of its subsystems. But in most organisations it is situations rather than the organisation as a whole that needs changing, and the remaining systems approaches address that.

Hard systems thinking (HST) assumes we can know both the state of the current system and the state we desire. It sees the main challenge as identifying how to reach one from the other, preferably employing mathematical equations or propositional logic. Early ISD believed computers could facilitate this, especially for administrative or industrial processes. In ISD driven by HST it is assumed that the role of IT is to control or "objectify" the situation, removing uncertainty.

Checkland (1981) argued that HST is fundamentally unsuited to management decisions, in which different players appreciate the situation in different ways and see different things as problematic in it. Hence they cannot necessarily agree on what the relevant system is, let alone what the desired new state should be. He proposed the term "soft systems thinking" (SST) to differentiate what motivated his soft systems methodology (SSM) from HST. SST is more suited to "human activity systems". The focus in SST is to expose the diversity of perspectives, and, welcoming all, try to reach a consensus about what should be done, including what IS should be developed.

SST however has been criticised by Jackson (1991) and Ulrich (1994), as being isolationist, assuming consensus rather than conflict, and taking participation for granted, unable to handle social power structures and not critically self-reflective. They have developed two strands of what is known as critical systems thinking (CST), both of which see themselves as in a line of progress: HST – SST – CST. Using Habermas' social theory, especially from (1972), CST believes there to be three types of rationality and interest, the instrumental interest of the empirical-analytic sciences, which characterises HST, interpretive rationality of historical-hermeneutic sciences, which characterises SST, and emancipatory interest of the critical sciences, which characterises itself. In emancipation—for example from oppressive work conditions common in the 1970s or from unconscious compulsions—CST recognises a transcending normativity in systems design.

The influence of HST may be traced in the lifecycle and structured methods in the list of ISD generations, SST in the next three and CST in the last two.

Jackson's more recent work (2006) on "creative holism" makes two additions to this picture. He now differentiates HST, SST, CST on the type of relationship between participants: unitary, pluralist, and coercive. (CST is now called emancipatory systems thinking because he now reserves the name "critical systems thinking" for an over-arching framework that combines all the others, but here CST will continue to be used as in his earlier work (1991) because other authors have referred to it.) He extends to a second dimension, of degree of complexity, arguing that HST, SST, CST deal with simple situations, while system dynamics, complexity theory, and "postmodern systems thinking" are complex, the first two being unitary, the third, coercive. His new approach is somewhat less convincing than his 1991 one, in that there is no complex + pluralistic, it is difficult to believe SST is restricted to "simple" situations, his argument that both emancipatory and postmodern are coercive is unconvincing, and postmodern systems thinking emerges from only one source. Because of this, and because he makes no reference to the Burrell-Morgan model, which is our current topic, the following discussion relies on his earlier work (1991).

6.1.2.2 Direct Use of the Burrell-Morgan Model

Instead of going via systems approaches, Hirschheim, Klein, and Lyytinen (1995) have applied Burrell and Morgan's model directly to ISD. After examining the ontological, epistemological, and value assumptions of the four paradigms, they discuss the impact each paradigm would have on ISD as such (including role of the IS designer, nature of IS application, objectives for IS design and use, legitimation of the objectives, and deficiencies of each paradigm), on ISD functions (including preferred metaphor for defining information and for framing ISD, problem finding and

Table 6-1. Impact of paradigms on some aspects of ISD

ISD aspect	Functionalist	Social Relativism	Radical Structuralism	Neohumanism
Role of IS designer	Expert (master engineer)	Catalyst	Warrior	Emancipator
Information seen as:	Product, made available and traded	Reflective journey with partner	Means of manipulation; weapon in ideological struggle	Means of control, sense-making, argumentation
Raison d'être	Maximizing savings, minimising costs, improving competitive advantage	Improving creativity, sense-making	Reducing alienation by giving the workforce the control of the productive resources	Emancipation from unwarranted constraints (physical, social) by improved control and understanding

formulation, analysis, logical design, "physical" design and technical implementation, organisational implementation, and maintenance), and on aspects of the developed system (including technology architecture, kind of information flows, control of users, control of systems development, access to information, error handling, training, and *raison d'être*). A selection of their analysis is presented in Table 6-1.

Hirschheim et al.'s use of the Burrell-Morgan model is more directly relevant to ISD than the systems approach, but both exhibit problems rooted in the model itself.

6.1.3 Practical Critique of Paradigms

The Burrell-Morgan model has thus been used, both directly and indirectly, as a framework to understand ISD, and still is today.

It has received some criticism as being over-simplified (Hirschheim et al., 1995, p. 49). For example, it seems to overly narrow one's view, making it difficult to be open to the diversity of everyday experience. For example, the roles Hirschheim et al. recognise for IS designers—expert, catalyst, warrior, and emancipator—do not exhaust all the roles the designer might adopt. Even though there might be situations when it is valid for each of these roles to predominate, there are many others when other roles might seem more appropriate. For example, the roles which this author has taken on in his experience of ISD include many that go beyond these four. They are listed in Table 6-2. (Explorer means that ideas were explored by means of ISD, stimulant, that the client's views were to be stimulated, butler, that he was serving the needs of a project but had considerable responsibility, but was not in the role of expert, artist, that he tried to generate something beautiful (teacher) that he taught others about ISD and IS.)

Likewise, information can be seen in other ways than product, journey, weapon, and means of control or argumentation, and the *raison d'être* for the IS can extend into many aspects, as discussed in Chapter IV.

Table 6-2. ISD roles assumed by the author

Project	Reference	ISD role
C.A.D.	Basden & Nichols (1973)	Explorer
Business data	-	Hired hand
CLINICS	Basden & Clark (1980)	Butler (responsible servant)
AusCor	Basden & Hines (1985)	Stimulant/teacher, Explorer
Wheat Counsellor	Jones & Crates (1984)	Explorer, Expert, a bit of Warrior
Elsie	Brandon, Basden, Hamilton, Stockley (1988)	Explorer, Teacher
Istar	Basden & Hibberd (1996)	Emancipator, Explorer
KgSvr	Basden (2000)	Butler
IRKit	-	Explorer, Artist
'New View' web site	www.basden.demon.co.uk/xn/nv/	Warrior, Stimulant
BHG web site	www.basden.demon.co.uk/xn/bhg/	Stimulant

Moreover, these paradigms do not capture the experience of the technology as such with which IS developers must engage (which is discussed in Chapter VII).

Real-life ISD cannot be compartmentalised into HST, SST, or CST, especially if it is of good quality. The good developer does take some things about the domain of application to be given (reflecting an aspect of HST) but does try to sensitive to a wide range of interpretations (SST) and does question the status quo (CST). It is often not entrapment by one of these systems approaches that prevents this so much as an attitude such as laziness. Just as with Burrell and Morgan's four paradigms, so these three systems approaches do not adequately represent fully the everyday approach taken in ISD.

6.1.4 Everyday Experience in ISD

Vignette 3 in the Preface gave only part of the author's story in ISD. More is seen in Table 6-2. The following additional details will be referred to later.

In 1970 he began programming in machine code to build data-structuring resources for the computer-aided design (CAD) system described in Basden and Nichols (1973), which he programmed in FORTRAN. In this project he was exploring suggestions by others that computers could lay out electronic circuits topologically rather than geometrically, which required highly flexible data structures. It was fulfilling work, especially as he sought to respect human ways of designing. See Basden (1975) for the full description.

He then joined a pharmaceutical company as programmer of business data and experienced the frustration and fun of working in a team with a poor supervisor. In 1975 he joined a small team to develop a database system for general practice,

CLINICS, comprising around 30 programs written in the COBOL language. Medical information, he discovered is complex, error-prone and subject to multiple interpretations! Combining the roles of programmer, analyst, planner, and software consultant, he also learned the importance both of good comments in program code and good relationships with users.

Joining the corporate laboratory of a large chemical company in 1980, he began developing knowledge based systems for corrosion and crop protection. His experiences of knowledge elicitation are described in Vignette 3 including his discovery of the importance of separating "understanding" from contextual knowledge and of developing a close relationship with the domain experts. A particular satisfaction lay in enabling others to become active in KBS, rather than protecting his own power-base, and especially in helping a corrosion expert to build his own KBSs. Another satisfaction lay in using KBS to challenge the status quo: agronomists at the time assumed industrial farming, but the author led them to admit the possibility of low-input farming, and built that into the Wheat Counsellor system.

These experiences led him to the project that produced the *Elsie* system described in Chapter IV. In addition to programming, knowledge elicitation, and representation, he took more strategic responsibility for KBS projects as a whole. As described in Vignette 3, the success of *Elsie* led to a 3-year research project to develop a "Client Centred Approach" (Basden, Watson, & Brandon, 1995), based not on technical nor social aspects of ISD but on responsibility.

Later, he returned to programming, in machine code and the C language, to develop IRKit as a basic data-structuring resource, which expanded his ideas of the 1970s. This not only enabled the development of the innovative proximal KBS development tool, Istar (Basden & Brown, 1996), but enabled him to begin serious testing of his ideas about appropriate "aspects of knowledge" outlined in Vignette 1 and taken further in Chapter VII.

More recently, his main development effort has been of Web sites in which there is tension between the fluid ongoing development of idea and the need for clear ideas for the reader.

There is, of course, much more, and some of the experience is expanded elsewhere in this chapter and Chapters IV and VII. But the author's account portrays a great variety and at the same time a substantial coherence, in which all are interwoven into a single fabric that is ISD. The author believes his account is by no means unrepresentative of the everyday experience in ISD, but the reader must judge how representative it is. The main lack he is aware of is that he has never been involved in a huge ISD project, neither as pawn nor as manager. However, he would argue that, with the possible exception of career ladders and rivalries in large teams, he has experienced most of what happens there.

The difficulty we face is that most accounts of ISD have already adopted one of the above frameworks for understanding and couch them in its terms. So accounts like that of the author, of the interwoven diverse facets, are rare in the academic discourse in ISD. In the light of this variety none of these theoretical frameworks, and especially the Burrell-Morgan model, seem adequate to help us understand ISD; they hardly seem even relevant!

6.1.5 Philosophical Critique of the Paradigms

Hirschheim et al. (1995, p. 49) cite over-simplification and artificiality as criticisms of the Burrell-Morgan model, though they saw it as the best available at the time for ISD. However, the reason why the Burrell-Morgan model is inadequate may be understood philosophically. Eriksson (2006) has examined both the systems approaches and the Burrell-Morgan model through the lens of Dooyeweerd's theory of ground-motives.

He argues that HST, SST, and CST all presuppose the Nature-Freedom ground-motive (§2.3.1.4). HST is self-evidently of the Nature pole, SST is of the Freedom pole and "Compared with HST and SST, CST makes a serious attempt at representing the complete Nature-Freedom ground-motive. It not only articulates explicitly the two realms but also attempts to provide a link between the two." This is shown in Table 6-3 [which is adapted from Eriksson (2006, p. 226)].

But, according to Eriksson (2006, p. 226), CST "has not succeeded in solving the very fundamental tension between the realm of nature and that of freedom, which is: How can man maintain his autonomous freedom in a mechanistically determined world?" This is because it is rooted in Kant, whom Dooyeweerd criticised for not being critical enough. (See §2.4.5; Eriksson also provides a cogent summary of Dooyeweerd's critique of Kant.)

MST, multimodal systems thinking (De Raadt, 1991), conjoins Dooyeweerd's aspects to Beer's (1979) viable systems model. Its use of aspects provides diversity of norms to act as practical guide. It claims to be founded on the Creation-Fall-Redemption ground-motive, but Eriksson criticises it for not taking seriously enough issues of Biblical interpretation and suggests that, because VSM presupposes the NFGM, so does MST. MST does indeed betray strong influences of the Nature pole. Nevertheless, De Raadt is to be applauded for the first attempt to define a systems approach based on Dooyeweerd. It is still in development and used in practical analysis.

Table 6-3. The ground-motive commitments of varieties of systems thinking

	Systems Thinking Framework			
	HST	SST	CST	MST / DST
Dominating Religious Ground-Motive	Nature	Freedom	Nature - Freedom	Creation - Fall - Redemption

DST, disclosive systems thinking (Strijbos, 2006), seeks to develop a systems approach based on some parts of Dooyeweerd's thinking without the help of others. Its central idea is that of diverse intrinsic normativity (§3.1.4.9) and sees ISD (and other business analyses) as a process of "disclosing" the innate normativity of a situation. It has links with Schuurman's (1980) "liberating vision for technology" discussed in Chapter VIII. It does not make use of Dooyeweerd's aspects, and has yet to be extensively applied to practical situations; Eriksson does little more than mention it.

Following his analysis of HST, SST, CST, MST, Eriksson discusses Jackson's use of the Burrell-Morgan model. He argues (2006, pp. 231-2) the model is based on the NFGM (both dimensions are expression of Nature v. Freedom) and thus "This both articulates and forces the investigated systems thinking paradigms into the unbridgeable tension of dualism, founded on the assumption of autonomous reason." As a result, "it does not allow the detection of problems in the very ground-motive that governs many of the systems thinking paradigms," "it mis-conceptualizes systems thinking paradigms that are not based on the Nature-Freedom ground-motive" (e.g., MST) and "does not inquire explicitly the sources of norms of these paradigms." The problem, Eriksson argues, is the Nature-Freedom ground-motive itself.

A new framework, based on a different ground-motive, CFR, is explored here, using Dooyeweerd's ideas, but not quite in the way either MST or DST does so.

6.1.6 Toward a Different Framework for Understanding

Each of Hirschheim et al.'s generations, being a historical response to certain perceived problems (often brought about by previous generations), is likely to point to something important in the pursuit of ISD. The issue addressed in each generation

Table 6-4. Main aspects of ISD generations

Generation	Concerns	Aspect
Lifecycle	"to control .. standardization"	Formative
Structured approaches	"to increase productivity"	Economic
Evolutionary, Prototyping	"concerned .. to get the right system"	Juridical
Socio-technical	"to ensure participation .. resolve conflicts"	Social
Sense-making and problem-formulation	"sense-making" "consensus among multiple perspectives"	Analytic Aesthetic
Trades union	"that workers' rights .. should prevail" "that .. democracy should prevail"	Juridical Social
Emancipatory	"barriers to communication" "questioning of dominant forms of thinking"	Lingual Pistic

becomes meaningful and problematic with reference to a particular aspect or two; Table 6-4 shows this author's analysis of them.

The shalom principle (§3.4.3) implies that focus on one aspect to the exclusion of others will inevitably give rise to problems, associated with one or more of the ignored aspects. This can account for why each generation led to others. But it also suggests that for high quality ISD all aspects need to be taken into account in ISD. This does not, however, indicate that we should simply amalgamate all the generations, because the ethical aspect is under-represented and the concerns shown do not always align with the kernel meaning of the aspect, but sometimes only with one small part thereof (e.g., Trades Union approach is concerned with juridical "what is due" but only for a small subset of stakeholders).

The current ways in which ISD is understood have already distorted the portrayal of its everyday experience. Therefore ISD will be reinterpreted from Dooyeweerd's perspective of CFR and multiple aspects.

6.2   ISD as Multi-Aspectual Human Activity

The central proposal of this chapter is that ISD may be understood as multi-aspectual human functioning (§3.4.1), in which every aspect is important and deserving of attention (§3.4.3). It is tempting because it is concerned with a technology and development, to jump to the conclusion that IS development is qualified by the formative aspect. But such an approach too quickly narrows the focus, robbing us of an everyday perspective. DST exhibits this tendency. A Dooyeweerdian approach that retains the lifeworld perspective is to begin not with a hidden normativity to be disclosed but with the very visible multi-aspectual human functioning that is ISD. The central focus is on cosmic meaning and law as they are actualised in the everyday life of ISD (§2.4.2-4).

Aspectual normativity is as important in ISD as it was in human use of computers (Chapter IV), but it is important in a different way. When considering HUC, the normativity of aspects was important to allow us to differentiate beneficial from detrimental impacts for purposes of evaluation and looks to the past. In ISD normativity is important as guidance for methodology and looks to the future.

6.2.1   Several Multi-Aspectual Functionings

Following the approach in Chapter IV, several multi-aspectual human functionings may be discerned in the lifeworld of ISD, which are interlaced with each other. Four will be considered:

- The overall ISD process
- Anticipating use: how possible use impacts on design of the IS and vice versa
- The creation and crafting of the IS: technical artefact and its context of use
- Encapsulation of domain meaning

These are by no means foreign to extant accounts of ISD. For example "anticipating use" is "user requirements analysis" in lifecycle approaches and is central to sense-making and emancipatory approaches. "Creation of the IS" is design, implementation, and testing in lifecycle approaches and is central in unstructured, structured, and evolutionary approaches. But Dooyeweerd leads us to emphasise things differently, and allows all four to take place in parallel. The advent of iterative and agile methods has shown this is correct.

In most extant approaches, the first is seen as a whole of which the others are parts, but assuming a part-whole relationship tends to separate out the various functionings into sequential stages (as in the waterfall model), and deprives the supposed parts of meaning. But each of them can be meaningful without the overall ISD process (for example knowledge elicitation may be seen as analysis unconnected with ISD), and each has a different qualifying aspect. So the relationship between them is a structural enkaptic relationship among wholes rather than a sequential relationship among parts.

The qualifying aspects of these four are, respectively, the aesthetic with social (orchestrating the whole process), juridical (responsibility for how the IS can be used), formative with lingual, and analytic (identifying relevant concepts). The reasons are given when the four are examined in more detail.

The author's colleague, Gareth Jones, has been undertaking exciting action research which has explored in detail how Dooyeweerd can be used in all these processes and which has generated knowledge based systems for assessing environmental sustainability. The work will be published in due course, but an early portion of his work, using Dooyeweerd to identify stakeholders, may be found in Jones and Basden (2004).

6.3 The Overall ISD Process

If the shalom principle is valid, every aspect is important in the overall process of ISD. But in which order should they be considered? The aesthetic aspect (harmony) is central to the coherence of the project. In ISD a number of people are involved—team members, participants or other stakeholders—and they must be involved not

Table 6-5. Aspects of the ISD process as a whole

Aspect	.. of overall ISD process	ISD Approach
Biotic / Organic	Health of members of development team.	
Psychic / Sensitive	Pain, pleasure experienced by the ISD team	
Analytic	Clear objectives, goal.	SI boundary ctq
Formative	Planning, History. Development of IS.	Iterative development Spiral and W/F models
Lingual	Knowledge representation. Documentation, Archives. Open dialogue. Seeking information.	EISD, SSM
Social	Authority in the ISD team. Respect for views. Friendships	Structure of team
Economic	Budget, deadlines. Access to expertise. Technical limitations.	Structured development
Aesthetic	**Orchestrating the project** Fun	Unstructured development MultiView, XP
Juridical	Contract to deliver. Responsibility.	CMMI. CCA
Ethical	Attitude of self-giving.	Egoless programming
Pistic	Vision for project. Loyalty to project.	SSM *Weltanschauung* SI 'sacred'

Creating the I.S.
Quantitative
Spatial
Kinematic
Physical
Biotic
Psychic
Analytic
Formative
Lingual
Social
Economic
Aesthetic
Juridical
Ethical
Pistic

Quantitative
Spatial
Kinematic
Physical
Biotic
Psychic
Analytic
Formative
Lingual
Social
Economic
Aesthetic
Juridical
Ethical
Pistic
Anticipate usage

as individuals but as a cohesive group, so the social aspect is of central importance. It is therefore useful, though not essential, to consider these first. The aspects of the overall process are summarised in Table 6-5, where the horizontal arrows indicate enkaptic relationships to other multi-aspectual functionings.

6.3.1 The Aesthetic Aspect

The ISD project may be likened to a symphony, the ISD team being the orchestra. There are main players—strings, wind, percussion, brass sections—with the occasional emphasis on one instrument type or another. Solo pieces come in from the outside, as it were. Each plays to the best of their ability (anticipating the juridical aspect) but each, including soloists, subordinating themselves to serving and supporting the others rather than seeking aggrandisement (anticipating the ethical). The music expresses something (lingual). The playing has a certain overall plan but there is much improvisation (formative), which is seen not as unfortunate deviation from plan but as making the whole even more meaningful. And yet improvisation is the minor element within the context of the major plan, and is done with economy (economic). Moreover, the symphony does not stand alone in the concert but beautifies and enhances its surrounding pieces, bringing out their own beauty and integrity.

One would never outsource the playing of sections of the piece! That unduly elevates the economic aspect in ISD. One of the strengths of the original unstructured approach is that, when done well, it facilitates an aesthetic holism. (But when done badly, it is destructive.) Avison and Wood-Harper's (1990) Multiview emphasises the aesthetic aspect. So might Beck's (2000) Extreme Programming with its slogan "embrace change", if change is seen not just as dynamics but as surprise.

6.3.2 The Social Aspect

The norm of the social aspect is respect, but what form should this "respect" take? Hirschheim and Klein's (1994) Emancipatory ISD (EISD) is one approach in which the social aspect is central and emphasises power-relationships. But the lifeworld of ISD involves many social relationships that are not based on power—from "social" events to helping each other out. To what extent should control be exercised, and to what extent should the team comprise unstructured social relationships?

Dooyeweerd's (1986) theory of social institutions can help us understand and manage power-relationships among stakeholders in ISD. He differentiated intracommunal, intercommunal and interpersonal relationships, exhibited in ISD, respectively, in the need to make a cohesive team (and hence often a formal structure), the interests of the various external stakeholders, and in the social friendships of all participants. Dooyeweerd suggested we should treat all three differently and beware of assuming the norms for one type carry over into others.

Power-relationships are valid only in the first type. Intracommunal relationships, that is within a true social institution, are dependent upon relationships of "authority and subordination", as Kalsbeek (1975, pp. 199-200) calls them. Whereas, under NFGM, authority and subordination are seen as limiting, to Dooyeweerd they are enabling, and should have no negative connotation. Moreover, they should be tempered by other aspects, especially the juridical and ethical.

But no such power structures should be imposed in the interaction between communities or individuals. As Kalsbeek (1975, p.200) points out, "we never find an authority structure in intercommunal and interpersonal relationships." It is not valid for one stakeholder's interest to subordinate that of another. However, there is also respect for special expertise, which may be seen as intercommunal in that the expert, as expert, is member of another community. Kalsbeek continued: "this does not deter certain people or groups or classes from exerting a considerable influence outside of their area of authority because they have special gifts or capital at their disposal." Table 6-6 summarises how the norm of respect is different for each type and might be useful in avoiding inappropriate social functionings in practice.

Table 6-6. Norms for different social relationships in ISD

Type of Social Relationship	Type of Respect
Intracommunal	Authority + Deference (tempered) Assumption of role
Intercommunal	Giving due importance to each interest and view Respect for expertise
Personal	Friendship, Consideration

6.3.3 Pre-Social Aspects

The lingual aspect of ISD overall is manifested in communication within the team and with all other stakeholders and also in such things as seeking information and reporting. Open dialogue, central to EISD and to Checkland's (1981) soft systems methodology (SSM), is what this refers to. (The lingual aspect of the representation and coding of knowledge is of a different multi-aspectual functioning, dealt with later.)

The formative aspect is manifested here in planning of the ISD process and in the history of the project. The difference between formal structure [e.g., waterfall model and Boehm's (1988) spiral model], and informal (iterative development), is visible at the formative aspect. The formative aspect is important in creating the IS itself, which is discussed later.

Functioning in the analytic aspect involves making distinctions relating to use, which include such things as between who are stakeholders and who are not, between relevant and irrelevant issues, and between what the IS should be called upon to do and what it should not. These are all versions of what Midgley (2000) calls boundary critique in his systemic intervention (SI). Perhaps one of the most important analytic functionings is the identification of secondary users, as discussed in Chapter IV.

The sensitive-psychic aspect of ISD is manifested in the emotions and feelings of team members and other stakeholders, and the biotic is their health. Few extant methodologies give these prominence.

6.3.4 Post-Social Aspects

Of the post-social aspects, the economic is exhibited in management of the project, and in the limited resources of time, budget, expertise, personnel, access to participants, and the like, which the structured approach emphasises. It has received considerable attention in research and practice, so little more need be said here.

The juridical aspect is important, not just because of implied or actual contracts of delivery between the parties, emphasised in CMMI for example, but more importantly because of responsibility for repercussions when the system is in use, and thus toward all stakeholders, whether acknowledged or not, emphasised in the author's (1995) client centred approach. This responsibility is the main reason why ISD should be orientated so as always to critically and sensitively anticipate use.

Dooyeweerd's ethical aspect is specifically centred on self-giving—in contrast to traditional approaches to ethics—implying a norm of generosity. This is recognised in Gerald Weinberg's (1999) notion of egoless programming.

The pistic aspect is concerned with faith, loyalty, and vision of who we are as part of the ISD project. It is manifested, for example, in treating some stakeholders and issues "sacred" and others "profane" (Midgley, 2000). But it is more centrally manifested in our commitment to a vision of what the meaning of the IS is that we are developing, what eventual use will mean, and in loyalty to the project. This may be different for different stakeholders, who hold deeply-held beliefs and perspectives that Checkland (1981) calls (*Weltanschauungen*).

Variety in such deep, pistically-held perspective can result in conflict. The more intractable conflicts arise from undue absolutization either of some concrete thing, such as one's own preferred solution, or an aspect. Such conflicts cannot be resolved by open dialogue, partly because their holding is pistic functioning, and partly because each aspect provides a distinct rationality (§3.1.5.2). Habermas' notion of ideal dialogue, which is at the centre of EISD, perhaps proves less than useful to us, even as a counterfactual ideal, because there is no logical link between what makes sense in different aspects. Holding disparate visions together involves pistic functioning, which itself depends, if Dooyeweerd is correct, on good aesthetic and ethical functioning. Assuming such an attitude, the Dooyeweerdian notion of aspectually centred perspectives, as discussed in §3.3.4, can help.

6.3.5 All Aspects Together

The aspectual approach to normativity, adoption of a suite of aspects like Dooyeweerd's, and the shalom principle give us a precise and yet flexible basis on which to understand the quality of the ISD process. According to this, ISD projects will tend to run well to the extent that those involved fulfil the underlying norms of all the aspects. Seriously ignoring any of the norms can jeopardise the overall success of the project. But it must be reiterated that since aspectual norms can never be fully known by theoretical thought, nor even defined, it is dangerous to try to make them into a method or set of rules. But Table 6-5 can help point to the kinds of normative issues that demand attention during an ISD project.

The validity of this multi-aspectual approach can be neither proved nor disproved scientifically because it concerns the lifeworld of ISD. However, support for it may be found in the records or the honest recollections of many ISD projects. In the author's experience of ISD project failure or difficulty:

- In one project, a key player left half way through, and the project could never recover. (pistic loyalty)
- In many student group projects there is lack of commitment. (pistic)
- In another, a competitive, self-seeking attitude pervaded the team. (ethical)
- In yet another, insufficient attention was given to the core knowledge and too much to the usability features, so the knowledge base was never completed. (juridical, aesthetic)
- Yet another project was too fragmented. (aesthetic)
- Many projects exceed time or cost budgets. (economic)
- Another project was jeopardised by animosity between team members. (social)
- Some projects fail technically. (formative)
- Some projects lack clear objectives. (analytic)

In such an analysis, are we merely filling slots? This common self-criticism in Dooyeweerdian application is discussed in Chapter IX. But even if it is, using the aspects as slots to fill can provide useful methodological guidance in ISD because it stimulates the developer to think of things often overlooked—as long as the suite of aspects used broadly covers the wide range of issues in the lifeworld of ISD. This is precisely what is claimed for Dooyeweerd's suite (§3.1.6).

6.4 Anticipating Use

Since 1970s user requirements analysis (URA) has probably been the most common method for anticipating use. Most versions of URA suffer because users seldom fully express what they need, are not aware of all the potential benefits and innovations resulting from IS use, nor the limitations, and change their minds and ways of working in unanticipated ways when they begin to use the IS (see Chapter IV). There have been many attempts to overcome such problems, of which Checkland's (1981) soft systems methodology (discussed later) is an early example. More recent proposals have questioned the very elements that URA assumes. For example, Patnaik and Becker (1999) urge a focus on "needs" rather than requirements, because needs are more stable and can be met by a variety of solutions. While most methods focus on problems to be solved (the negative) Norum (2001) advocates appreciating the positive rather than problems to be solved because this leads to more "life-giving" changes.

It would be useful to understand where these and other methods fit in an overall picture and whether there are any parts not yet covered. The generality of Dooyeweerd's philosophy might help provide this.

Whereas in Chapter IV, the focus is mainly on present and past actuality, in ISD we look forward, anticipating future possibilities for such use, which guide the creative process of design and development. Even IS maintenance is directed to the future. Thinking philosophically, the notion of law-based functioning gave Dooyeweerd a means of understanding anticipation of future possibility (Dooyeweerd, 1984, I, p. 105):

Everything that has real existence has many more potentialities than are actualized. Potentiality itself resides in the factual subject-side; its principle, on the contrary, in the cosmonomic-side (law-side) of time. The factual subject-side is always connected with individuality (actual as well as potential), which can never be reduced to a general rule. But it remains bound to its structural laws, which determine its margin or latitude of possibilities.

This latitude of possibilities is governed by the normativity of the aspects, each aspect providing latitude of a different kind.

Whereas development itself might be formative, possibility implies responsibility. That is, the IS developers are responsible for what they develop and, though never solely so, for the uses to which the IS is put and the impacts thereof because these are influenced by its design. For this reason we may see the qualifying aspect of anticipating use as the juridical.

To fulfil this juridical norm every stakeholder should be identified and involved in the ISD process as far as possible. Both the client centred approach (CCA) (Basden, Watson, & Brandon, 1995), Midgley's (2000) systemic intervention emphasises this aspect. But neither give guidance on how to ensure all the stakeholders are identified. Dooyeweerd's suite of aspects can, however, help us do so very practically (Jones & Basden, 2004).

- Asking of each aspect "What roles are there connected with this aspect in the situation of use?" will identify those who should participate.
- Asking "What repercussions will use of the system have in this aspect, and on what or whom will it have repercussions?" can help identify other stakeholders who would be affected even though they have no role connected with it—such as animals, the public, society, environment. A host of likely repercussions can be exposed, and thus help distinguish the important from the trivial.

In practice, it can be useful to consider the three types of use we identified in Chapter IV, HLC, ERC, HCI (human living with computers, engaging with represented content, human-computer interaction). Here is a selection of questions that developer might ask, mainly geared to HLC.

Analytic aspect: HLC: When they use our artefact, users will make all kinds of distinctions relevant to their living or work that we have not considered. To do so appropriately they must understand it clearly: build in transparency so they can understand the structures of the system and how it works.

Formative: HLC: Will this be used for purposes we cannot envisage (c.f. *Elsie* in Chapter IV), because their context is not ours? So build in flexibility and robustness.

Lingual: HLC: Will users be able to explain or communicate better to their colleagues, or keep more salient records because of the IS we deliver? HCI and ERC: Users must interpret what we put on screen and give input. To ensure they will understand what we intend, we must carefully design wording and explanations.

Social: HLC: Will use of this make people less or more socially active in healthy ways? HCI: Cultural connotations may cause problems if it will be used globally (e.g., a Web page).

Economic: HLC: As a result of using this, will the way resources are managed (e.g. raw materials, paper, users' time) change? Ensure this always tends toward more frugality rather than waste or superfluity. Remember climate change: not only energy consumed by a computer left on for hours when not used, but will use of it result in more flights, road journeys, etc.? Though SSM (Checkland, 1981) deals with "environmental constraints" it does not adequately recognise the flexible responsibility involved; see later.

Aesthetic: HLC: Users should find all they do with your system interesting and stimulating, and the IS should fit harmoniously with its lifeworld context(s) of use, yet provoking new thinking therein.

Juridical: HLC: Ensure that, as a result of using this system, all stakeholders will be given their due. This is more important even than ensuring adherence to national and international law, which may be seen as (deficient) attempts to define "due".

The notion of emancipation in Hirschheim and Klein's (1994) EISD is mainly juridical.

Ethical: HLC: Will use of our system make users more selfish and competitive or more self-giving, generous, collaborative? This was one reason why *Elsie* (Chapter IV) was a success. ERC: Will users come to "love" the IS?

Pistic: HLC: It might surprise us how important is the pistic aspect. Churchman (1971) suggested:

If we look at faith from the design point of view, we ask whether a faithful inquiring system is better than a faithless one. No matter how slight the chance, the gambler must in some sense have faith in the one possibility that is favourable to him. (p. 240)

The "gambler" may be the IS developer. But they can also be the user, who is using the IS in new ways. ERC: Will its users and all other stakeholders trust it, or might it let them down? HLC: Is the use of this system in line with the vision of the users and their organisation? If so, is that vision appropriate? When using it, will users be stimulated toward questioning deficient visions? Will "good faith" be encouraged or will it be hindered [see §4.6.2 and Walsham (2001)]?

There is much more to be said in each aspect, and the reader is left to fill the gaps. Many of these impacts in HLC will be unexpected, indirect, or long-term, and it is the IS developer's responsibility to design for these as far as possible. Of course, many such impacts cannot be predicted, but an aspectual analysis can help indicate the general kinds of issues that might have been overlooked.

One example: As developer of the Wheat Counsellor system (Jones & Crates, 1985), which advised farmers on use of fungicides, the author took on himself a responsibility of a pistic nature. Realising that British agriculture was then too heavily dependent on chemicals, and that the pendulum would soon begin to swing away from this state, he probed the source agronomists for their expertise about what they would advise if a farmer wished to use fewer chemicals. After some initial resistance, they divulged the advice quite readily, and Wheat Counsellor became not only a better product but one that was more trusted. (This is why, in Table 6-2, he had a bit of a Warrior role.)

The IS developer has a (God-given?) responsibility to ensure they are sensitive to a wide range of aspectual issues. Functionalistic "user requirements analysis" is a pale shadow of such multi-aspectual analysis anticipating use, and no prior specification can adequately capture its richness, nor can the usual spiral or "iterative" methods.

Patnaik and Becker's (1999) focus on needs rather than requirements may be seen as beginning to escape subjectivism in ISD and to recognise there is something that transcends the users. Norum's (2001) focus on the positive may be seen as opening the door to considering *shalom* rather than merely reacting to presenting problems. But these still provide only meagre guidance on appreciating richness of this *shalom* and still suffer from overlooking important taken-for-granted factors. Aspectual analysis could enrich them considerably, and stimulate a consideration of such factors. How it can enrich SSM is discussed later. This is perhaps what DST is aiming at, and it is a pity that the use of Dooyeweerd's aspects is eschewed. It is certainly what good developers already do, and what this author had been trying to achieve in most of his ISD projects listed in Table 6-2, most of which were undertaken long before he discovered the great benefit of Dooyeweerd's suite of aspects.

6.5 Creating the IS

Creation of the IS—and its on-going maintenance—may be seen philosophically in terms of becoming and change, and Dooyeweerd can throw light on this (see §3.2.4). Though maintenance should perhaps be considered separately, it will be merged with IS creation here because most of the important issues apply to both.

One of these is that creating (or maintaining) the IS is not just about constructing a technical artefact or system. Creating the IS involves shaping also the user's knowledge and the human and organisational context of use (see Structure of ERC in Chapter IV).

That both knowledge in the artefact and that instilled in the context of use are important is indicated by the Dooyeweerdian belief, expounded in Chapters IV and V, that the knowledge represented "in" the computer is nothing without the user's subject-functioning; it is the user's knowledge which is the more important. This echoes West's (1992) view that, even in the case of KBS technology, in which the knowledge represented in the artefact reaches its peak of sophistication, "it may not be necessary to convert this information ... for inclusion within the expert system [KBS]" but to rely on it being active in the users. This involves training users—if you like, domain knowledge must be "put into" both the computer system and the users!

6.5.1 Aspects of Creating the IS

The shaping of both is a multi-aspectual human activity (§3.4.1). Though the aspects of ISD overall and anticipating use were each discussed, here only a few indicative

Table 6-7. Aspects of creating the information system

Aspect	... of shaping context	.. of creating artefact
Analytic	Who may and may not use it; What they must know	Clarifying concepts, Relevant v. irrelevant knowledge, concepts.
Formative	Purpose it serves (HLC) Planning the context	Structuring data, Designing algorithms, Getting it working.
Lingual	Training users	Representing knowledge as the program. Writing documentation, tutorials, help systems, etc.
Social	Organisational structures.	Relationship with all involved in ISD, esp. domain experts.
Economic	Management of IS use.	Program efficiency, Use of limited screen area. Scaling up.
Aesthetic	Making use enjoyable. Artefact fits context.	Style of UI. Beauty of program,
Juridical	Ensuring use is appropriate to context.	Doing justice to information and knowledge.
Ethical		Loving the program.
Pistic	Shaping visions of users.	Impact of religious views on how we program.

Kg Elic^n Quantitative, Spatial, Kinematic, Physical, Biotic, Psychic, Analytic, Formative, Lingual, Social, Economic, Aesthetic, Juridical, Ethical, Pistic

Kg Repr^n

Chore

Delight, Satisfaction

examples are given, of aspects of artefact creation and of shaping the context of use. Table 6-7 shows these; the knowledgeable reader is encouraged to criticise this and make their own more detailed analysis.

The main (qualifying) aspect of creating the IS is the formative—planning and shaping both the artefact and its context of use. The ISD community as a whole has considerable experience of the analytic to lingual aspects, and books on, and methodologies about, these abound. This involves both the chore of functioning in the earlier aspects and the delight that is functioning in the later ones; all the aspects are important to being fully human in ISD (§3.4.2).

Creating the IS links strongly with knowledge elicitation and representation, via the analytic, formative, and lingual aspects as shown, and this is explored later in this chapter, and further in Chapter VII.

6.5.2 The "Chores" of Creating the IS

It is well known that the task of creating the IS is hard work, and most methodologies designed to guide it concern themselves with the analytic to economic aspects in varying degrees. Central, of course, is the discipline of actual programming (or database creation, OO class creation, etc.)—good program structure, meaningful variable names, comments that explain "why" as well as "what; this is all covered by the cosmic normativity of the lingual aspect. But it has long been recognised that many other aspects are also relevant, such as the economic aspect of keeping to time, the formative aspect of planning the IS, and so on.

Agile SDMs have returned the focus to IS creation, and each seems to address a particular problem. For example the Crystal series of ASDMs (Cockburn, 2005) is aimed at overcoming barriers to communication, while Extreme Programming (Beck, 2000) has the slogan "Embrace change". ASDMs might be seen as a return to the everyday life of programming—as is evidenced by the wide range of human aspects given importance by the 15 principles that support Extreme Programming's values; see Table 6-8. Such an aspectual analysis of an SDM can indicate the degree to which it is open to everyday reality.

["Embracing change" was difficult to assign to an aspect. The most obvious aspect is the formative (deliberate shaping) but this principle expresses rather the welcoming of change that occurs, hence the aesthetic aspect of harmony. It might have been the ethical aspect (self-giving: willingness to bend to the will of others), which is otherwise missing.]

Table 6-8. Aspects emphasised in Extreme Programming

XP Principle	Aspect
Assuming simplicity	Aesthetic
Incremental change	Formative
Embracing change	Aesthetic
Quality work	Juridical
Teaching learning	Lingual
Small initial investments	Economic
Playing to win	Pistic
Concrete experimentation	Formative
Open, honest communication	Lingual
Working with people's instincts	Social
Accepting responsibility	Juridical
Local adaptation	Social
Travelling light	Economic
Honest measurement	Juridical
Rapid feedback	Lingual

6.5.3 The Delight that is Creating IS

But other aspects are also important, including some rather surprising ones. Yet, in *Things a Computer Scientist Rarely Speaks Talks About*, Donald Knuth (2001), the designer of T$_E$X, recalls (p. 130):

I got hold of a program from IBM called SOAP, written by Stan Poley. That program was absolutely beautiful. Reading it was just like hearing a symphony, because every instruction was sort of doing two things and everything came together gracefully. I also read the code of a compiler that was written by ...: that code was plodding and excruciating to read, because it just didn't possess any wit whatsoever. It got the job done, but its use of the computer was very disappointing. So I was encouraged to rewrite that program in a way that would approach the style of Stan Poley. In fact, that's how I got into software.

What Knuth refers to is the aesthetic aspect of programming. Pacey (1996, pp. 80-81) likewise refers to the "existential joy" in technology. One can "love" a program, giving oneself for it (ethical aspect), do justice to all the knowledge and information represented in it (juridical), and Knuth is also quite open about how his religious beliefs have impacted his programming. What is it that makes creation of the artefact satisfying and a delight, rather than a chore? (I am not aware that this question has ever been addressed.) It is these four post-economic aspects—which, sadly, receive little attention. But seeing creation of the IS as multi-aspectual, in which the shalom principle is important, can help ensure such issues are given due, but not undue, importance.

6.6 Encapsulating Knowledge

Businesses today pride themselves on being "knowledge-led" and seek to build and maintain "knowledge assets", in which useful knowledge is held electronically. Much of this is held in sophisticated databases, but even the humblest program can be seen as a "knowledge asset" insofar as it runs according to knowledge about its domain that is encapsulated inside it. For example a calculator encapsulates knowledge about numbers, laws of arithmetic, and so forth. A word processor encapsulates knowledge about words, paragraphs, and the laws of the lingual aspect such as that most words should be in a vocabulary or that some need special emphasis in use as italics or headings. The IS that is a word processor situated in a particular context of use encapsulates even more knowledge—about accepted writing styles and layouts and about which words and phrases are well known, and so on. In a KBS like *Elsie* (Chapter IV) the knowledge encapsulated is complex and multi-aspectual.

To develop such an IS—both the technical artefact and its context of use—involves eliciting such knowledge and representing it: (a) using a programming or knowledge representation language to construct the artefact and (b) in documents, videos, conversations, stories, and the like to develop the context of use.

But what knowledge must be elicited? Many knowledge management (KM) texts, such as Collinson and Parcell's practical handbook (2001), emphasise the capturing and encapsulation of specific knowledge to build an organisation-wide "knowledge asset"; they include stories, incidents, lessons learned, results of actions, sample presentations, and so on. But such specific information is of little value as a guide to future action unless it is generalised. So they advocate distilling the key messages, to move toward good practice. So it is usually generic knowledge which is encapsulated within programs. (This is why it was suggested in Chapter V that a program may be seen as a constructed, virtual law side.)

6.6.1 Knowledge Elicitation and Representation

The terms "knowledge elicitation" and "knowledge representation" are taken from the KBS community, though they are equivalent to analysis and design, implementation, and testing in structured SDMs. But "elicitation" and "representation" are used here because it was there they reached their pinnacle of sophistication and responsiveness to the lifeworld of the application, because in KBS it was important to elicit and represent very complex, highly nuanced, meaning of the domain, and current methods owe a lot to KBS research and practice. Knowledge elicitation is concerned with identifying relevant "knowledge" of the domain and conceptualizing it, and knowledge representation is concerned with expressing that in computer-readable symbolic form.

Knowledge elicitation has traditionally been seen in mining metaphors: extracting nuggets of "knowledge" from those who are expert in the domain, which are subsequently represented in propositional form. But this metaphor was critiqued in the early 1990s on the grounds that some knowledge is un-extractable, some takes the form of "stuff" rather than nuggets, while yet other knowledge is generated by the very process of elicitation and representation [see Basden and Hibberd (1996) for discussion of this]. Fiol and Huff (1992), and many others, found that the very act of representation stimulates elicitation rather than being merely a mechanical expression of what has been elicited.

But under Dooyeweerd, to think of knowledge as either extractable or un-extractable, or nuggets or stuff, is misleading, because it presupposes a distance between knower and known (a Kantian gulf between thought and thing) which Dooyeweerd rejects (§3.3.2). Rather, it is better to focus on the multi-aspectual human activity that is knowledge elicitation-and-representation. It concerns itself with cosmic and concrete meaning relevant to the domain of application, involving the analytic aspect of distinguishing what is relevant meaning from what is less so, and conceptualizing it, formative aspect, of structuring and relating it, and lingual aspect, of expressing it in a computer-readable knowledge representation language (the type of which is discussed in Chapter VII). These are the arrows shown in Table 6-7. The qualifying aspect of elicitation on its own is the analytic, and that of representation is the lingual, but it is better to see all three as intertwined.

But all other aspects are important too. For example, the social aspect is important in knowledge elicitation because those with relevant knowledge might withhold it if not treated with respect. Trust and friendship become extremely important for developing KBS, as mentioned in Vignette 3. The harmony of all the knowledge encapsulated (aesthetic aspect) is vital. But of especial importance is the juridical aspect of doing justice to the meaning of the domain.

6.6.2 Doing Justice to Domain Meaning

In Chapter IV it was shown how the norm that should guide ERC (engagement with represented content) is that of doing justice to domain meaning. One technique that has emerged to assist this is the domain ontology, a computer-processable statement of, as Guarino, Masolo and Verere (1999), put it, "a set of things whose existence is acknowledged by a particular theory or system of thought." But such a definition presupposes a theoretical rather than lifeworld attitude towards knowledge elicitation. Perhaps better is Gruber's (1995) definition, "a formal explicit specification of a shared conceptualization" because while it acknowledges the need for formal specification (readable by computer as a data model) it seems to leave open the possibility of this attempting to represent the lifeworld of the domain. But there is a hidden presupposition, that it is clear what a shared conceptualisation is or that it is straightforward for the developer to know it, and also that knowing is necessarily analytical in nature, which Dooyeweerd questioned (§3.3.3).

Because (in general) the IS will be part of the everyday life of the user, the knowledge that is represented in it must relate naturally to everyday use. It is thus multi-aspectual. This means that the analytic aspect, formative and lingual aspects of knowledge encapsulation must reach out to all the aspects, as already mentioned and indicated in Table 6-7. The knowledge the developer seeks is neither "objective" nor purely personal (so-called subjective) because Dooyeweerd goes beyond both (see §3.3.1). The challenge is, therefore, to ensure:

- That every relevant aspect of the domain is recognised;
- That each of these aspects is appropriately understood in terms of its kernel meaning; and
- That all relevant concepts within each aspect are elicited (including things and their types, properties, constraints, relationships, operations, and so on).

The first two refer to the law side of reality, the third to the entity or fact side. Traditionally, only the third has been recognised, leading to the erroneous notion of "complete and accurate knowledge".

6.6.3 Virtual Reality

In the case of systems that provide the users with a virtual world, however, these challenges must be modified because the virtual world might be imaginary rather than real, especially in computer games. As discussed in Chapter IV the quality of virtual world is governed more by its (virtual, represented) law side than by its subject side, so the third challenge discussed earlier is even less important in relation to the other two. Table 4-7 showed how the virtual world in the game *ZAngband* invokes nearly every aspect; all these must be represented explicitly to form that software.

The virtual environment community differentiates believability from plausibility. Both refer to the virtual law side that is the program. Believability may be seen as arising from faithful representation of the laws of all aspects (e.g., there is gravity, social grouping), and plausibility as minor modification of these laws (e.g., gravity strength varies with time), enabling the exploration of interesting fictional possibilities.

6.6.4 Everyday Experience and Understanding

Knowledge elicitation has undergone two shifts in emphasis. The first was from computer models, which modelled theories (especially physical), but in the 1970s it became clear that such "book knowledge" could not form the basis of KBS (then called "expert systems") that could provide useful advice or solve real-life problems. So the possibility of encapsulating actual problem-solving experience, as rules of thumb or heuristic rules, was explored. Two classic KBSs of this era were *MYCIN* (Shortliffe, 1976) and *Prospector* (Hart & Duda, 1977).

But basing a knowledge base on experience led to "brittle" systems, whose knowledge bases were opaque and for which the explanation facilities were poor. The author's own experience, mentioned in Vignette 3 of the Preface, for example, recommended that what he called "understanding" should be sought instead. It could be extracted from experience by separating out context-dependent problem-solving knowledge (Attarwala & Basden, 1985). "Understanding" was very definitely not a return to theoretical knowledge, but included "everyday" understanding. Incorporating everyday understanding was the second step, and it became known as "second generation expert systems" by Steels (1985) and Weilinga and Breuker (1986). Pat Hayes' (1978, 1985) *Naive Physics Manifesto* was an early visionary contributor to this movement. Systems incorporating such understanding would "degrade gracefully".

But, at the time, neither Steels nor this author had any philosophical justification for believing there was "understanding" to be sought and separated; it just seemed to work. However, it may now be justified by Dooyeweerd's distinction between law- and subject-side knowledge. Heuristics, as expressions of concrete experience, are subject-side knowledge and therefore cannot express that which is generally so, while understanding is law-side and can do so.

If this is so, then there is a different type of understanding for each aspect. So-called causal networks were devised at the time, on the assumption of a universal causality, but lack of insight into the difference between kernel causality (physical aspect) and its analogical echoes (§3.1.4.7) led to problems. What this implies, and will be explored in Chapter VII, is that it may be worthwhile to prepare a knowledge representation facility in which the different type of understanding in each of the aspects is made available.

6.6.5 Tacit and Explicit Knowledge

The issue of tacit knowledge was first highlighted by Collins (1974), as a problem for knowledge elicitation. "Knowledge" is conveyed lingually, but speaker and hearer might assume subtly different meanings in the words used; Collins' example is what "short" means in reference to a wire in an early laser. During the 1980s and 1990s especially interest in tacit knowledge mushroomed, first because of the challenges in building KBS, then because of misunderstandings of knowledge stored in organisational repositories. Such misunderstandings can be both hidden because each assumes that others understand things in the same way as they do, and dangerous, because the stored knowledge can mislead without any warning.

The challenge is to make tacit knowledge explicit, but without destroying its life-world characteristic. Polanyi's monograph, *The Tacit Dimension* (1967), has been a core reference in this discourse, but he maintained that tacit knowledge can never be made explicit; most of his examples are of sensory-motor knowledge. But many, including the author of this work, have found otherwise for much taken-for-granted conceptual knowledge; such knowledge, long forgotten and now apparently second-nature in operation, can often be made explicit by judicious interview techniques, such as Winfield's (2000) MAKE. Meanwhile, Baumard (1999) discusses "tacit knowledge in organisations", which takes on a metaphorical meaning: knowledge can be "tacit" to the organisation-treated-as-person because it is not recorded, even though someone in the organisation might have the knowledge required in explicit form.

The Dooyeweerdian notion of multi-aspectual knowing can provide some insight to distinguish several reasons why knowledge might be "tacit":

- Intuitive grasp of aspectual meaning (§3.1.4.10, §3.3.5) fundamentally cannot be explicated fully, so it might be more fruitful to rely on its influence on user's activity, by judicious design of the artefact, rather than to try to represent it explicitly therein.

- The different aspectual ways of knowing (§3.3.3) implies tacit knowing in several ways. Knowing in the pre-analytic aspects (e.g., our experience of colour, sound, or muscular feeling, our knowledge of how to ride a bicycle) is likely not to be explicated without significant distortion because such things are continuous rather than crisply distinguishable. Therefore it is usually unwise to attempt precise and full description of this, and wiser to capitalise on the user's functioning.

- The non-absoluteness of the lingual aspect (§3.1.4.4) implies that language can never fully carry the meaning intended, so misunderstandings can arise.

- However, the lingual aspect does have considerable power to carry meaning, and a lot can be explicated usefully. Such knowledge is tacit by virtue of being taken for granted, which is often due to cultural reasons and is often possible to explicate to some degree by stimulating the expert's memory, either by listening to stories or by seeking understanding as Attarwala and Basden's (1985) approach sought to do. In Collins' (1974) example what "short" means was misunderstood in this way, but eventually explicated. Careful attention to the levels of abstraction (§3.3.7), especially lower-level, can help to make knowledge explicit without overly distorting it or severing its relationships with its context. This is used in Winfield's approach outlined below.

- Baumard's (1999) "tacit knowledge in organisations" is socially rather than lingually or analytically tacit and hence might be explicated by functioning in the social rather than merely lingual aspect, which encourages individuals holding the required knowledge to release it into the public space.

Such an understanding of tacit or taken-for-granted knowledge can usefully guide the handling thereof.

This has implications for knowledge elicitation. The limits to knowing discussed in §3.3.9 should make us always cautious about the claims we make for elicited knowledge, but on the other hand we should not despise our intuitive grasp of domain meaning during knowledge elicitation. Rather, we should allow it to speak to us about the domain of application and explore what it tells us. Some of the following practical devices help us do this.

6.7 Practical Devices

6.7.1 Aspectual Analysis

As in Chapter IV, aspectual analysis is useful. Here it has been used to ensure quality (shalom) of the overall ISD process and creation of the IS and to stimulate the consideration of possibilities in anticipating use and identifying responsibilities. But it is perhaps less useful in those simple forms for knowledge elicitation. Instead, for this, a "multi-aspectual knowledge elicitation" method has been developed.

6.7.2 Multi-Aspectual Knowledge Elicitation: MAKE

Winfield (2000) has done some very interesting work, in devising a sophisticated methodology for analysis of domain knowledge, that is centred on Dooyeweerd's notion of aspects: multi-aspectual knowledge elicitation (MAKE). MAKE combines the identification of aspects of an application with more detailed analysis of the individual issues, concepts, laws, and so forth, that are important for the application, and is useful for generating domain ontologies and for making some types of tacit knowledge explicit.

His approach is to start by asking the experts in the application domain to identify a couple of the aspects they deem most important, and to grow a recognition of the relevance of others aspects by a gradual process. The participants first identify a few whole aspects, then they start to identify concepts and laws, and so forth, within aspects. In the process, concepts come to mind that do not fit well within currently identified aspects, and the participants are thus led in a very natural manner to identify other relevant aspects. The steps of the MAKE process may be seen as:

1.   Introduction (e.g., explanation of kernel meanings of aspects, and obtain statement of requirements).

2.   Identify a few (e.g., a couple) important aspects.

3.   Focus on one of these aspects and specify any laws, axioms, data, definitions, and constraints that apply to the domain.

4. Identify as many concepts as possible that lie in this aspect (Note: May need to check the concepts at a later stage to ensure they fall within the correct aspect).

5. Apply Low Level Abstraction to each concept, which needs, or is thought to need exploding.

6. Repeat steps 3-6 as necessary.

7. Use the aspectual template to identify any new aspects, which may apply to the concepts specified (build bridges between concepts and aspects), and return to step 3.

Low Level Abstraction was a concept that Winfield developed from the 1991 edition of Clouser (2005) and refers to becoming aware of the various aspectual properties of things yet without isolating them from the things; see §3.3.7.

An aspect map is drawn as the analysis proceeds, an example shown in Figure 6-1, derived from Winfield (2000) for a veterinary practice. It shows concepts placed within aspects, with various relationships between them, such as "Desk acts as friend" and "Desk handles cash" (though their labels are not shown). The aspect map

Figure 6-1. Aspect map generated in multi-aspectual Knowledge Elicitation

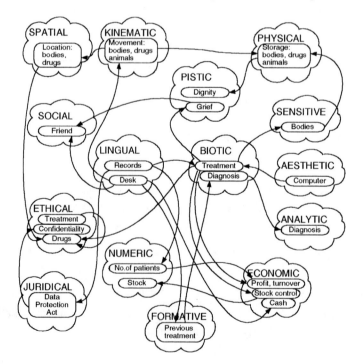

was generated during dialogue between the veterinary surgeon and Mike Winfield with some modification after time for reflection. The surgeon's views always had priority.

Several things may be noticed. "Drugs" appears several times, showing it to be of multi-aspectual importance. Some things seem to us to be in the "wrong" aspect (we might expect grief to be in the sensitive aspect) but the interviewee's view is given priority because of the particular meaning they give the concept. Certain aspects have more connections than others, suggesting they are of greater immediate importance. It is clear that the veterinary surgeon had little difficulty in understanding Dooyeweerd's suite of aspects and using them in analysis.

Winfield refined and tested MAKE on eight case studies, mostly with participants who had never heard of Dooyeweerd or the aspects before, and found consistently that MAKE was easy to learn (both by experts and Winfield's students who carried out some of the case studies) it was not difficult for participants to grasp enough understanding of the aspects in order to undertake this process, that nearly every aspect (typically 13 out of 15) was identified in each of the case studies, that it is useful over a wide variety of domains (tree planting, sustainability, veterinary practice, Islamic food laws, youth advice and management of a local housing business unit).

6.7.3 Characteristics of MAKE

MAKE has a flexible structure that places the interviewed expert very much at the heart of the knowledge elicitation process, and works even when participants understand aspectual meaning differently from the way Dooyeweerd did. It was found that MAKE improved on current methods of knowledge elicitation in a number of ways, by:

- Improving the breadth of knowledge elicited;
- Aiding the elicitation of (certain types of) tacit knowledge;
- Facilitating multiple views of knowledge;
- Encouraging reflection by the expert; and
- Eliciting the underlying "theories" relating to the domain.

If the aspectual law-and-meaning pertains across all cultures, as claimed in §3.1.4.1, and to the extent that participants' intuitive grasp of aspectual meaning is not too distorted, then MAKE might be particularly useful in inter-cultural or cross-cultural analysis. The interviewee in the case of Islamic food laws said he found it extremely useful.

One general benefit of Dooyeweerd's approach based on law rather than entity is that it is more tolerant of novel ideas. For example, in considering virtual organisations, an entity-centred perspective would see that all organisations (as entities) have involved the spatial aspect, suggesting that virtual organisations could not work, whereas if we considered the aspects, we can see that the social and spatial aspects are distinct and thus virtual organisations might work.

MAKE is a simple and effective method for analysing a domain. But we can also see an example of how a method of analysis can arise directly out of a Dooyeweerdian framework for understanding knowledge domains. The use of MAKE has confirmed empirically a number of things:

- That applications knowledge is multi-aspectual;
- That the aspects are distinct and thus deliberately thought about;
- That most aspects can expect to be relevant;
- That each aspect gives rise to concepts, known laws, axioms, constraints, and the like;
- That these are all interconnected, that such interconnections include connections with other aspects (e.g., inter-aspectual echoes); and
- That the kernel meanings of the aspects may be intuitively grasped.

We cannot say that MAKE was derived deductively from the framework, but rather that it was inspired by a knowledge of Dooyeweerd. The development of MAKE exemplifies how those working in the various of areas of information systems should be able to develop methods, theories, classifications, and the like on the basis of frameworks for understanding inspired by Dooyeweerdian philosophy.

6.8 Enriching SSM

Here we briefly review how Dooyeweerd has been used by several thinkers to engage with an existing framework for understanding ISD. The framework is Checkland's (1981) soft systems methodology (SSM). In ISD, SSM has been used mainly to help identify appropriate purposes for the IS, bring to light main factors that should influence its design. This discussion should not be seen as a full critique of SSM, but rather as a demonstration of some ways in which Dooyeweerd's thought can engage with existing frameworks, not to supplant them but to support, underpin, and enrich them.

Whereas SST, CST, MST, and DST have presented themselves as replacing or absorbing earlier paradigms, several thinkers have attempted to enrich Checkland's (1981) SSM using Dooyeweerd explicitly, in effect transplanting it as it is or wishes to be from the Nature-Freedom to the Creation-Fall-Redemption ground-motive.

Bergvall-Kåreborn (2001) characterises the aim of SSM as "to improve real-world situations by orchestrating changes of appreciation through a cyclic learning process." This speaks of a diversity of norms, relationships, and perspectives among those involved in, or affected by, the situation. Such diversity must be not only acknowledged but "orchestrated", so that all those involved or affected may receive their due, and do so in a coherent way. "Changes" speaks of dynamically bringing about new insights into the situation rather than merely gathering existing perspectives together. SSM has four stages: of finding out about a situation that needs to be improved, modelling, comparison and taking action, all usually undertaken by a group of participants. Finding out yields a rich picture, modelling yields root definitions and conceptual models of systems that might effect improvements, comparison is between these these models and the real situation, and taking action should result. Checkland (1981, pp. 224-5) suggested that a good root definition and conceptual model will specify at least six elements:

- **C** – customers: "Beneficiaries or victims affected by the system's activities"
- **A** – actors: "Agents who carry out, or cause to be carried out, the main activities of the system, especially its main transformation"
- **T** – transformation process: "The means by which defined inputs are transformed into defined output" (where input is current situation and output is desired situation)
- **W** – Weltanschauung: "An outlook, framework or image that makes this particular root definition meaningful"
- **O** – ownership of the system: "Some agency having a prime concern for the system and the ultimate power to cause the system to cease to exist"
- **E** – environmental constraints: "Features of the system's environments and/or wider systems which it has to take as 'given'"

which has since then been known as "CATWOE". T is the desired improvement, but the W concept is, as Checkland (1981, p. 18) points out, "the most important one in the methodology." By making Ws (perspectives) explicit during analysis new insights can be generated that might, as Bergvall-Kåreborn (2001) puts it, "break away from self-imposed constraints and frames of mind." Though SSM has been in use for well over 20 years, almost in its original form, there are a number of problems, which are discussed by Bergvall-Kåreborn, Mirijamdotter and Basden (2004), and which have motivated various attempts to enrich SSM using Dooyeweerd.

Bergvall-Kåreborn (2006) sees a human activity system (such as an hospital) as functioning in all aspects, but with a primary purpose or meaning that is led by a qualifying aspect (§3.2.5). She shows that how a system functions is governed by what we treat as the qualifying aspect. The notion of qualifying aspect was embedded in SSM and this was tested by applying it to a case study that studied two programs aimed at creating new work opportunities in a small municipality in Sweden.

Whereas most Dooyeweerdian scholars assume, and try to identify, a single qualifying aspect of things, Bergvall-Kåreborn allowed the qualifying aspect to vary to reflect different perspectives. Asking "What if we treat the qualifying aspect of our system as X?," she found, helped clarify especially the meaning of W and T in SSM. It might stimulate fresh insights and overcome SSM's tendency to generate regulative solutions.

Mirijamdotter and Bergvall-Kåreborn (2006) carried out an "appreciative critique and refinement" of SSM as a whole. First they demonstrated, by case studies (of attitudes taken by young people to the area in which they live), how aspectual analysis can enrich SSM's rich picture, by being sensitive to the diversity of aspects and also be differentiating good from bad in each. In the design and comparing phases, five criteria are traditionally employed to judge quality of transformation. It is easy to see these are arbitrarily chosen (they all begin with E), but Mirijamdotter and Bergvall-Kåreborn showed how Dooyeweerd's aspects not only justified these five, but also extends the set of criteria. This then led them to devise useful evaluative questions. Finally, they show how, by the worked example of an analysis of the Estonia ferry disaster, aspectual analysis can help build a good conceptual model and CATWOE. Thus, in every phase of SSM, Dooyeweerd is able to enrich its activity.

Basden and Wood-Harper (2006) have undertaken a detailed critical reinterpretation of CATWOE using Dooyeweerd's ideas. Most of the elements may be understood in aspectual terms. To summarise (and over-simplify): T is multi-aspectual functioning. C is no longer individual people or organisation but repercussions in each aspect of T. E is primarily the intrinsic normativity of all the aspects, which are the law side, and only secondarily the constraints afforded by the subject side. A is multi-aspectual competences. W is perspectives centred on various aspects. O is responsibility for C and A in particular. These re-definitions overcome a number of extant problems in SSM.

They claim that in effect they have transplanted CATWOE from the sterile soil of NFGM into the more fertile soil of CFR, so that it can bear more fruit. They also discuss the validity of doing this. Though both they and De Raadt (1991) have brought Dooyeweerdian thought into contact with extant systems models, they have done so in different ways. De Raadt accepted Beer's viable systems model and its various concepts largely as they were given. Though he criticised the narrow biotic foundation of VSM, he did not criticise its structure but merely suggested multi-aspectual versions of each element. He also treated Ashby's notion of variety in the same uncritical way. By contrast, Basden and Wood-Harper critically reinterpreted the elements of CATWOE in the light of Dooyeweerd's Meaning- and Law-centred approach and accounted for why each is necessary to the model. They also justified doing this, especially on the basis that if Checkland's writing is examined, Meaning proves to be much more important than Being, and so his attempt to ground SSM in systems theory, which presupposes Being, may be seen as mistaken and also unnecessary. Dooyeweerd could, they claim, provide a much sounder philosophical foundation for SSM. In this way they might have overcome SSM's subjectivism without robbing it of its sensitivity to human perspectives.

6.9 Conclusion

The starting point for formulating a philosophical framework for understanding IS development is to see it not as a technical, nor even socio-technical, operation but as a human activity that exhibits a strong everyday flavour even though theories may be made about it and methodologies created to guide it.

Upon briefly reviewing the history of the area that is IS development (including recent developments like agile system development methods), and examining paradigms that pertain in ISD suggested by the time-honoured Burrell-Morgan model, it was found that it would be useful to attempt to understand ISD from the standpoint of the CFR ground-motive. The review was brief and limited, intended merely to indicate the usefulness and value of CFR (as understood by Dooyeweerd), rather than to demonstrate conclusively that CFR is superior to the others. Attempting such a demonstration is left to another occasion, and would have to take into account approaches not based on the Burrell-Morgan model.

However, one reason why CFR is likely to be useful is because of its sensitivity to everyday experience, especially its ability to address complexity (cohering diversity) and normativity. ISD was reinterpreted:

- ISD is multi-aspectual human functioning, with especial focus on the normativity of the aspects.

This is similar to the framework formulated for computer use, but there is a difference. In computer use, aspectual normativity is directed toward the past and present, and to evaluation thereof, whereas ...

- In ISD aspectual normativity is directed to the future, and to guidance toward it. Spheres of law are seen as enabling possibility.

Approaching ISD as everyday experience revealed, as with use:

- ISD is constituted of several different multi-aspectual functionings, each of which needs to be treated in a different way, but which interweave with each other in an enkaptic relationship.

These include the overall ISD project or process, anticipating use, creating the IS, and knowledge elicitation and representation. Each was then examined, to yield the following understanding of them:

- The overall project is to be guided by the shalom principle, but the aesthetic and social aspects are key, the aesthetic aspect in its focus on harmony to achieve a coherent project as its qualifying aspect, and the social because ISD is teamwork. An insight from Dooyeweerd about the difference between types of social institutions was helpful in establishing the proper place of power and non-power relationships, including authority, subordination, respect, friendship, and intimate trust. Aspects of the overall project reach out to the other three human activities examined.

- Anticipating use is qualified by the juridical aspect of responsibility for all outcomes in future use. Aspectual normativity as possibility is a key insight.

- Creating the IS is likewise multi-aspectual functioning, and the analytic to economic aspects thereof have been long recognised, but, less recognised, it can be a delight as well as a chore if the four latest aspects are given their due.

- Knowledge elicitation and representation are qualified by the analytic, formative, and lingual aspects and are best seen as enkaptically interwoven with each other rather than separate, in that each stimulates and depends on the other. Each of these aspects reaches out to the diverse aspectual meaning of the domain of application, which should be respected.

For the latter, Dooyeweerd's theory of knowing was important, including the insights this offered about the nature of tacit or taken-for-granted knowledge, which is currently recognised as a major challenge.

The fact that only certain of the aspects have been used as inter-process links, as shown in Table 6-5, suggests there might be other processes, but the discussion in this chapter indicates the ways by which different processes can be understood and how they can be integrated into the whole of ISD.

Two practical devices were offered, simple aspectual analysis as was found useful in the area of computer use but with an orientation to future possibilities and to guidance of the ISD process, and Multi-Aspectual Knowledge Elicitation, developed by Winfield (2000), which has proven useful in developing domain ontologies and making some kinds of taken-for-granted knowledge explicit.

It was also shown how, instead of constructing a purely Dooyeweerdian methodology such as MAKE is, extant methodology can be critiqued, supported, and enriched by Dooyeweerd. How this might be achieved was illustrated by briefly examining the work of several authors who have applied Dooyeweerd in this way to Checkland's soft systems methodology. The reader is referred to the cited literature to find out more.

The benefits of this framework, if developed into guidelines, are that the overall ISD process should be healthier and more satisfying to all involved, as well as more efficient, because it attends to all aspects of the process. The frequency of developing the wrong IS should reduce because the framework provides a way of imagining the potential of the domain more clearly, in all its aspects. Artefact creation should become a more human, and yet more efficient, process. Knowledge elicitation and representation should be more complete because of the encouragement to explicitly consider more aspects of the domain meaning.

One question has not been considered in this chapter: what information technology resources do IS need to have at their disposal? A house is built from bricks, timber, nails, copper piping, and the like. From what basic building blocks should the IS (at least its technical component) be constructed? What tools are appropriate? This is the topic of the next chapter.

References

Attarwala, F. T., & Basden, A. (1985). A methodology for constructing Expert Systems. *R&D Management, 15*(2), 141-149.

Avison, D. E., & Wood-Harper, A. T. (1990). *Multiview: An exploration in information systems development.* Henley-on-Thames, UK: Alfred Walker.

Basden, A. (1975). *An interactive system for layout of small printed circuit boards.* Unpublished doctoral thesis, University of Southampton, UK.

Basden, A. (2000). Some technical and non-technical issues in implementing a knowledge server. *Software Practice and Experience, 30*(10), 1127-1164.

Basden, A., & Brown, A. J. (1996). Istar: A tool for creative design of knowledge bases. *Expert Systems, 13*(4), 259-276.

Basden, A., & Clark, E. M. (1980). Data integrity in a general practice computer system (CLINICS). *Int. J. Bio-Medical Computing, 11*, 511-519.

Basden, A., & Hibberd, P. R. (1996). User interface issues raised by knowledge refinement. *International Journal of Human-Computer Studies, 45*, 135-155.

Basden, A., & Nichols, K. G. (1973). New topological method for laying out printed circuits. *Proc. IEE, 120*(3), 325-328.

Basden, A., Watson, I. D., & Brandon, P. S. (1995). *Client centred: An approach to developing knowledge based systems.* Chilton, UK: Council for the Central Laboratory of the Research Councils.

Basden, A., & Wood-Harper, A. T. (2006). A philosophical discussion of the root definition in soft systems thinking: An enrichment of CATWOE. *Systems Research and Behavioral Science, 23,* 61-87.

Baumard, P. (1999). *Tacit knowledge in organisations.* London: Sage.

Beck, K. (2000). *Extreme programming explained: Embrace change.* Boston: Addison-Wesley.

Beer, S. (1979). *The heart of the enterprise.* New York: Wiley.

Bergvall-Kåreborn, B. (2001). The role of the qualifying function concept in systems design. *Systemic Practice and Action Research, 14*(1), 79-93.

Bergvall-Kåreborn, B. (2006). Reflecting on the use of the concept of qualifying function in system design. In S. Strijbos & A. Basden (Eds.), *In search of an integrative vision for technology: Interdisciplinary studies in information systems* (pp. 39-62). New York: Springer.

Bergvall-Kåreborn, B., Mirijamdotter, A., & Basden, A. (2004). Basic principles of SSM modeling: An examination of CATWOE from a soft perspective. *Systemic Practice and Action Research, 17*(2), 55-73.

Boehm, B. W. (1988, May). A spiral model of software development and enhancement. *IEEE Computer, 21*(5), 61-72.

Brandon, P. S., Basden, A., Hamilton, I., & Stockley, J. (1988). *Application of expert systems to quantity surveying.* London: Royal Institution of Chartered Surveyors.

Burrell, G., & Morgan, G. (1979). *Sociological paradigms and organizational analysis.* London: Heinemann.

Checkland, P. (1981). *Systems thinking, systems practice.* New York: Wiley.

Churchman, C. W. (Ed.). (1971). The religion of inquiring systems. In *The design of inquiring systems: Basic concepts of systems and organization* (pp. 237-246). New York: Basic Books.

Clouser, R. (2005). *The myth of religious neutrality: An essay on the hidden role of religious belief in theories* (2nd ed.). Notre Dame, IN: University of Notre Dame Press.

Cockburn, A. (2005). *Crystal clear: A human-powered methodology for small teams.* Boston: Addison-Wesley.

Collins, H. M. (1974). The TEA-Set: Tacit knowledge and scientific networks. *Science Studies, 4,* 165-186.

Collinson, C., & Parcell, G. (2001). *Learning to fly: Practical lessons from one of the world's leading knowledge companies.* Oxford, UK: Capstone.

De Raadt, J. D. R. (1991). *Information and managerial wisdom.* Pocatello, ID: Paradigm Publications.

Dooyeweerd, H. (1984). *A new critique of theoretical thought* (Vols. 1-4). Jordan Station, Ontario, Canada: Paideia Press. (Original work published 1953-1958)

Dooyeweerd, H. (1986). *A Christian theory of social institutions* (M. Verbrugge, Trans.). La Jolla, CA: Herman Dooyeweerd Foundation.

Eriksson, D. M. (2006). Normative sources of systems thinking: An inquiry into religious ground-motives of systems thinking paradigms. In S. Strijbos & A. Basden (Eds.), *In search of an integrative vision for technology: Interdisciplinary studies in information systems* (pp. 217-232). New York: Springer.

Fiol, C. M., & Huff, A. S. (1992). Maps for managers: Where are we, where do we go from here? *Journal of Management Studies, 29*, 267-286.

Gruber, T. R. (1995). Toward principles for the design of ontologies used for knowledge sharing. *International Journal of Human-Computer Studies, 43*, 907-928.

Guarino, N., Masolo, C., & Verere, G. (1999). OntoSeek: Content-based access to the Web. *IEEE Intelligent Systems, 14*(3), 70-80.

Habermas, J. (1972). *Knowledge and human interests* (J. J. Shapiro, Trans.). London: Heinemann.

Hart, P. E., & Duda, R. O. (1977). *Prospector: A computer based consultation system for mineral exploration*. Menlo Park, CA: SRI.

Hayes, P. J. (1978). The naive physics manifesto. In D. Michie (Ed.), *Expert systems in the micro electronic age* (pp. 242-270). Edinburgh, Scotland: Edinburgh University Press.

Hayes, P. J. (1985). The second naive physics manifesto. In R. J. Brachman & H. J. Levesque (Eds.), *Readings in knowledge representation* (pp. 467-486). Los Altos, CA: Morgan Kaufmann.

Hines, J. G., & Basden, A. (1986). Experience with use of computers to handle corrosion knowledge. *British Corrosion Journal, 21*(3), 151-156.

Hirschheim, R., & Klein, H. K. (1994). Realizing emancipatory principles in information systems development: The case for ETHICS. *MIS Quarterly, 18*(1), 85-109.

Hirschheim, R., Klein, H. K., & Lyytinen, K. (1995). *Information systems development and data modeling: Conceptual and philosophical foundations*. London: Cambridge University Press.

Jackson, M. C. (1991). *System methodology for the management sciences*. New York: Plenum Press.

Jackson, M. C. (2006). Creative holism: A critical systems approach to complex problem situations. *Systems Research and Behavioral Science, 23*(5), 647-657.

Jones, G. O., & Basden, A. (2004, April 19-24). Using Dooyeweerd's philosophy to guide the process of stakeholder engagement in ISD. In M. J. de Vries, B. Bergvall-Kåreborn, & S. Strijbos (Eds.), *Interdisciplinarity and the integration of knowledge: Proceedings of the 10th Annual Working Conference of CPTS* (pp. 1-19). Amersfoort, Netherlands: Centre for Philosophy, Technology and Social Systems.

Jones, M. J., & Crates, D. T. (1985). Expert systems and videotext: An application in the marketing of agrochemicals. In M. A. Bramer (Ed.), *Research and development in expert systems: Proceedings of the Fourth Technical Conference of the British Computer Society Specialist Group for Expert Systems*. London: Cambridge University Press.

Kalsbeek, L. (1975). *Contours of a Christian philosophy*. Toronto, Ontario, Canada: Wedge.

Knuth, D. (2001). *Things a computer scientist rarely talks about*. Stanford, CA: CSLI.

Midgley, G. (2000). *Systemic intervention: Philosophy, methodology and practice*. New York: Kluwer/Plenum.

Mirijamdotter, A., & Bergvall-Kåreborn, B. (2006). An appreciative critique and refinement of Checkland's soft systems methodology. In S. Strijbos & A. Basden (Eds.), *In search of an integrative vision for technology: Interdisciplinary studies in information systems* (pp. 79-102). New York: Springer.

Norum, K. E. (2001). Appreciative design. *Systems Research and Behavioral Science, 18,* 323-333.

Pacey, A. (1996). *The culture of technology*. Cambridge, MA: MIT Press.

Patnaik, D., & Becker, R. (1999). Needfinding: The why and how of uncovering people's needs. *Design Management Journal, 10*(2), 37-43.

Polanyi, M. (1967). *The tacit dimension*. London: Routledge and Kegan Paul.

Schuurman, E. (1980). *Technology and the future: A philosophical challenge*. Toronto, Ontario, Canada: Wedge.

Shortliffe, E. H. (1976). *Computer-based medical consultations: MYCIN*. New York: Elsevier.

Steels, L. (1985). Second generation expert systems. *Future Generation Computer Systems, 1,* 213-221.

Strijbos, S. (2006). Towards a "disclosive systems thinking'. In S. Strijbos & A. Basden (Eds.), *In search of an integrative vision for technology: Interdisciplinary studies in information systems* (pp. 235-256). New York: Springer.

Ulrich, W. (1994). *Critical heuristics of social planning: A new approach to practical philosophy*. Chichester, UK: Wiley.

Walsham, G. (2001). *Making a world of difference: IT in a global context*. Chichester, UK: Wiley.

Weinberg, G. M. (1999). Egoless programming. *IEEE Software, 16*(1), 118-120.

West, D. (1992). Knowledge elicitation as an inquiring system: Towards a "subjective" knowledge elicitation methodology. *Journal of Information Systems, 2,* 31-44.

Wielinga, B. J., & Breuker, J. A. (1986, July). Models of expertise. *Proceedings 7th European Conference on Artificial Intelligence* (pp. 306-318).

Winfield, M. (2000). *Multi-aspectual knowledge elicitation*. Unpublished doctoral thesis, University of Salford, UK.

Chapter VII

A Framework for Understanding Information Technology Resources

Information systems developers make use of information technology resources. IT resources might include, for example, tools like editors, compilers, linkers, and debuggers (usually now integrated into development environment software), the languages in which knowledge may be represented (programming or knowledge representation (KR) languages), protocols for inter-program communication (which includes file protocols for various purposes), and building blocks such as ready-made algorithms and libraries of software. Whereas ISD is concerned with specific types of application, IT resources and tools aspire to be valid regardless of type of application.

The communities that have researched and practised in this area include computer science, software engineering, systems engineering, software architecture, programming language design, KR language (KRL) design, and also the data modelling, object-orientation, logic programming, knowledge based systems, and some artificial intelligence communities. There is, of course, considerable interaction and collaboration with IS developers since the latter influence the shape of resources delivered to them. (Often the same person works in both areas.)

Such resources must, of course, be designed and developed, and the methodological issues discussed in Chapter VI are relevant to this. But this chapter is not concerned with methodology for developing them. It is concerned with their quality and appropriateness when employed by IS developers. The central philosophical question addressed in this chapter is: on what basis can or should we understand the need for, and quality of, such resources? The central practical question is: what ranges of resources should we aim for? The very nature of IT resources will be reinterpreted in the light of the everyday experience that is preparing them for IS developers (as recommended by Dooyeweerd in §2.4.2). But what is "everyday" experience in this area? To assess that, it is necessary to review some of the things that have influenced the design of IT resources.

7.1 Influences on Design of IT Resources

At least four influences on the design of such resources can be traced, especially of programming and KR languages, the first three of which make it a specialist task.

1. **The way machines work:** Computers work by obeying instructions held in memory in sequence (some now employ several parallel sequences), sometimes jumping to a sequence beginning elsewhere in memory. Most of the instructions store bit patterns in memory cells or retrieve from them for processing. The way computers work is reflected in assembler languages, which provide convenient languages and tokens to express the sequences of instructions. It is also reflected in some of what were called high-level languages in the 1960s. For example in FORTRAN IV programs, though at a higher level of granularity, were still composed of a sequence of individual instructions, with many "GOTO" statements with numeric labels to alter the sequence. Program variables were seen as symbol level versions of memory cells. BASIC and COBOL also reflect this sequential working.

2. **Convenience to programmers:** But that was not very convenient to programmers. The GOTO construct, for example, was notoriously blamed for making programs difficult to understand, debug and modify, and was replaced by other control structures like REPEAT and DO. The long sequence of instructions was replaced by short syntactic units called blocks, which could be nested. Memory cells were replaced by program variables bound together in structures, and the content of these variables was expected to be processed as pieces of data rather than bit patterns, which led to the need to define their data type (integer, character, etc.). Complex mathematical formulae could be written declaratively (this latter was found in FORTRAN: its name is shortened from

"Formula Translation"). ALGOL, PASCAL, BCPL, C, C++ and JAVA are languages that reflect this motivation in their design.

3. **Theories of how to represent knowledge:** There are problems with such "procedural" languages, and "declarative" languages began to appear in response. Rather than specifying individual instructions ("what to do"), these declare "what is so" and depend on some simple internal software engine to work out what to do. Each reflects a theory of how knowledge might be formalised and represented as a program: a knowledge representation theory. LISP reflects the theory that nested lists, with some being interpreted as (mathematical or logical) functions, can constitute a program; it was an early example of a "functional language". PROLOG reflects the theory that the world can be formalised and represented as statements in first order predicate logic, and is perhaps the best known "logic programming" language. The relational data model (RDM) as originally conceived by Codd (1970) reflected the theory that the world can be represented as points in multi-dimensional space and its processing by set-theoretic operations. The object-oriented model (OO) reflects the theory that the world can be represented as active objects that fall into predefined classes. Wand and Weber (1995) have attempted a comprehensive KR proposal, grounded in the philosophical theory of Bunge. RDM, OO, and Wand-Weber are discussed later. OO is perhaps the predominant KR formalism in use today, but there is a move to a fourth approach.

4. **The structure of real-life meaning:** Throughout these developments, another motivation is reflected in some languages: to recognise what is meaningful in certain application areas. Usually this is made available to the IS developers as ready-made features in a language. For example, COBOL, though designed in the 1950s, has many features that embody some of the things found meaningful in business, including records with fields, dates, currencies, and so forth. APL embodies some things found important in mathematical domains. During the 1970s, Stamper's (1977) LEGOL may be seen as an attempt to recognise what is meaningful in the legal domain, including agents and norms. Though not a language as conventionally understood, geographic information systems (GIS) may be seen as an attempt to recognise spatial things and operations. Many generalised versions of domain ontologies built in the 1990s may be seen as an attempt to formalise and generalise specific spheres of meaning found in applications on top of a KR language like OO. Design patterns, an attempt in the field of architecture to design buildings and towns that reflect the diversity of human living (Alexander et al., 1977) has been proposed as an approach to designing the resources required for software designers; this is discussed later.

There is some overlap between the third and fourth. The theory-based formalisms, LISP, PROLOG, and OO could be seen as offering capability to handle lists, logic and active things as found in real life respectively. But this is their hidden pitfall: making it easy and convenient to handle such things, they privilege seeing the whole of the diverse meaning of applications in those terms. This is often inappropriate and constitutes a reduction. This echoes Dooyeweerd's contention that theory as such is not incompatible with everyday life, and can even enrich it (§3.4.5); it is the theoretical attitude which causes problems, by which we try to view all the diversity of life through a single construct.

A framework for understanding KR resources must include some discussion of the link between the lingual activity of representing knowledge and how computing machines work at the bit level, and this is discussed later. Of the other three only the last will be discussed in depth here. This is so for two reasons. One is philosophical: we seek an understanding that is sensitive to the everyday, so neither theories of how knowledge may be represented nor the convenience of IS developers should circumscribe our discussion. The other is practical, as expressed in the call for "KR to the people".

7.1.1 "KR to the People"

Reflecting on the experience of the 1980s, the decade in which KR had become mature as a discipline and before it had been constrained to become "knowledge management", Brachman (1990) suggested what would become likely scenarios and what issues would particularly need research. Many are still relevant today, especially one that stood out as relating KR to the wider context of IS development: "KR to the people":

It is likely that by the millen[n]ium "knowledge systems" will be a common commercial concept. This has important implications for the future of KR. Among other things, KR components will increasingly find themselves in the hands of non-experts, raising a novel set of issues. (Brachman, 1990, p. 1090)

KR had become a specialist field from which "the people" were excluded, mainly because, as mentioned earlier, it had become treated in one of the first three ways mentioned. But it is "the people" who should be the ones to use a KR language and other resources to construct knowledge (or information) systems because it is they who are immersed in the everyday meaning of the application, including all its nuances, cultural meanings, and taken-for-granted assumptions and norms. The need for "KR to the people" is clear in knowledge management (KM), for example much entry of complex knowledge (as opposed to simple data) is undertaken by "the people" (e.g., managers).

But, though Brachman highlighted the issue of "KR to the people", he made no attempt to address it. "KR to the people" has yet to materialise. This is the main issue depicted in Vignette 1 in the Preface, where this author tried to gain "appropriateness" by distinguishing four "aspects of knowledge".

7.1.2 Appropriateness

What criteria should a KR language or toolkit meet in order to facilitate "KR to the people"? The conventionally recognised criteria that emerged in the KR community are sufficiency, efficiency, and expressive power (Minsky, 1981; Levesque & Brachman, 1985). But these are criteria by which the various theories of KR could be evaluated, and are not sufficient for a lifeworld approach. Basden (1993) argued that another is more important: appropriateness. Appropriateness is when the mapping from the meaning of the domain to the available KR resources is natural: what is primitive to our intuition in the domain is matched by a primitive building block or language construct, leaving aggregation of building blocks to be needed only for what is complex in the domain.

In that proposal, which had its roots in the author's experience of representing knowledge relating to the laying out of electronic circuit boards (Basden & Nichols, 1973), to data in general medical practice in the 1970s (Basden & Clark, 1979), to the chemical industry (Basden & Hines, 1986; Hines & Basden, 1986; Jones & Crates, 1985), and to the surveying profession in the 1980s (Brandon, Basden, Hamilton, & Stockley, 1988), the author developed the notion of "aspects of knowledge" (Basden, 1993) that had the characteristic of being irreducible to each other, long before he had encountered Dooyeweerd's notion of aspects (and he had only a hazy notion of irreducibility, §3.1.4.2). In retrospect, some of these can be aligned with some of Dooyeweerd's aspects:

- **Items:** Analytic aspect (distinct concepts)
- **Relationships:** Formative aspect (formed structure)
- **Values:** Quantitative aspect (discrete amount)
- **Spatial:** Spatial aspect (continuous extension)

with a fifth that was suggested later

- **Text:** Lingual aspect (symbolic signification)

This concerns, not the syntax of a KR language as much as what Chomsky (1965) referred to as the deep structure of languages—the broad kinds of meaning that its tokens have to convey or signify, regardless of syntactic form. (Reference to Chomsky does not imply acceptance of his theory.)

7.1.3 Extant KR Languages

In terms of appropriateness to the full range of meaning found in everyday experience, extant KR formalisms (KRFs) do not fulfil even that modest proposal. Table 7-1 indicates the degree to which procedural, functional, logic-programming and OO formalisms, COBOL and GIS, and Basden's (1993) proposal, facilitate representation of the meaning of various aspects of applications.

Several things stand out. The first is the large number of aspects for which no direct support is offered, and thus the general poverty of extant KRFs. The second is that apart from Basden's yet-to-be-implemented ideal there are no instances of full support (denoted by "*****"). For example, though most support integers and continuous numbers, only a very few languages support ratios as such (including, for example, removing common factors by Euclid's algorithm). For example, support for the important formative notion of structural relationships is inadequate by the

Table 7-1. Aspectual capability of extant KRLs

Aspect	Proc	F.P.	L.P.	OO	COBOL	GIS	B1993
Quant'ive	***	****	*	**	**	*	*****
Spatial						****	*****
Kinematic							
Physical							
Organic							
Psychic							
Analytic	*	*	****	**	**	*	*****
Formative	***	**	**	****	**	**	*****
Lingual	*	**	*	*	**		*****
Social							
Economic					*		
Aesthetic							
Juridical							
Ethical							
Pistic							

extant KR approach (for the full richness of relationships as we experience them). (An early approach, similar to OO, Quillian's (1967) semantic nets, did have a richer notion of relationships, but sadly this was not taken up by later KRLs.)

Since 1993 a number of other authors have raised similar issues, including Gennart, Tu, Rothenfluh, and Musen (1994), Stephens and Chen (1996), and Wand and Weber (1995). Greeno (1994) discussed a similar notion to appropriateness in user interface design: affordance, which originated in biology with Gibson (1977). But, with the exception of Wand and Weber, few of these have attempted to ground their discussions in philosophy.

7.1.4 A New Approach

From this analysis, the whole picture is rather disappointing despite several decades of research and development in technological resources. There still remains considerable opportunity for innovative and ground-breaking research. Yet this is as unlikely to happen now as it has been for 30 years. The reasons might be structural—financial investment in current resources, legal frameworks and the habits of developers and researchers, to name but three. But underlying these are presuppositions.

It is presupposed that all cosmic meaning should be able, in principle, to be represented using a single basic KR concept like lists, logic or active things, as long as the chosen one is sufficient, efficient, and has good expressive power. For now it is enough to note that this is a presupposition, rather than a self-evident truth; its untenability is shown later.

Likewise it is presupposed that, because both involve some expression of meaning using a computer language, the development of IS is essentially the same as the creation of IT resources, and thus this area and that of ISD should be treated as the same. The untenability of this lingering presupposition was demonstrated by the discovery that the software engineering methods used to create general resources have proven largely inappropriate for development of IS artefacts for use in human everyday life. The kind of software entities of interest here are not shaped to any particular application, but seem more general.

The questioning of both presuppositions suggests that it might be useful to explore another avenue not based on them. Dooyeweerd's ideas allow us to reinterpret the whole notion of IT resources, starting from everyday life and cosmic meaning (§2.4.2, 2.4.3).

7.2 Semi-Manufactured Products

The difference between an IS and the IT resources it is made of is not unlike that between a house and the bricks out of which the house is built, or between a piece of furniture and the blocks of timber and nails out of which it is fashioned.

7.2.1 The Notion of Semi-Manufactured Products

"In modern life," said Dooyeweerd (1984, III, p. 129), "materials are technically formed into semi-manufactured products, before they are again formed into utensils." Trees are formed into planks before they become a chair (pp. 131-132). The IS artefact or system, such as a traffic route-finder, may be seen as utensil; the IT resources, such as the GD graphics library, from which it is made may be seen as semi-manufactured products (SMPs).

Can this intuitively-recognised difference be accounted for philosophically? Dooyeweerd suggested that the leading aspect of utensils is internal while that of SMPs is external. The meaning of SMPs awaits fulfilment, in the construction of something else. Likewise, the IS has an internal leading aspect (governed by ERC; see Chapter IV) but IT resources have an external leading aspect, because their meaning awaits fulfilment. [Dooyeweerd did not finalise the idea (and there is evidence that it was still under development) (Dooyeweerd, 1984, III, p. 132). Indeed, consideration of IT resources might provide fruitful ground for discussion of this issue by Dooyeweerdian scholars.]

Semi-manufactured products have the formative as their founding aspect (Dooyeweerd, 1984, III, p. 131). Whereas nails and planks are a mechanical technology, shaped by the need to control and distribute physical forces (physical aspect), our resources are information technology, shaped by the need to represent (lingual aspect) diverse fragments of reality. For example graphics algorithms give the IS developer the ready-made language tokens and functions with which to represent and implement some of spatial reality.

7.2.2 The Proposal

Even though both creation of IT resources and IS development involve design and programming, the requirements that guide them should differ because they have very different responsibilities. Whereas IS developers bear some responsibility for possible repercussions on stakeholders in specific situations (see "Anticipating Use" in Chapter VI), the creators of resources do not. The norm of ERC, discussed in Chapter IV, was that the user should be enabled and encouraged, by the represented content, to do justice to the meaning of the domain. If the IS developer—especially "the people"—is to provide represented content that does this, they must find the process of, and resources for, representing knowledge easy and natural. It is the responsibility of those working in this area to create such resources as make it possible for IS developers to represent all the reality they wish to easily and naturally. As has been made clear in Chapter VI, this could, in principle, involve every aspect.

So, the responsibility of IT resource creators is to aspectual meaning as such. While IS developers are oriented to the concrete situations and possibilities of the subject side, the creators of IT resources are oriented to the law side. This responsibility pervades the notion of appropriateness outlined previously and is also alluded to in our use of aspects to evaluate extant KRLs in Table 7-1, but here it is made explicit.

The tokens of the KR language express and implement primitive entities, properties, processes, and so forth, that are meaningful in given spheres (aspects). Meaning of the domain might be of any aspect. If the IS developer is to be facilitated in representing all the diverse meaning of the domain, then the tokens and other resources should reflect the meaning in every aspect. That is the central proposal of this chapter. It is not an onerous responsibility; it is a joyful mission.

7.2.3 Aspectual Capability of KRLs

To assess the aspectual capability of any KR or programming language properly requires examination of the tokens and language constructs it offers, including its standard set of functions. The tokens, constructs, and functions are used by the developer to represent fragments of meaning, including what the developer wants the computer to do and what stuff this is to be done to (e.g., test two text-strings for equality). These fragments are usually most meaningful by reference to one aspect (e.g., test-for-equality is analytic, text-string is lingual).

The C language is usually considered to be a programming rather than a KR language, but it nevertheless is used to represent the programmer's knowledge of what they want the computer to do. Table 7-2 shows a number of tokens, constructs, and standard functions offered by the C language, along with fragments of meaning for which each is used, and the main aspect in which such fragments are meaningful.

The main support C gives is for things whose main meaning is in the quantitative, psychic, analytic, formative, and lingual aspects. A few C functions support things whose main meaning lies in other aspects, such as spatial, economic, juridical. But they are few, and some (juridical and economic) support aspects of the program itself which is being developed rather than of the application, and hence are in brackets.

But even in the well-represented aspects, support is very patchy. For example, to represent quantitative meaning, tokens are offered to represent amounts, addition

Table 7-2. Aspects of elements of C language

C tokens	Function in C language	Aspect
int	Amount	Quantitative
long, short	Determine how many bits to use	Psychic
char	Character	Lingual
"..."	Text string	Lingual
>, <	Numeric comparison	Quantitative
==	Identity comparison	Analytic
=	Assignment of value to variable	Formative
++, --, +=, -=	Increment, decrement	Quantitative
&, |, >>, <<	Bitwise manipulation	Psychic
for	Apply same procedure multiple times	Formative
if	Make a distinction	Analytic
{ ... }	Block of code or structure	Formative
struct	Define a structure	Formative
a->b	Pointer	Formative
a.b	Identify part of a structure	Analytic
*var	Define pointer	Formative
typedef	Define type of data	Analytic

Function	Function in C language	Aspect
strstr()	Find substring in string	Lingual
assert()	Check program validity	(Juridical)
sin(), cos()	Trigonometric calculations	Spatial
qsort()	Sort a data array	Formative
remove()	Remove a file	(Economic)

and other arithmetic, comparison, and structurings to represent arithmetic expressions and equations, and so forth. But also found in everyday quantitative life are ratios, fractions, proportions, means, standard deviations and other statistical things, approximations, and infinity—and most of these are absent from C, as from most extant KRLs. To represent lingual meaning, the meaningful resources include words, phrases, sentences, paragraphs, vocabularies, thesauri, emphasis, cross references, and the like—but all that is offered in C is the text string and a few character-based string manipulation functions: the necessary resources for the lingual aspect are mostly missing.

(One exception to this for the lingual aspect is the HTML protocol, which includes tags for citing, headings, bullet lists, tables, cross references, etc., as well as all the standard "physical" ones like italics.)

Similar analysis of most extant KRLs shows a similar picture. This lack of support for most aspects might be explained by their historical focus on reflecting theories about how knowledge may be represented, rather than being orientated to the everyday lifeworld meaning of domains. It might also be explained by the presupposition mentioned previously, that "application" meaning does not need to be represented directly but may be implemented indirectly via a single aspectual notion like lists, logic, or active entities.

7.2.4 Problems of Missing Aspects

The ramifications of missing aspects are enormous. If facilities for aspects are missing then IS developers will find it difficult and unnatural, or even impossible, to represent meaning of that aspect. This can prevent them from doing justice to the full meaning of the domain. Or else, they must find ways to implement it in available aspects. This generates a number of problems, which to this day plague IS developers.

- **Specialist knowledge:** They need specialist knowledge of how to implement the missing aspectual things as data structures and algorithms. For example an arbitrarily complex shape (which happens to represent the boundary fence of a piece of woodland) might be implemented as a list of coordinates that define its boundary and the type of curves or straight lines that join them (Figure 7-1a), so they need skills in using linked-list structures. Worse, to expand the shape by a given amount (Figure 7-1b) requires esoteric knowledge of complex trigonometry. But the notion of complex shape in everyday experience is spatially primitive and not esoteric.

- **Oversimplifications:** Even worse, the result of such expansion might be a shape with a hole in it (Figure 7-2), in which case a single list of coordinates is insufficient. The IS developer needs to be aware of such exceptions to the ordinary assumptions they might make. Frequently, such complexities are not discovered until long after the basic data structures have been embedded in place for some time, and it proves very costly and dangerous to change them. A very

Figure 7-1. A complex shape and its expansion

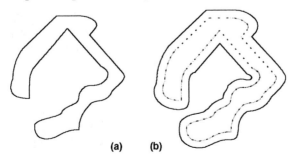

(a)   (b)

Figure 7-2. A complex shape with hole

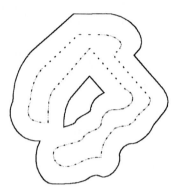

common oversimplification is relationships, which are often implemented as pointers or database keys, even though, in everyday life, relationships involve inverses, have attributes and might even themselves make relationships. Such oversimplifications might be valid at first, but can cause problems later when the system is expanded for new usage contexts because they have become embedded.

- **Inner workings:** Sometimes the inner workings of the resources are exposed to the danger of interference, whether deliberate or unwitting.

- **Hidden side-effects:** In some toolkits, the resources offered have hidden side-effects, which are meaningless to the pure form of the language but implemented nevertheless. For example, the ASSERT predicate in PROLOG, which, to the logical engine merely returns TRUE, but which, as a side effect, creates a new proposition or predicate. Such side-effects are dangerous, esoteric, and make the IS difficult to maintain or upgrade.

- Sometimes, what Wand and Weber (1995) call **"redundant" variables** are offered, which can confuse "the people" and make the IS difficult to maintain or upgrade in the future.

- Sometimes there is undue **ambiguity** about the meaning of a resource, which leads to confusion, a reduction in interoperability between systems and in misunderstandings between people, whether developers, maintainers, or users. The primitives should, as far as possible, accord with informed intuition of each aspect of knowledge; this will enhance intersubjectivity between and among developers and users.

These problems imply that the set of IT resources offered to the IS developer by what we will call a KR toolkit should cover the entire range of aspects, and should, together, do justice to the kernel meaning of each aspect.

7.3 Aspectual Design of IT Resources

How do we do justice to each aspect? One proposal is that a KR toolkit should provide primitive or basic facilities that are meaningful in every aspect, which "the people" (or other IS developers) can employ in construction of artefacts and computer systems for human life.

7.3.1 Philosophical Roles of Aspects to Indicate Primitives and Tokens

In everyday life, "the people" experience entities, properties, relationships, processes, norms, rationalities, and so on of each aspect. The meaning of each aspect is manifested in the cosmos by means of its philosophical roles. Therefore, our proposal here is that basic facilities and tokens can be defined for most philosophical roles of each aspect (refer to the relevant subparts of §3.1.5):

- Aspect as mode of being indicates types of "things" to make provision for (for example the sentence in the lingual aspect).
- Aspect as ways of functioning suggests activity to provide as procedures (for example, Soundex searching, translation).
- Aspect as basic type of property suggests attribute types to cater for (for example, emphasis, pronunciation, correctness of syntax).
- Aspect as ways of relating suggests type of relationship and interacting (for example, cross reference, synonym).
- Aspectual rationality indicates inferences to be built in (for example, if two sentences follow each other then it is likely they are about the same topic).
- Aspectual law suggests constraints that would be meaningful (for example, vocabulary and rules of syntax).
- Aspect as way of describing suggests the style of tokens of the "language" by which this meaning might be expressed.

This can lead to a very practical proposal.

7.3.2 A Practical Proposal: Aspectual Modules

Thus, a module could be created for each aspect to implement these roles of each aspect: things, activity, properties, ways of relating, inferences, constraints, and language tokens, and style suited to that aspect. "Language" is not assumed to be solely textual, and the tokens might be any input or output symbols; either as output to the user via screen, speakers, and so forth, or as input, via mouse or keyboard gestures (for example, for the quantitative aspect, not only digits but also sliders and bars of varying length are common "tokens").

Here are examples of what aspectual modules might look like. organised in accordance with seven roles that aspects fulfil in an application (O and I refer to output and input style for expressing meaning of this aspect). All the lists give merely a few examples to illustrate or stimulate ideas, and they should be severely criticised; much more research is needed to identify proper lists.

This proposal could be used in at least two ways: to suggest new lines for KR research, and, when more fully developed, to provide a yardstick by which extant KR toolkits may be evaluated.

Quantitative Aspect (discrete amount)

Things: integers, ratios, fractions, proportions, etc.; also types that anticipate later aspects such as 'real numbers' for the spatial aspect

Actions: incrementing, scaling, statistical functions, etc.	**Properties:** accuracy, approximation
Inferences: arithmetic	**Relatings:** greater and less than, sets, etc.
Constraints: e.g. a given quantity remains that quantity until changed	**O:** Digits, Bar length, Contour lines **I:** Hit keys, Drag bar, Drag contour

---

Spatial Aspect (Continuous extension)

Things: space itself, shapes, lines (straight or curved), areas, regions, dimensional axes, etc.

Actions: join, split, stretch, deform, rotate, overlap, expand, etc.	**Properties:** size, orientation, distance, side (in, out, left, right), etc.
Inferences: those of geometry and topology	**Relatings:** spatial alignments and arrangements, touching, crossing, overlapping, surrounding, topology, etc.
Constraints: e.g. boundaries should not have gaps	**O:** Shapes, spatial arrangements **I:** Drag to draw, modify

---

Kinematic Aspect (smooth movement)

Things: movement, path, flow, centre of rotation, etc.

Actions: start, stop, rotate, follow a path, etc.

Properties: velocity, speed, direction, divergence, curl, duration of movement

Inferences: e.g. s = v * t -- those often found in the field of mechanics

Relatings: faster/slower, forward/back, travel together, etc.

Constraints: the Hare does beat the Tortoise

O: Animation
I: Joystick/keys to give direction, speed

Physical Aspect (Energy, mass, etc.)

Things: waves, particles, forces, fields, causality, impacts; also mechanical things, chemicals, solutions, liquids, gases, crystals, materials, etc.

Actions: physical interaction, expanding a field by inverse square law, dissolving, chemical reacting, etc.

Properties: mass, energy, charge, frequency, force, field strength, Newton-power, etc.

Inferences: various energy functions, etc.

Relatings: causes, attracts/repels, etc.

Constraints: conservation of mass / energy / momentum, laws of thermodynamics, etc.

O: 3D ray-traced perspective view
I: Haptic devices

Organic (Biotic) Aspect (Integrity of organism)

Things: organism, organ, system boundary, tissue, air, food, life, population, environment, dysfunctions; checksums, etc.

Actions: regulate, grow, ingest, excrete, reproduce, repair, die, etc.

Properties: health, stamina, age, etc. (c.f. the 'stats' in role playing games)

Inferences: e.g. parent implies child

Relatings: parent/child/mate, food chains, symbiosis, system-environment, etc.

Constraints: need for sustenance and benign environment, etc.

O: Fractal 3D views
I: 'Soft' haptic device

Psychic Aspect (Sensing, feeling)

Things: signals (sounds, sights, etc.), channels, states (esp. emotional), memories, motor actions, etc.

Actions: respond, remember, forget, feel, push, etc.

Properties: colour (hue, saturation, value), pitch, volume, etc.; angry, happy, etc.

Inferences:

Relatings: e.g. stimulus-response

Constraints: sensitivity ranges of sense organs, etc.

O: Colour, sound
I: Linear sliders (e.g. HSV for colour)

Analytic Aspect (Distinction)

Things: distinct concepts, objects, labels to identify things, etc.

Actions: e.g. distinguish, deduce	**Properties:** truth values, difference and sameness, etc.
Inferences: those of logic, etc.	**Relatings:** contradiction, logical entailment, identity, etc.
Constraints: e.g. principle of non-contradiction, entity integrity (as in relational databases)	**O:** Icons, Menus, Tick boxes **I:** Click to select

---

Formative Aspect (Formative power)

Things: structuring, relationships, modifications, plans, means and ends, goals, intentions, power, etc.

Actions: form, compose, relate, revise, undo, seek, effect a meaningful change (change a state), plan, etc.	**Properties:** feasibility, efficacy, version, strength (as of a relationship), etc.
Inferences: graph searching, synthesis activity, etc.	**Relatings:** means and ends, the purpose of something, sequence of operations (history), part-whole, etc.
Constraints: e.g. referential integrity	**O:** Box-and-arrows graph; Buttons **I:** Drag boxes, arrows; Click to activate

---

Lingual Aspect (symbolic signification)

Things: nouns, verbs, etc.; words, sentences, etc.; bullet lists, headings, cross references, quotations, etc.; word roots, languages

Actions: write, draw, understand, send message, text search, find equivalent meaning, translate, etc.	**Properties:** tense, case, emphasis, cultural connotation, etc.
Inferences: those of syntax, semantics, etc.	**Relatings:** synonyms, antonyms, opposites, cross references, rhymes, thesaurus relationships, etc.
Constraints: spelling, grammar, pragmatic context, etc.	**O:** Text **I:** Alpha-numeric characters (keyboard)

Social Aspect (social interaction, institutions, keeping company)

Things: person, group, role, institution, title, name, nickname, etc.

Actions: communicate, befriend, adopt a role, give respect, etc.	**Properties:** status, leadership, formality and informality, address (postal, phone, email), etc.
Inferences: e.g. how to address someone	**Relatings:** friendship, acquaintance, respect for, membership, organizational structure, hierarchies, etc.
Constraints:	**O:** Organisation charts, etc. **I:** As analytical+lingual?

Economic Aspect (Frugality, limited resources, managing)

Things: resource, limit (complex), supplier, consumer, exchange, market, human resources, etc.

Actions: distribute resources, allocate price, etc.

Properties: limits, prices (values), etc.

Inferences: e.g. management forecasting

Relatings: supplier-consumer, relationship with resource limits, inter-currency, etc.

Constraints: e.g. no net loss of resources except via defined inputs and outputs

O: e.g. Tables of figures
I: As analytic+quantitative

Aesthetic Aspect (Harmony, enjoyment)

Things: nuances, harmonies, surprises, humour, fun, leisure, sport, etc. plus all the beings found in the various arts

Actions: harmonize e.g. music, play with, etc.

Properties: situatedness, harmony, surprisingness, paradox, interesting/boring, etc.

Inferences:

Relatings: nuance, echoing, counterpoint/complementarity, etc.

Constraints: "Less is more in art" [C.S. Lewis]

O: Decoration + accompanying music
I: As psychic with fine control
Both e.g. Colour circle device

Juridical Aspect ("to each their due")

Things: dues, responsibilities, rights, coded laws, policies, contracts, security measures, owners, policies, (in)justice, etc.

Actions: make contract, decide the essence of a case, judge, make retribution or recompense, etc.

Properties: security ratings, equity, proportionality, appropriateness, etc.

Inferences: e.g. consider evidence

Relatings: retribution, ownership, etc. Many cross-references between clauses

Constraints: laws of land, idea of what is due to each type of thing, ensure consistency, etc.

O: Text with cross references
I: As lingual?

Ethical Aspect (Self-giving love)

Things: attitudes, gifts, sacrifices, etc.

Actions: give (without expectation of reward), forgive, etc.

Properties: generosity, etc.

Inferences:

Relatings: Buber's I-Thou relationship, marriage/troth, etc.

Constraints: self-giving must be genuine, not for gain, etc.

O: As lingual+aesthetic?
I: As lingual+aesthetic?

```
Pistic Aspect  (vision, faith, committing)

Things: commitments, beliefs, trust, creeds, rituals, etc.


Actions: make a commitment (after,      Properties: degree of certainty,
maybe, weighing up the evidence), trust, trustworthiness, etc.
worship, etc.

Inferences:                             Relatings: committed-to, believe-in,
                                        trust, etc.


Constraints: commitments should be      O: As lingual+aesthetic?
kept, etc.                              I: As lingual+aesthetic?
```

This proposal is an initial one, which has yet to be worked out.

7.3.3 Implementation at the Bit Level

How may these be implemented? In terms discussed in Chapter V, these basic facilities are the raw pieces of data, each being of a particular type depending on the aspect. But pieces of data (analytic aspect) are implemented in things meaningful in the psychic aspect, such as bit patterns, memory address adjacency, machine code, and digital signals. So there must be a mapping between bit patterns to basic types of data, and between machine code and the valid manipulations for these types of data.

But, owing to the irreducibility of the aspects, these mappings are never given *a priori*, but must be designed by us. In principle, aspectual basic type of data requires its own different mapping. Table 7-3 shows some of the mappings (or "coding") for each aspect. (FPB refers to "fixed point binary", a quantity that increases using binary coding from 0, represented by all bits 0000..0000, to 1.0 or 100%, all bits 1111..1111, however many bits are used. Note the important word "contiguous" in the bit column, which refers to the bit patterns running contiguously in memory, which is a phenomenon meaningful in the psychic aspect as it is used in Chapter V.)

It may be noticed that most of the defined bit codings are for the early aspects. Because of foundational dependency between aspects (§3.1.4.6), we can expect that many types of data in the later aspects make use of types in earlier aspects. For example, economic price, double-entry book-keeping, or transfer of resources usually make use of quantitative amount, analytic distinction, and formative relating. Nevertheless, there seems to be still some meaningful facility in the later aspects which cannot be seen in terms of facilities of earlier aspects, an example being the economic notion of data compression such as by zip coding, which yields, not amounts or relationships but merely a long bit pattern.

Table 7-3. Bit pattern codings for selected aspectual resources

Aspect	Basic facility	Bit pattern coding
Quantitative	Integer Ratio Proportion, Probability 'Real' number	Binary, BCD Two contiguous bins FPB 00..00-11..11 FPB mantissa + Bin exponent
Spatial	2-D field (e.g. Funt) (x,y) complex number Direction, angle	Bitmap (grid of bits) Two contiguous Man+exp Circular FPB
Kinematic	Movement	Pen colours alterations Sequence of bitmaps by pointers
Physical	3-D grid Collision detection	Array of reals 'AND' bitmaps (Amiga)
Organic-Biotic	Integrity of data	Checksums
Psychic / Sensitive	Colour Picture Sound waveform	3x8 bits each FPB for RGB Grid of colour cells Contiguous array of FPBs
Analytic	Distinct concept (datum) Truth value	Allocated contiguous memory Single bit (0, 1)
Formative	Structure Relationship	Contiguous memory. Memory address (pointer), Linked list
Lingual	Text characters Emphasis (bold etc)	ISO, EBCDIC ANSI Escape sequences, or bits to flag style
Social	URL / email address	Four FPBs
Economic	Data compression	e.g. Zip coding
Aesthetic		
Juridical	Data protection	128-bit encryption
Ethical		
Pistic		

7.3.4 Reflections on the Proposal

These lists are the kind of thing the author was aiming for in his "aspects of knowledge" in Vignette 1 in the Preface. Despite the serious flaws in these lists, they exceed those supported by most extant KR approaches. Furthermore, the reader is likely to agree that the things mentioned are not esoteric, but are features encountered in everyday living.

Some of the benefits that might be expected if such a proposal was actualised include the separating out of different characteristics of things (e.g., spatial things are continuous rather than objects), easing of the task of IS developers, fewer errors during development, enhanced reliability of software, enhanced ability to extend and upgrade it as requirements change, especially unforeseen ventures into new uses.

But is the ambitious proposal feasible? What is proposed here has not yet been implemented in full, but the author's Istar KBS software (Basden & Brown, 1996) started to be developed along these lines. One reason to believe it might be feasible (in the long term) is that there is real-life software qualified by each aspect, which offers some of the basic facilities mentioned. Table 4-5 showed some of the extant software that provide some of the facilities related to each aspect. It demonstrates clearly that at least some meaning in every aspect has been represented in software, and thus that it is at least possible to do this. Also domain ontologies generated for practical applications could contribute knowledge of aspectual meaning.

Moreover, many things seen as frills or specialised cases today are the aspectual primitives of tomorrow. Take the case of cascading delete, for example. Deletion of an entity is an operation meaningful in the analytic aspect, but cascading delete (where the delete operation "cascades" through a network of relationships to delete entities related to the main one) presupposes structure, which is of the formative aspect. Initially, cascading delete was seen as a special case; nowadays it is often the norm. Likewise the notion of database transaction (commit, rollback, etc.) is meaningful in the formative but not the analytic aspect.

That the ethical and pistic cells are empty, in both Tables 4-5 and 7-3 means that little explicitly ethic- or pistic-qualified software has been produced and that the need for specific bit codings has yet to emerge. This might be due to the down-playing of these aspects in modern life for which IT has been designed. It may be that if IT had been developed, not by Western thought, but in sub-Saharan Africa, where generosity is a way of life and the spiritual is not divorced from the physical, that these cells would have been full. Such a possibility is taken up in Chapter VIII.

7.4 Integration

The distinctness of aspectual modules is guaranteed by aspectual irreducibility. But how are these to be integrated so that the IS developer can harness the philosophical roles of all aspects relevant to their needs to construct their IT artefacts or systems? This may be done by providing features that reflect the inter-aspect relationships of dependency, analogy, and reaching out (§3.1.4.6-8). Some of this is found in current practice, though it is seldom recognised as such, and a Dooyeweerdian view might help clarify issues and stimulate new directions for research.

7.4.1 Foundational Inter-Aspect Dependency

Inter-aspect dependency in the foundational direction (see §3.1.4.5) implies that a module for any aspect will require the facilities offered by a module of earlier aspects. Inter-aspect dependency can be direct or indirect.

Figure 7-3. Directed acyclic graph of some foundational dependencies

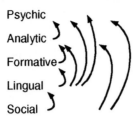

For example, a lingual module would require to handle concepts like structuring (formative), distinctness (analytic), colours for display of text and sounds for speech (psychic), layout on screen (spatial) and such things as word count, average word length (quantitative), and thus depend on those aspectual modules. Then a social module would depend on the lingual and thus indirectly on all those, but would also depend directly on the analytical distinguishing of who is important in a social situation, and the psychic aspect of emotion toward others. But none of these earlier aspects depend foundationally on the social (though *knowledge of them* does).

Inter-aspect dependency implies a cosmic order among the aspects, but this is not simply a linear sequence. Rather, it forms a directed acyclic graph that, as illustrated in Figure 7-3. In principle the graph is fully connected, but in practice some links are latent rather than actual; see Dooyeweerd (1984, II, p. 164) for a discussion of foundational dependency.

This type of inter-module link has been well-known almost since computers were invented (the subroutine concept) but perhaps in a rather arbitrary manner. It is also manifest in OO classes. The main contribution Dooyeweerd might make here is to provide strategic clarity to the designer of a suite of modules. Such clarity is particularly important to meet the challenge of complexity.

Implementing this is relatively straightforward because the designer of a module will know what facilities from earlier modules will be needed. But anticipatory dependency and analogy are more challenging.

7.4.2 Anticipatory Inter-Aspect Dependency

Dependency in the anticipatory direction looks toward what an aspect facilitates rather than what facilitates it. It is concerned with cosmic possibility, much of which has yet to be opened up by creative human endeavour (see §3.4.6). In IS anticipation is important for building future-proof computer system architectures. But there is a difficulty.

To Dooyeweerd, the possibility of facilitation of later meaning is within the aspect from the start. But irreducibility of aspects implies that the way in which an earlier aspect facilitates a later cannot be determined *a priori*. This means that when we design a module for an early aspect we might find, a considerable time later, that the way we have designed and implemented it is insufficient to support the new meaning, or at least clumsy in doing so.

For example, what should result from adding 270 and 260? Obviously 530. But if these figures are angles in degrees, the result should be 170. But degrees and circular addition are only meaningful in the spatial aspect, and not in the quantitative.

The designers of an aspectual module can be expected to have expertise in that aspect but perhaps not in others that it anticipates. The designers of the quantitative module cannot always be expected to know aforetime all the subtly different types of quantities and behaviours needed. Nor, similarly, for all other aspectual modules.

This is pernicious because usually the modules that implement the earlier aspects are developed first and come into use early, and become embedded in many applications. So they should be correct in all their anticipations right from the start. If they are not, we stand the risk of overly constraining our later possibilities because of untoward assumptions made. For example, we implement the resources of the lingual aspect to support lexics, syntax, and semantics—and fail to cater for the complexity of linguistic pragmatics which anticipates the social and later aspects.

Failure to anticipate can be a particular problem with off-the-shelf modules unless one knows that its designer has an attitude of openness to later cosmic meaning.

Therefore each module should be designed in such a way that it anticipates later modules without needing to be rewritten when they arrive. The challenge is how to cater for unforeseen new types of things, functions, properties, constraints, and so on, providing a clean way of enabling a module to function in a different way when unanticipated cosmic meaning presents itself. This is not the same as the challenge to the IS developer of altering a business application system when the business requirements change; that type of change usually reflects a change in the subject side. The challenge we face here is a change in (our knowledge of) the law side. It is more fundamental and requires more careful consideration.

Call-back hooks is one way of catering for unforeseen expansions, but usually they are provided grudgingly and without much careful thought. The OO community's emphasis on polymorphism might constitute a partial recognition of anticipatory dependency, but their penchant for encapsulation works against making it possible. As far as this author is aware, research into this issue of law-side anticipatory dependency is long overdue.

7.4.3 Inter-Aspect Analogy

Inter-aspect analogy, in both directions, is as difficult as anticipatory dependency to predict and cater for.

Take the case of causality, discussed in relation to KR by Nilsson (1998, p. 326ff). Causality is seen, not as physical, but as an intuitive linking of "causes" and "effects". As a result of knowledge elicitation, a "causal network" of nodes and arcs may be built, throughout which effects may be propagated from "first causes" in order to simulate the domain or make predictions. For each aspect a slightly different "causality" algorithm is likely to be needed. Nilsson argues that the Bayesian algorithm for accumulation of evidence is appropriate for such causal links, which makes probabilistic rather than precise calculations of effects from causes. This notion of causality has become established within the KR community so that Bayesian networks furnish the IS developer with a kind of general purpose resource for representing causality. The Bayesian algorithm is very useful because its ability to approximate can overcome many minor discrepancies between different aspectual repercussions, it can cope with non-determinate repercussions, and it is, in effect, an acknowledgement that some antecedent parameters have been omitted from the represented content. But its usefulness can mislead. Basden, Ball, and Chadwick (2001) found (without any reference to Dooyeweerd) when assessing the amount of trust one can place in Internet certification statements, which involved echoes of causality in the juridical, pistic, and other aspects, that the algorithms required are very different.

Perhaps extensibility features similar to those required for anticipatory dependency will be adequate, but since their exact form is likely to be different, it would be wise to keep them separate. There are differences from anticipatory dependency, such as that analogy lacks the necessity found in dependency, and, by its nature, analogy is often more difficult to clearly define.

What is novel in this proposal is not that we can, for example, use causal networks for pistic software, but that we can guide the development of such general purpose facilities according to aspectual analogies, rather than in an ad-hoc manner.

7.4.4 Implementing Aspectual Reach-Out

Aspectual "reach-out" might be rather easier to cater for. It is not esoteric. For example:

• Quantitative reach-out implies a different kind of amount in each reached-to aspect. This implies a need for units associated with each aspect (feet, metres, pounds, kg, currency, etc.), the ability to convert between units at the will of the user, and to add new units when necessary.

- Analytic reach-out implies a different type of distinction in each aspect and different types of inferences. Social logic differs from physical (Winch, 1958); see §3.1.5.2. This implies different types of identification and deduction for each aspect—as was realised quite early by Stamper (1977), who was struggling with how to store legal data.

- Lingual reach-out. Dooyeweerd's aspects could point to new syntactic and semantic structures and laws to be incorporated into lingual software (such as verse).

Many examples of such reach-out features have been added to mature real-life software qualified by different aspects in order to make life easier for developers or users. But these accretions have been largely ad-hoc, left to IS developers. Dooyeweerd's aspects could perhaps provide some systematisation of this and stimulate new directions in research.

7.4.5 Reflection

The vision of an integrated, multi-aspectual KR toolkit presented here is a long-term one and will require considerable research effort. Before it can be properly judged, we should attempt to define comprehensive modules for every aspect and how to integrate them. There is evidence that this may be possible and even desirable because features are becoming important in practical software even today, as indicated in Table 4-5, even from later aspects—such as the juridical feature of copyright notices and authentication checks, and extant research into many of these issues continues apace, being forced upon us by our everyday experience.

So it may be that the main contribution of this proposal is not the vision of a grand multi-aspectual toolkit so much as that it provides a basis on which the diverse areas of research may be integrated as part of a wider picture.

The vision of a multi-aspectual toolkit designed in this way opens up an intriguing possibility: is it possible to insert new imaginary virtual law-spheres among the given ones, or perhaps modify the meaning of existing ones, and then create virtual worlds with these and see how well they run? This could, in principle, forge an interesting test for Dooyeweerd's suite of aspects against others.

However, there are several problems with our proposal that need to be addressed over the longer term. First, it is not always clear what shape a toolkit might take for the later aspects, such as the juridical. It might be that (as currently) it is sufficient to express all the post-lingual aspects in language. But there is reason to suspect that, as application in later aspects matures, we will find we need something that cannot be written in language but must be implemented directly at the bit level (assembly language) as a primitive symbolic signification of aspectual meaning; this is not for efficiency reasons but to do justice to its meaning. Or, alternatively, we must fall back on domain meaning being inscribed into the context of use (the user's own knowledge and lifeworld experience) rather than represented in the technical artefact as such (see "Creating the IS" in Chapter VI).

Second, if a complete set of facilities were implemented for every aspectual module would not the resulting software package be unwieldy? Though this has not been researched yet, there are two reasons for believing this might not be entirely so. One is that the aspects are readily learned, and intuitively grasped; as Winfield (2000) and Lombardi (2001) have found, it becomes second-nature to consider them in any situation. The other is that since the aspects have irreducible meaning, then clear separation should be possible between resources of each aspect. The saying, "A solution should be as simple as possible, but no simpler" Budgen (2003, p. 75) attributes to Albert Einstein. The reality which IS developers encounter and represent is of an aspectual complexity that should not be unduly simplified, but rather supported and made explicit and understandable. Good modularity involves "separation of concerns", and this is precisely what this proposal offers.

7.5 Relating to Extant Discourse

The earlier text has outlined a long-term project for a KR toolkit inspired by Dooyeweerd. Here we discuss how a Dooyeweerdian approach might be used in three different ways to critique extant proposals for data models or KR languages, and its relationship to Alexander's design patterns.

7.5.1 Dooyeweerdian Critique of Extant Data Models

It is possible to use Dooyeweerdian philosophy to analyse data models and KR approaches. There are two ways. We could set up a Dooyeweerdian proposal, as outlined previously, as a yardstick against which to measure the data models; for example, we could find out in which aspects they are strong or weak (as with Tables 7-1 and 7-2). Alternatively, we can look at the problems each has experienced in everyday life and explore how Dooyeweerd might explain those problems and perhaps propose a solution. The latter will be used. The approach will be to expose root presuppositions or aspectually-inspired world-views.

Three brief analyses will be presented to illustrate the kind of approach that might be taken. Comprehensive critiques must wait for another occasion.

7.5.1.1 The Relational Data Model

The relational data model (RDM) was defined by Codd (1970) as a reference model to structure and search data in databases. The popular version, available in current software, sees the world as tables containing records, which are composed of sets of attributes that contain values; some attributes may be keys that point to other records, thus forming relationships. But Codd's original data model ("pure" RDM) treated data as points in multi-dimensional spaces. Each table ("relation") is a multi-dimensional space, in which each axis of the space represents an attribute in which all its possible values are mapped onto positive integers, allowing infinitely many possible values in any attribute if desired. Each record in the table (tuple of such integers) is a point in the space. All operations are on sets of tuples and generate other sets. The latter allows operations to be chained together to provide very sophisticated overall operations.

But pure RDM gives problems. If two tuples have the same set of attributes (e.g., two men of the same name) they are the same point and so have no separate existence in the relation (cannot be stored as separate records); but in real life this restriction is onerous even if rarely encountered. Second, if we allow keys to form relationships, we introduce the problem of referential disintegrity. Third, the set of operations that immediately offer themselves do not match the operations we want in everyday life (for example, the cartesian product is almost useless while the more useful relational join is absent and must be assembled from basic operations). Fourth, many people misapply the relational join because they keep forgetting to make two key-attributes equal to each other. Fifth, ordering of records according to the user's wishes is not supported in pure RDM. (Practical relational models do allow duplicate records, sorting, etc., due to accretion of "dirty" features on top of "pure" RDM, as the language or data model was exposed to the everyday life of IS developers.)

The root of these problems becomes clear when we understand it aspectually. The analytic-formative notion of entities and relationships is reduced to the quantitative-spatial notion of points in space. Whereas the analytic aspect allows distinction of two things with same properties, the spatial does not. Treating a relationship, which is a formative thing, as quantitative forces "the people" to handle artificial attributes, leading to the second and fourth problems. Ordering according to wishes is a formative notion and has no meaning in the spatial aspect (the quantitative aspect's ordering is fixed). This can explain the accretion of features foreign to the original reference model.

7.5.1.2 Object-Orientation and Subject-Orientation

The object-oriented (OO) KR approach sees the world as objects that possess a number of attributes and operations, as dictated by pre-defined classes, which are very easy to define by inheriting properties from other classes. The design ethos emphasises reusability (an economic norm) by means of polymorphism and encapsulation. Much of the work of OO programming consists of defining classes of objects, in terms of their attributes and operations. For example (example from Harrison and Ossher, 1993):

Class: Tree

Attributes: Height, Weight, CellCount, LeafMass

Operations: Grow, Photosynthesize

There are many standard texts on OO, a classic being Booch (1991).

Though it has become the premier KR approach of today, OO exhibits problems. While many arise from foibles of specific implementations and variations (such as whether multiple inheritance is allowed), some are more fundamental. Harrison and Ossher discuss one of these: the notion of object as of pre-defined type is not only philosophically suspect but also problematic in practice. In particular, other attributes and operations, not in the class, might be meaningful to different subjects. To a tax assessor, the tree's meaningful attributes include AssessedValue and meaningful operations include EstimateValue and ComputeTax, to a forester, these would include SalePrice, TimeToCut, ComputeProfit, while to an eagle, they would include FoodValue, BuildNest (the subjects do not have to be human). Such issues become important especially when developing suites of cooperating application programs.

Might this diversity of attributes be met by simply defining the class with as many attributes as possible and then filtering out those that are not needed by each program? Harrison and Ossher think not, on the philosophical grounds that "subjective perception is more than just a view filtering some objective reality. The perception adds to and transforms the reality" (p. 413).

How can a Dooyeweerdian view contribute to such a debate (of which this is only a tiny fraction)? First, we may seek to identify several ways in which the Dooyeweerdian view might affirm OO:

- Classes, with their inheritable properties, may be seen as a simple attempt to implement type laws (see §3.2.5).
- The class-subclass hierarchy is commensurable with Dooyeweerd's notion of types, subtypes, and so forth.

- The notion of "object" as a thing that acts and possesses properties by virtue of classes is at least commensurable with the Dooyeweerdian notion of being as something that is active by virtue of law.

- Multiple inheritance (where allowed) recognises the irreducibly distinct spheres of law-and-meaning.

- That objects can override what is inherited from classes reflects the plasticity of type laws.

- Polymorphism reflects the transcendence of the aspects (§3.1.4.1).

Then we can use Dooyeweerd to reveal problems and perhaps provide a way to overcome them.

- Encapsulation assumes a part-whole relationship, but Dooyeweerd draws attention to other types, like enkapsis. In practice, sometimes the developer or even user wishes to see or manipulate some "hidden" property, such as efficiency. This could be allowed by restricting encapsulation to genuine part-whole relationships and allowing aspects of the object that tend to become hidden, to be always visible in principle.

- The hierarchical nature of the inheritance relationship betrays a universalistic notion of the totality of things. Dooyeweerd criticises this and replaces it, not with an individualistic notion, but with enkaptic interlacements. For example:

  o **Hermit crab and shell:** Objects linked by subject-object enkapsis
  o **Data, algorithms, bit level implementation thereof:** Objects linked by foundational enkapsis
  o **City and orchestra, football team, university:** Objects linked by territorial enkapsis
  o **Fauna and flora living in habitat:** Objects linked by correlative enkapsis.

  But OO can find it difficult, or at least inappropriate, to cater for these using only the part-whole and inheritance relationships alone.

- The notion of object itself is problematic, because it presupposes the primacy of existence over meaning. Harrison and Ossher's notion of subject-orientation comes closer in that the nature of the object (as defined by its properties) depends on what it means to various subjects, and this is dynamic.

Harrison and Ossher use the example of a tree having many subject-relevant properties. So, coincidentally, does Dooyeweerd:

The tree's structure seems at first to be simple, but on deeper theoretical analysis it proves to be highly complex because this structure appears to be possible only in the universal inter-structural coherence. ... Here this natural thing proves to be included in an extremely complex interwovenness with the structures of temporal human society [such as the roles of woodcutter and tax assessor]. (1984, III, p. 833)

But Harrison and Ossher's proposal might also exhibit weaknesses, in line with the current vogue toward subjectivism. It is in danger of dissolving the object, such as the tree, into amorphous nothingness and of giving no fundamental basis for differentiating types of object. But, as Dooyeweerd realised, everyday experience does not allow this but things present themselves to us with their own natures even while these can be of variable meaning to different subjects. His solution was type laws, which are founded on the more fundamental laws of aspects and yet are of immense plasticity, which might offer the stability their proposal lacks. (However, this author is not attracted to that idea.) Even without type laws, if Dooyeweerd's notion of diverse law spheres (aspects) were to be implemented in a way that transcends all classes and objects, allowing them to respond to any such laws without requiring the developer to explicitly specify in advance which are relevant to their objects or classes, then it would be relatively straightforward to allow unforeseen properties to be added to objects in a non-cumbersome manner.

7.5.1.3 The Wand-Weber KR Ontology

Wand and Weber (1995) sought to define the building blocks for a comprehensive KR approach. They outline three models: a state-tracking model, concerned with how a computer system responds to and keeps track with the real world; a decomposition model, concerned with decomposing a system into subsystems; and a representational model, concerned with the grammatical constructs that the KRL should offer, and how they relate to the ontological constructs that should be offered, if they are to be both complete and clear. Their representational model is the most detailed and is philosophically grounded in Bunge's (1977) *Ontology 1: The Furniture of the World*. They were introduced to Bunge almost by accident, when they began seeking for a suitable philosophical ontology. They used Bunge because they believed, "it deals directly with concepts relevant to the information systems and computer science domains" and "Bunge's ontology is better developed and better formalized than any others we have encountered" (Wand & Weber, 1995, p. 209). The grammatical constructs map one-to-one to the following ontological constructs:

- **Things:** Properties, states (stable and unstable), events (external and internal, well- and poorly-defined), transformations, histories, couplings, systems, classes, and kinds
- **Laws that pertain to these:** State laws, (lawful) conceivable state spaces, (lawful) event spaces, lawful transformations
- **Related to system:** System compositions, system environments, system structure, subsystems, system decompositions, level structures

This goes further toward appropriateness than other extant KR approaches do and they show how it may be used to evaluate the entity-relationship data model. But it becomes clear there are deficiencies which impact KR practice.

For example, event is defined as "A change of state of a thing." This may be satisfactory when considering the values held by attributes, but not so appropriate when considering the creation and deletion of a thing, spatial changes or changes to the global context. To say, in extremis, that we could treat items, relationships and properties all as the states of bits held in computer memory makes a category error, confusing the bit level and symbol level.

While the wording of the definition of event can be changed to include entity creation and deletion, it betrays a presupposition that the aspect of formative power may be reduced to the quantitative with no loss of meaning. This presupposition is found in Bunge.

More fundamentally, their ontology begins with "thing". This would preclude many domains in which "things" are not meaningful such as continuous space (Funt, 1980), liquids or aesthetic nuances. This, again, comes from Bunge.

Perhaps more important for "the people" is that the WW ontology is in some ways "unnatural" without needing to be. For example, Wand and Weber differentiate internal from external events on the basis of whether an event links to changes in the external world. But to "the people" a more important distinction is between those events that are meaningful to the domain of application, and those that are not. Many of what W-W call internal events, such as the calculation of secondary results and explanations, might be meaningful.

Likewise a coupling is defined solely in terms of how one thing acts on another. This makes it clumsy to represent relationships of meaning rather than action, such as ownership of a field, or which fields surround a piece of woodland. These both betray a presupposition that divorces meaning from being/action.

W-W's use of Bunge can also be criticised. They actually made use of only one part of Bunge's ontology, *The Furniture of the World* (1977), and only part of that, related to the notions of thing and change, and not those related to substance, form, possibility and spacetime. But Bunge's second ontology, *A World of Systems* (1979), which explores different systems genera—physical, chemical, biological, social, technical—they made no use of at all.

That could be seen, not as a true criticism, but merely as unfinished work, except that W-W give no reason for excluding systems genera nor any indication how they could incorporate these genera. Wyssusek (2006) criticises W-W for using Bunge in a way Bunge would not have intended. Specifically, while Bunge made ontological commitments which gave ontological content to the terms he used, Wand and Weber did not but merely employed the terms without Bunge's intended content. (W-W explicitly state they are agnostic about whether the ontology is of an objective or a socially-constructed reality.) Whether this is a weakness or a strength in W-W is debatable because the Bungean foundation itself is shaky.

Wyssusek shows how Bunge's ontology, though seemingly logical, fails to match its own claims because of its presupposition of dialectical materialism and adoption of a scientific world-view. As a result, Bunge cannot provide an adequate treatment of either meaning or normativity. W-W make many references to meaning but these cannot be integrated with the rest of their proposal. This is not surprising because Bunge focused on the worlds of science and paid very little attention to the lifeworld.

Moreover, Bunge's system genera, which W-W might wish to use to finish their work, may be questioned. Whereas physical, chemical, biological form a linear sequence, social and technical are placed side by side, but no justification is given for this, nor even any explanation. And where is the psychological level, which appears in most similar attempts to define levels (such as Hartmann)? The answer Bunge gives (1979, p. 247) is "We might have distinguished a system genus between biosystems and sociosystems, namely psychosystems. We have refrained from doing so from *fear* [our emphasis] of encouraging the myth of disembodied minds." Is "fear" a proper philosophical reason? The problem is that Bunge ultimately fails to offer a sound basis for differentiating systems genera.

Despite the briefness of this critique (which merely selectively summarises a fuller one) it indicates problems in W-W's proposal, even though it is superior to most others on offer.

These are the kinds of problems that Dooyeweerd predicted would always emerge from the immanence standpoint (§2.3.3): the divorcing of meaning from reality, "unmethodical treatment", and absolutization (reductionism). Instead, Dooyeweerd offers an alternative rendering of these issues, based on a transcendence standpoint (§2.3.4):

•   As has been demonstrated, Dooyeweerd provides a philosophically integrated account of meaning and normativity in relation to being, rationality and functioning which can lead to a practical KR proposal. Using Dooyeweerd would enable W-W's references to meaning to be incorporated as a systematic part of their proposal.

- Aspects provide a sound basis for both differentiating and ordering systems genera, and reasons are given for the quality of Dooyeweerd's suite (§3.1.6). Indeed, Bunge's systems genera is a subset of Dooyeweerd's (see Table 3-2).

- It is drawn from lifeworld and not just the worlds of science (§2.4.2).

- It does not arbitrarily include furniture of the world and exclude system genera. Rather, both are incorporated into a single integrated approach.

As a result, Dooyeweerd might yield a better foundation for the W-W proposal than Bunge can. It is a pity, perhaps, that Wand and Weber were not presented with Dooyeweerd rather than Bunge when they sought their philosophical foundation for KR.

7.5.1.4 Three Types of Critique of Data Models

We have used Dooyeweerd in three different ways. The critique of Codd's relational data model showed how Dooyeweerd can expose aspectual reduction, of analytic-formative to quantitative-spatial, and thereby explain the problems that have arisen in practice. The critique of OO showed how Dooyeweerd can lay bare the presuppositions underlying the central notion of "object". While acknowledging strengths in the notion, it also exposed the nature of a fundamental weakness already detected by Harrison and Ossher (1993). Unlike Harrison and Ossher, however, Dooyeweerd frees us from the subjectivist, anti-objectivist reaction, to acknowledge and integrate the insights of both sides. The critique of Wand and Weber's ontology showed how Dooyeweerd can be used to address specific issues and trace the root of specific problems to a variety of philosophical presuppositions. It also suggested how Dooyeweerd could replace Bunge as the philosophical foundation for Wand and Weber.

What comes through clearly from all critiques is the influence of the immanence standpoint in our data models and KR approaches. It has forced their designers toward reduction, to presupposing the self-dependence of things and/or events and away from taking meaning—the all-important application meaning to users—seriously as a systematically integral part of their proposals. Meaning, where it is acknowledged at all, is assumed to be reducible to epistemology. All such proposals must, of necessity, concern themselves with ontology rather than only with epistemology, and yet the immanence standpoint has driven Western thought into opposing the two and pushing ontology out of fashion in most IS communities today. But Dooyeweerd, taking the transcendence standpoint, is able to bridge such gulfs, taking ontology seriously and allowing us to systematically include meaning (and also normativity) as part of that ontology. This is what our earlier proposal was able to achieve.

These three critiques have been brief and only indicative, but they have shown that further work in the directions indicated is likely to be fruitful. For our final application of Dooyeweerd we turn to what is an overall approach and ethos rather than a precisely defined data model, that based on Alexander's design patterns.

7.5.2 Dooyeweerd and Alexander

Information systems developers face challenges not unlike those found in architecture—the design and putting together of a complex whole with which human beings will engage as they live, using generic ideas and components applied to the specific situations. So it is no surprise to find that approaches devised in architecture are being brought into service here. Both require an interdisciplinary approach, which in turn demands a diversity of types of components that are "reusable" and yet flexible. Design patterns is a notion borrowed from the architect, Christopher Alexander (Alexander et al., 1977), and adapted to software by, for example, Gamma, Helm, Johnson, and Vlissides (1995).

7.5.2.1 Alexander's Vision

The vision of Alexander and his team, expressed in their two books *The Timeless Way of Building* (1979) and *A Pattern Language* (1977), is to make "towns and buildings ... come alive" which will not happen "unless they are made by all the people in society, and unless these people share a common pattern language, within which to make these buildings, and unless the common pattern language is alive itself" (1977, p. x). This strongly echoes "KR to the people" and the Dooyeweerdian proposal discussed earlier for a way of representing the diversity of everyday meaning.

They present a "language" composed of "patterns", each of which "describes a problem which occurs over and over again in our environment, and then describes the core of the solution to that problem, in such a way that you can use this solution a million times over, without ever doing it the same way twice." To construct the pattern language they:

... have also tried to penetrate, as deep as we are able, into the nature of things in the environment: and hope that a great part of this language, which we print here, will be a core of any sensible human pattern language, which any person constructs for himself, in his own mind. In this sense, at least a part of this language we have

presented here, is the archetypal core of all possible pattern languages, which can make people feel alive and human. (ibid., p. xvii)

Their hopes for this language are expressed in interesting phraseology: "that when a person uses it, he will be so impressed by its power, and so joyful in its use, that he will understand again, what it means to have a living language of this kind" (ibid., p. xvii).

Two hundred and fifty three patterns are defined, patterns 1-94 concern towns, 95-204 concern buildings and 205-253 concern construction—the wider context of buildings, buildings in use, and buildings as products to develop.

7.5.2.2 Design Patterns in Information Systems Design

In ISD, likewise, problems occur over and over again and the core solution to one problem is often applicable to others, though with some appropriate modification. So, partly in response to a rigidity that characterises the extant KR languages, and partly because the diversity of problems is not catered for in them, design patterns has been used as a model to generate better KR "languages" based on a patterns approach that could respond to the needs of software reuse and maintenance in a changing usage environment.

Gamma et al. (1995) have detailed 23 patterns for use in object-oriented design and development, five creational patterns, seven structural patterns, and eleven behavioural patterns. For example the Observer behavioural pattern defines, for a given object, which objects need to updated or informed when it changes. It is surprising how few design patterns they believe are necessary (compared with Alexander's 253). This might be because their only case study is of designing a document editor. For example, while they have a Composite pattern to generate part-whole hierarchies, they do not recognise other types of relationship. Curiously they do not treat lists as patterns but rather as "foundation classes" that patterns may call upon. But they do invite comment and extension.

Several writers have made a critique of the patterns approach, such as Budgen (2003). But, unfortunately—a personal observation—while the structural and methodological elements of design patterns have been adopted and the software engineering community does seem to have taken seriously Alexander's desire that pattern languages should be created according to the needs of the discipline, the original vision of joy and life seems missing.

The roots of this approach are in practical and "intuitive" notions—which makes it important in our study as a lifeworld-oriented perspective. But it seems to lack any proper philosophical foundation or underpinning. As a result, the discipline is at the mercy of unprincipled application or development, in response to undue focus on specific classes of problem; one example is the Flyweight pattern, made necessary by a problem arising from an inappropriate implementation of text (§7.5.2.4).

7.5.2.3 Dooyeweerdian Analysis of Design Patterns

We can see, or perhaps feel, a degree of affinity between Alexander's design patterns and Dooyeweerd, in their motivation, orientation and outworking. Both thinkers "tried to penetrate ... into the nature of things" (Alexander et al., 1977, p. xvii). Both were courageous enough to attempt complete coverage, as Alexander's "all possible pattern languages" and Dooyeweerd's suite of aspects do. Both recognise richness. Alexander's orientation to the idea of a core of reality, a "nature of things in the environment" with a recognition that every person holds a different view, is not unlike Dooyeweerd's acknowledgement of a reality that transcends us alongside the freedom of human knowing and believing. We can see a deep similarity in their beliefs about language as something individual yet social, and alive yet responsible.

Especially in words like "alive" and "joyful" we can immediately feel the affinity between Alexander and Dooyeweerd in their outworking. "Alive" is close to Dooyeweerd's notion that all things are subjects rather than merely passive objects (§2.4.4). "Joyful" is similar to the Dooyeweerdian notion of shalom as full, healthy, positive functioning in diverse aspects (§3.4.3). Each pattern is related to a "problem" that makes it meaningful, but also to its "context" within wider patterns and to "smaller" patterns, which are not seen as parts but as important in their own right. This echoes the Dooyeweerdian notions of primacy of meaning and enkaptic interlacement.

Asking, of the 253 patterns, "What is this pattern trying to achieve? What (aspectual) normativity makes this a problem to be addressed?," to see which aspect is most meaningful in each, gives the results shown in Table 7-4. For this analysis, a differentiation was made in the social aspect between groups of people and cultural matters, and in the aesthetic aspect between harmony (integration, balance, and coherence) on the one hand, and style (beauty, fun, and rest) on the other. There is a row for patterns that could not be clearly placed, and one for multi-aspectual patterns.

Such an aspectual analysis of design patterns may reveal a number of things, if we take Dooyeweerd's suite to be reasonably comprehensive and well-founded. Four considerations confirm expectations of the quality of Alexander's work:

- Aspectual spread can indicate to what extent the creators of the set of patterns are sensitive to the diversity found in the lifeworld (§3.4.5), as opposed to focusing on certain aspects currently deemed fashionable by researchers or

Table 7-4. Aspectual spread of Alexander's patterns

Aspect	Towns	Buildings	Construction
Quantitative		96	231
Spatial	21 37	167 195	
Kinematic	20 23 34 49 56	120 131	
Physical			211 212 213 215 218 219 225 227 234 236
Biotic	4 47 65 70 72	118 169 170 175 177	
Sensitive	54 55 60 62 71 74 82 92 93 94	97 98 105 109 114 115 119 121 125 132 137 138 142 161 163 164 173 176 180 182 190 192 196 197 199 201 202 117 203	223 230 235 237 250 233
Analytic	14 15 53 57	102 110 111	229
Formative	16 50 73 78 83	104 108 146 171	208 240
Lingual	12 18 43		
Social	2 13 27 30 31 33 36 41 44 45 48 61 67 68 75 76 77 86 89 91	95 100 101 122 123 124 127 129 139 141 143 147 148 149 151 152 179 185 186 188 193	242
Soc: Culture	6 8 40	ʼ	253
Economic	11 19 22 32 39	103 106 145 150 153 162 178 198 200 204	206 228
Aesth: Style	28 38 46 58 59 63 69 87 88 90	112 113 128 134 135 174 191	216 217 221 224 226 232 238 239 244 245 247 249 251 252
Aesth: Hmny	3 5 9 17 29 35 42 51 52	99 107 116 126 130 133 140 156 157 158 160 166 168 181 194	214 222 241 246 248
Juridical	1 7 10 79	172 183 184	205 207 209 210 243
Ethical		136 187	
Pistic	24 66 80 81 84	154 155	
Multiple	26		220
Unplaced	25 64 85	144 159 165 189	

developers. That every aspect is represented in Alexander's set suggests he and Dooyeweerd have a similarly wide recognition of meaning.

- Every thinker is part of a community and both community and individual stress certain things. Aspectual analysis can reveal which. That many patterns are related to the sensitive and aesthetic (style) aspects is to be expected in most architects, while the abundance of social and aesthetic (harmony) patterns reflects Alexander's desire that buildings should enhance community and harmonise.

- A different aspectual profile would be expected in the three columns, in that Construction is a physical and technical activity, in Buildings the human user is central, and Towns concern society and its (pistic) vision. In Alexander we find largely what we expect; if we did not, we could question their treatment of these issues.

- The low number of unplaced and multi-aspectual patterns can indicate that the patterns are clearly thought out as to their meaning. A high number would indicate either a confused understanding, or that Dooyeweerd's suite needs modification.

But one analysis shows a curious anomaly:

- Considering under-represented aspects can also provide insight. In Alexander, for example, why is there so little of the lingual aspect? In the three columns, respectively, it could cover public signage and communication, deliberate signification in and around the buildings, and communication during the construction process.

Other sparse aspects might be treated similarly. It is only an aspectual analysis of this kind that can expose the omission of whole spheres of meaning.

If we perform a similar aspectual analysis on the patterns oriented to software in Gamma et al. (1995), we find a very different picture; Table 7-5, where patterns are identified by page number, and are grouped into patterns that guide creation, structure, and behaviour of objects in the program.

With only 23 patterns our analysis must be more cautious. Since nearly half the patterns are primarily formative, we can clearly detect a heavy emphasis on techni-

Table 7-5. Gamma's design patterns for software

Aspect	Creation	Structure	Behaviour
Quantitative	127		
Spatial			
Kinematic			
Physical			
Biotic			
Sensitive			
Analytic			229
Formative	107 117	151 163 185	233 273 283 305 325 331
Lingual	97		243,
Social			
Economic		195 207	
Aesth: Style	87		
Aesth: Hmny		139	293
Juridical		175	223 315
Ethical			
Pistic			

cal matters. That the juridical aspect makes a reasonable appearance suggests that Gamma et al. recognise the need to do justice to the diversity of the world to be represented, and this is supported by examining their text.

But eight aspects are missing. With so few patterns, we should look at groups of, rather than single, under-represented aspects. Where, for example, are the aspects which are important to the user interface: the sensitive, spatial and kinematic? Where are the human and social aspects? Examining their text we find that most of the post-social aspects concern, not human use nor the wider social context of use, but the "objects" and classes of which the software is composed. Gamma et al. almost wholly focus on what Alexander called construction and display little awareness of software's use and wider context. This analysis might suggest that Gamma et. al's set of patterns is in danger of directing the developer's attention away from the all-important human and social issues of use of the software in context to mere technical issues, and of failing to support the sensitive, spatial and kinematic aspects of UI.

7.5.2.4 Reflection on Aspectual Analysis

Two types of aspectual analysis have been demonstrated here. The analysis of Alexander's patterns generated indications of what they deemed important that are reasonably supportable. But aspectual entries in Gamma et al. were too sparse for that. Instead, the presence and absence of entries were used to suggest issues that deserve further examination in the text, and it is this that exposes gaps or over-emphases in their approach.

It can also be helpful to subject single patterns, especially problematic ones, to Dooyeweerdian scrutiny. For example the Flyweight pattern (Gamma et al., 1995, p. 195) introduces clumsy complexity. It was motivated by the inefficiency of requiring each letter of text to hold the full set of attributes that text as such does (font, etc.), working to strip letters of such attributes. But it is an unsatisfactory solution because it misdiagnosed the problem, assuming that text may be seen as aggregations of letters into rows. Dooyeweerd's treatment of wholes (see Chapter III) urges us to ask what are the meaningful wholes of this lingual thing that is text. Our answer would presumably include such things as words, sentences, and headings, but not individual letters, since these cannot stand alone as lingual wholes. As wholes, letters are analytic rather than lingual, and so cannot be seen as part of words. In view of this, letters should never have been treated in the way text is, with all the lingually-relevant features. The root of the problem is a fundamental misunderstanding of the nature of part-whole relationships which Dooyeweerd could help clear up.

Sadly, such misunderstandings are very common in the OO community and indeed the whole systems community, because they have no basis for recognising true wholes nor enkaptic relationships.

Thus a Dooyeweerdian analysis of meaning can serve as a useful critique of individual patterns as well as of the whole approach. The root of the problem lies not so much in the individual pattern as in the OO paradigm, which is based on an existence-oriented presupposition that does not allow Being as such to be analysed and thus fails to reveal concepts like enkapsis (§3.2.2).

But Dooyeweerd can take us further than critique. If, as suggested in this chapter's main proposal, it is possible in principle to provide modules that implement the meaning of every aspect, then it should be possible to provide patterns related to every aspect and realise something of Alexander's vision in IS.

7.6 Conclusion

This area is concerned with the design and preparation of IT resources like knowledge representation languages, with their tokens, subroutines, and code libraries intended to be used by IS developers—equivalent to the design and preparation of things like bricks, planks of timber, nails, hammers, and so forth, to be used by builders of houses. As with the nature of computers (Chapter V), one might wonder to what extent the preparing of basic technological resources and building blocks can be viewed as everyday experience. It seems too technical a field. But a moment's reflection on the similarity to the construction industry will affirm that there is an everyday lifeworld.

7.6.1 Overview of Framework for Understanding

After a review of what has motivated the design of KR languages, it became clear that Brachman's notion of "KR to the people" was a key to the everyday lifeworld of this area. This is the normative proposal that "ordinary" people should be able to make use of these general technological resources in order to construct computer systems for themselves, without needing to call upon an expert programmer.

This led to the starting point that an important characteristic required of IT resources to achieve this is appropriateness—that the basic resources with which an IS might be constructed should be "natural". But, to understand what this means requires taking a step back to first understand the nature of these resources themselves. This is facilitated by Dooyeweerd's ontic analysis of similar resources in the construction industry:

- Basic technological resources are semi-manufactured products, the leading aspect of which is not internal but external; their very nature is to reach out to the diverse spheres of meaning of the domain of application.

Unlike planks, bricks, and nails, which are centred on the physical aspect, the operation of these resources is centred on the lingual.

Such resources have to be designed and implemented. But so do the technical artefacts of the IS developer; what is the difference? The key difference, it was suggested, is that what the IS developer designs for is concrete subject-side situations and requirements, while these basic resources embody the law side. Their generality is thus not just a matter of degree but different in kind. So the principle that should guide their design is:

- For each and every aspect, a set of basic facilities and KR language tokens should be available that express its cosmic meaning, as it may be actualised in the various philosophical roles that aspects have (things, properties, relationships, activity, inferences, constraints, etc.).

This is the root of appropriateness. It is based on the belief that aspects, as spheres of meaning and law, enable these very things (see §3.1.5). Thus this area makes use of Dooyeweerd's approach to ontology, as transcending us but based on meaning rather than being.

A practical proposal was made that a module that contains these things be prepared for each aspect. Evidence that this is feasible comes from the fact that software has been developed which is qualified by each aspect. The implementation of aspectual modules at the bit level was briefly discussed. But the modules must integrate and work with each other, and be open to future requirements, especially when we begin with modules only of the earlier aspects. So:

- The relationships between modules reflects that between aspects: the modules are irreducible to each other, but they are to be linked by foundational dependency, anticipatory dependency, analogy and reaching-out, all of which should be taken into account in the design of modules.

Anticipatory relationships are important for future-proofing the modules.

This proposal is only an initial one, and lacks much. Though protocols were included as a type of resource, these have not been discussed, except for an incidental mention of HTML. A comparison with the current practice of using OO-based domain ontologies to create facilities for later aspects need to be made. Technologies like neural nets should be brought into the discussion.

Nevertheless, some benefits accorded by this framework are that it provides fresh impetus to consider the diversity of basic, intuitive meaning that IS developers might encounter and not delegate the responsibility for implementing some of this to developers who might not be expert in it. Only thus will it be possible to provide the range of resources that IS developers actually need to cope with real-life complexity. This in turn will help bring "KR to the people".

Even though we can do a lot better than currently, achievement of this goal will never be complete because all aspectual functioning is non-absolute (§3.1.4.4). So this proposal might be seen as a grand vision for an all-encompassing KR toolkit. But, realistically, it might be more useful as a framework within which to situate extant work. For example, it could be seen as a counterfactual ideal against which to evaluate other proposals. Two examples were given by which current proposals were critiqued, that of Wand and Weber (1995), which represents one of the best-developed general KR ontologies to date, and the importation of the idea of design patterns from the field of architecture, which is arousing much excitement in the object-oriented community.

7.6.2 The Mission of Bringing Information Technologies into Being

We have been discussing how to bring small pieces of information technology into being, as building blocks and other resources for the IS developer to use. But should we bring information technologies into being? What should guide and motivate this process? A presumed inner dynamic of IT itself (technological determinism)? A social shaping of technology? Market forces of the IT industry? The whim or even the "existential joy" it brings to those who create it? And what about obsolescence?

Dooyeweerd held that humankind is called normatively to open up the potential of all aspects as spheres of meaning and law, and this for the overall blessing of the cosmos. This idea is examined in Chapter VIII, where the development of technology may be seen as a "meaning disclosure", and that of IT as a whole as the opening up of the lingual aspect of information and representation, and perhaps also the formative aspect (technology). So the general answer to the first question is "Yes", and it should be guided by the norms of all other aspects, especially post-lingual.

But more specifically, the coming into being of any particular piece of IT may be seen as an opening up of the aspect that is represented. The norms that should guide this progress is not that of the lingual or formative aspect themselves, but that of the represented aspect, followed closely by all other aspects, governed by the shalom principle (§3.4.3).

This leads us to think of obsolescence in a different way. A certain piece of technology might happen to no longer suit our current historical circumstances, but the meaning it discloses cannot be undisclosed. Therefore, in general, while we might cease to use it for a while, humanity has a mandate to protect its having-been-disclosed.

However, this disclosure of pieces of meaning is not the end of the story, because we are led outwards from the specific ones we develop to those our community develops to those society develops, and eventually to IT as a whole, as a global, historical, societal, religious phenomenon. This becomes the very environment in which we live and have our being, which affects our habits, aspirations, and expectations, and which is itself inscribed by by these very things, our world-view. This is the topic explored in the next chapter.

References

Alexander, C. (1979). *A timeless way of building.* Oxford, UK: Oxford University Press.

Alexander, C., Ishikawa, S., Silverstein, M., Jacobson, M., Fikidahl-King, I., & Angel, S. (1977). *A pattern language: Towns, buildings, construction.* New York: Oxford University Press.

Basden, A. (1993). Appropriateness. In M. A. Bramer & A. L. Macintosh (Eds.), *Research and development in Expert Systems X* (pp. 315-328). Cranfield, UK: BHR Group.

Basden, A., Ball, E., & Chadwick, D.W. (2001). Knowledge issues raised in modelling trust in a public key infrastructure. *Expert Systems, 18*(5), 233-249.

Basden, A., & Brown, A. J. (1996, November). Istar: A tool for creative design of knowledge bases. *Expert Systems, 13*(4), 259-276.

Basden, A., & Clark, E. M. (1979). Errors in a computerized medical records system (CLINICS). *Medical Informatics, 4*(4), 203-208.

Basden, A., & Hines, J. G. (1986). Implications of the relation between information and knowledge in use of computers to handle corrosion knowledge. *British Corrosion Journal, 21*(3), 157-162.

Basden, A., & Nichols, K. G. (1973). New topological method for laying out printed circuits. *Proceedings of the IEE, 120*(3), 325-328.

Booch, G. (1991). *Object-oriented design with applications.* Redwood City, CA: Benjamin-Cummings.

Brachman, R. J. (1990). The future of knowledge representation. In *American Association for Artificial Intelligence, AAAI-90: Proceedings of the Eighth National Conference on Artificial Intelligence* (pp. 1082-1092). Boston: AAAI.

Brandon, P. S., Basden, A., Hamilton, I., & Stockley, J. (1988). *Expert systems: Strategic planning of construction projects.* London: The Royal Institution of Chartered Surveyors.

Budgen, D. (2003). *Software design* (2nd ed.). Harlow, UK: Pearson Education/Addison-Wesley.

Bunge, M. (1977). *Treatise on basic philosophy, Vol. 3: Ontology 1: The furniture of the world.* Boston: Reidal.

Bunge, M. (1979). *Treatise on basic philosophy, Vol. 4: Ontology 2: A world of systems.* Boston: Reidal.

Chomsky, N. (1965). *Aspects of the theory of syntax.* Cambridge, MA: MIT Press.

Clouser, R. (2005). *The myth of religious neutrality: An essay on the hidden role of religious belief in theories* (2nd ed.). Notre Dame, IN: University of Notre Dame Press.

Codd, E. A. (1970). A relational model for large shared databanks. *Communications of the ACM, 13*(6), 377-387.

Dooyeweerd, H. (1984). *A new critique of theoretical thought* (Vols. 1-4). Jordan Station, Ontario, Canada: Paideia Press. (Original work published 1953-1958)

Funt, B. V. (1980). Problem-solving with diagrammatic representations. *Artificial Intelligence, 13*(3), 201-230.

Gamma, E., Helm, R., Johnson, R., & Vlissides, J. (1995). *Design patterns: Elements of reusable object-oriented software.* Reading, MA: Addison-Wesley.

Gennart, J. H., Tu, S. W., Rothenfluh, T. E., & Musen, M. A. (1994). Mapping domains to methods in support of reuse. *International Journal of Human-Computer Studies, 41*, 399-424.

Gibson, J. J. (1977). The theory of affordances. In R. Shaw & J. Bransford (Eds.), *Perceiving, acting and knowing* (pp. 67-82). Hillsdale, NJ: Erlbaum.

Greeno, J. (1994). Gibson's affordances. *Psychological Review, 101,* 336-342.

Harrison, W. H., & Ossher, H. (1993, September 26-October 1). Subject-oriented programming: A critique of pure objects. In Andreas Paepcke (Ed.), *Proceedings from Conference on Object-Oriented Programming Systems, Languages, and Applications (OOPSLA), Eighth Annual Conference*, Washington, DC (pp. 411-428). New York: ACM.

Hines, J. G., & Basden, A. (1986). Experience with the use of computers to handle corrosion knowledge. *British Corrosion Journal, 21*(3), 151-156.

Jones, M. J., & Crates, D. T. (1985). Expert systems and videotext: And application in the marketing of agrochemicals. In M. A. Bramer (Ed.), *Research and development in Expert Systems: Proceedings of the fourth Technical Conference of the British Computer Society Specialist Group for Expert Systems.* Cambridge, UK: Cambridge University Press.

Levesque, H. J., & Brachman, R. J. (1985). A fundamental tradeoff in knowledge representation and reasoning (Revised Version). In R. J. Brachman & H. J. Levesque (Eds.), *Readings in knowledge representation.* Los Altos, CA: Morgan Kaufmann.

Lombardi, P. L. (2001). Responsibilities towards the coming generations: Forming a new creed. *Urban Design Studies, 7,* 89-102.

Minsky, M. (1981). A framework for representing knowledge. In J. Haugeland (Ed.), *Mind design.* Montgomery, VT: Bradford Books/MIT Press.

Nilsson, N. J. (1998). *Artificial intelligence: A new synthesis.* San Francisco: Morgan Kaufmann.

Quillian, M. R. (1967). Word concepts: a theory and simulation of some basic semantic capabilities. *Behavioral Science, 12,* 410-430.

Stamper, R. (1977). Physical objects, human discourse and formal systems. In G. M. Nijssen (Ed.), *Architectures and models in database management systems* (pp. 293-311). Amsterdam: North Holland.

Stephens, L. W., & Chen, Y. F. (1996). Principles for organizing semantic relations in large knowledge bases. *IEEE Transactions on Knowledge and Data Engineering, 8,* 492-496.

Wand, Y., & Weber, R. (1995). On the deep structure of information systems. *Information Systems Journal, 5,* 203-224.

Winfield, M. (2000). *Multi-aspectual knowledge elicitation.* Unpublished doctoral thesis, University of Salford, UK.

Winch, P. (1958). *The idea of a social science and its relation to philosophy.* London: Routledge and Kegan Paul.

Wyssusek, B. (2006). On ontological foundations of conceptual modelling. *Scandinavian Journal of Information Systems, 18*(1), 63-80.

Chapter VIII

A Framework for Understanding Information Technology as Ecology

How do we bridge the "digital divide"? But do we even want to bridge the digital divide? IT as a whole seems to deliver the opposite of what it promises—think about the "paperless office" and about convenient, time-saving e-mail technology, which now consumes so much of our time.

But will not the Internet help to reduce climate change emissions by enabling us to "meet" without having to travel? Recently the author heard of four people who play bridge over the Internet and decided to meet for a "proper" game in Toronto. They flew from across the world, a total of something like 20,000 air-miles, causing as much climate change emissions as would 10,000 local "proper" games of bridge if all drove to them. The Internet was actually responsible for generating more climate change emissions.

Information technology (IT) deeply influences the way we live, our assumptions, aspirations, expectations, habits—our very life-and-world-views (LWVs). It has become an environment in which we live and work, a kind of ecology in which modern life takes place. We have created it, but it also creates, or at least deeply affects, us. This area is concerned with what Lyon (1988) called the information society and Castells (1996), the network society, and it is in this area that such things as Walsham's (2001) discussion of globalisation are considered. It addresses issues of the type found in Vignette 5 in the Preface.

The communities that contribute to research and practice in this area include those working in social theory, history of technology, philosophy of technology, society and technology studies, social shaping of technology, gender studies, globalisation and the environment, and many more. They are not so much concerned with a billion IT users in the world, a billion IS developments, a billion technologies that have been created, even averaged out, but rather with something deeper and more structural. This area is characterised by a very different type of relationship between human beings and IT. In the other areas, humanity is "outside" the technology, as its original conceiver (Chapter V), its creator (Chapter VII), its applier (Chapter VI), and its user (Chapter IV). But in this area the human being is "inside" the technology.

These questions relate to those in Western culture. Should the non-Western, developing, two-thirds world be encouraged to adopt IT? Western life-and-world-views and ways of living are spreading throughout the world, driven by globalisation on the one hand and pulled by aspirations of non-Western peoples on the other. Does this mean the destiny of humankind and indeed the whole world is to be forever controlled by IT?

Our everyday attitude stridently rejects such a possibility. But that might just be the content rather than the structure of the everyday attitude, and thus might be only culturally contingent (§2.4.2). So we cannot take that to be an answer to this question. Instead, we can break it down into three component questions:

- Is IT valid as an endeavour for humanity, or should we close down technological development and even use? If IT is valid, in what way is it valid?

If it is valid, and we do not want it to influence our lives, at least not in the controlling ways indicated earlier, then we need to ask a second question ...

- What the nature of the relationship between humanity and IT-as-ecology?

If we can understand the nature of such a relationship, then we need to understand why IT controls us in the way it currently does. Seeing the influence of IT on the way we live and work as a problem is not universal because what we see as problematic depends on the way we look at things (see Introduction). So we need a deeper view on ...

- Why have things gone (as we might see it) wrong? What is the root of this, and what should be done about it?

This chapter explores how Dooyeweerdian philosophy might throw light on these questions. It is to some extent a reinterpretation of the area. Unlike previous chapters, where other ideas were discussed first, this chapter will introduce Dooyeweerdian thinking immediately, making reference to other thoughts during the process.

8.1 On the Validity and Destiny of Information Technology

In formulating a framework for understanding this area, the first question to address is whether technological development is valid at all as a human endeavour. If not, then the rest of this chapter and indeed all four previous chapters are meaningless. It is not our primary purpose here to answer that question (though a positive answer will emerge), but rather to indicate how Dooyeweerd might help us address it.

This is one area where Dooyeweerd's awareness of the religious root, cosmic dimension, as well as his own exploration of the CFR ground-motive, are important. Throughout his writing one meets the notion of destiny: destiny and mandate of humankind, and destiny of the cosmos (§3.4.6). The religious root of the cosmos implies a destiny as such, and CFR implies a particular kind of destiny. Though much human life is now "inside" IT, human destiny transcends IT. Humanity is still responsible for the long-term direction that IT takes and what impact it has on human and other life as a whole.

Schuurman (1980) has made extensive use of Dooyeweerd to discuss this, and his conclusion is that it is valid but that current directions of development are wrong. He first examines the views of two sets of thinkers, of what he calls transcendentalists—Jünger, Heidegger, Ellul, Meyer—who were largely pessimistic about technology, and of what he calls the positivists—Wiener, Steinbuch, and Klaus, and to some extent, Habermas—whom he finds largely optimistic. After a lengthy and careful analysis of these views (to which the reader is referred for detail), he criticised both sets of thinkers (1980, p. 314):

The views of the transcendentalists and the positivists alike imply an unpromising, somber future. The reason for this, as I see it, is that both categories of thinkers espouse an autonomous philosophy or, as is the case with Ellul and Meyer, an attempted synthesis with autonomous thought in science and technology. In the autonomy of thought lies the ground of the tensions that pervade their philosophies, as well as the reason for their inability to indicate a way of escape from the real problems that modern technology entails.

In contrast to both the pessimistic and optimistic views of technology, Schuurman works out what he calls a "liberating vision for technology", and discusses what might "disrupt" this vision. Interwoven with a brief explanation of Schuurman's vision are comments on how it relates specifically to information and communication technology (ICT).

[Schuurman's later writings do not seem to add much of philosophical substance to his (1980). For example Schuurman (2003) is an updated re-statement of (1980) for a largely Christian audience, adding some theological considerations to a "liberating perspective" and "ethics of responsibility". So his earlier work (1980), which carefully established his foundations, will remain our source here.]

8.1.1 Schuurman's "Liberating Vision for Technology"

Schuurman's "liberating vision for technology" (1980) is an expanded version of Dooyeweerd's theory of progress as set out in §3.4.6, based on what he calls the "meaning-*dynamis*", which is his term for what we have called the aspectual law-framework in §2.4.4. His "liberating vision" has seven elements.

First, drawing on Dooyeweerd's theory of progress as the opening up of the aspects, he sets out a vision of what humankind's project of technology is: it should contribute to the opening up of the aspects. Technology is meaning-disclosure. Schuurman's discussion of Heidegger's notion of technicism as "revealment" sounds very similar.

Second, technological development of itself is utterly insufficient. It must be led, not by its own inner meaning and dynamic, but by that of other aspects. "A liberating perspective for technological development opens up when it is understood that the specific meaning of technology ought to be led by the normativity of the various postcultural aspects of our reality, namely, the lingual, social, economic, aesthetic, juridical, ethical, and pistical aspects" (Schuurman, 1980, p. 361). (This author would extend this to all aspects, especially in the light of climate change, because the three main components of climate change emissions—power generation, households, and transport—are all technological.) This emphasis on multiple aspects echoes that in every other area for which we have developed a framework for understanding.

Third, when this occurs, we can expect a deepening and enrichment of meaning as it is experienced within society and cosmos. Technology is no longer demonic, predatory, counter-productive, as the transcendentalists feared it is, but can be a blessing rather than a curse. The unexpected benefits accruing from *Elsie*, and the "existential joy" in technology, and the "joyful mission" of technology shaping, mentioned in Chapters IV, VI, and VII, may be seen as examples of such blessing. The reason it is often a curse is largely because we ignore or distort aspects of use, development, or shaping.

Fourth, "the future of technology is in fact not determined, but open" (ibid., p. 361). Since the postcultural aspects are normative (§3.1.4.9), "it follows that technological meaning-disclosure as deepening of meaning is not rigid in character." This implies a deep denial of the "technological imperative" and stresses that humanity has responsibility for the way technology is developed.

Fifth (ibid., p. 362), "the key to modern technological activity is to be found in designing—more particularly, in the engineer who does the designing. The engineer designs—or rather, projects—laws that are related to the fashioning of technological things and facts. This designing is also called the positivizing of laws or norms." (In §7.2.2 it was suggested that IT resources embody the law side.) Echoing Heidegger's stress that technicism is a verb rather than a noun, a process, and functioning rather than a thing, Schuurman takes this further by focusing on the human being who undertakes that process, a Dooyeweerdian subject responding to the law-framework.

Sixth, this implies a quality criterion for the engineer's technological activity: "The more finely attuned he is to the meaning-*dynamis* and the better he thus understands the normative principles which are given him to be worked out, the more adequately he unfolds this normativity" (ibid., p. 362). The importance of being sensitive to the aspectual law framework was stressed in previous chapters.

Seventh, the problems with technological activity are likewise explained with reference to the engineer's response to aspectual norms. "Should the engineer fail to attain a correct insight, either because he surrenders, self-sufficiently, to absolutizations and thus shuts the whole process off, or because of shortcomings or shortsightedness, then meaning-disruption will arise and, as we have observed, prevail" (ibid., p. 362). The discussion in previous chapters suggests, however, that it is not only the engineer's response that causes problems, but also that of users and IS developers.

8.1.2 Disruption of the Vision

Schuurman then outlines three absolutizations that have caused us problems—of technological-scientific thought, which "closes off all further technological meaning-disclosure," of social disclosure, as in Marxism, which results in a "leveling and stiffening of technological development," and of the economic aspect of technological development, as in capitalism that makes profit maximisation its key norm, "at the cost of overlooking the dangers to the environment" (Schuurman, 1980, pp. 362-3).

Later he expands on this, discussing five assumptions that disrupt this vision: technology seen as applied science, as handmaiden of economic life, as neutral instrument, as the will to power, and as technology for its own sake. All these disruptions are clearly visible in IT.

Schuurman traces a Dooyeweerdian-style religious root in the disruption that is common to all "technology as" absolutizations (ibid, p. 359):

The deepest *ground of the disruption of technological meaning is* religious *in character. Meaning-disturbance, which arises under the leading of various motives, is occasioned in the first instance by humanity's refusal to bow before the true meaning-*dynamis *of the creation, that is, by humanity's refusal to offer itself in freedom and responsibility.*

A concern of Weber [cited in Habermas (1987)] is that, in modernity, all meaning is lost to our experience. To Dooyeweerd (and Schuurman), this is because Meaning always refers beyond, first to other aspects, then ultimately to the Divine Origin of Meaning. So when something is absolutized, such as the formative aspect, its referring-beyond relationships are severed both from other aspects and from the root of all Meaning. Since it is relative and cannot be Meaning in itself, we experience loss of meaning.

8.1.3 Implications of Schuurman's Vision

If, as argued earlier, the leading aspect of information technology is the lingual, and if we take Dooyeweerd's view of progress as aspectual opening (§3.4.6), then we can see its development by humanity as an opening up of the lingual aspect. Whereas speech opened up the lingual aspect's propensity to communication, writing opened up the persistence of symbols, poetry and literature opened up its aesthetic potential, and print opened up its potential in the public sphere, information technology has begun to open up its dynamism and ability to respond meaningfully to each individual human situation. Thus the destiny of IS is bound up with the opening of the lingual aspect but guided by the norms of all other aspects, so that IT/IS can contribute to opening up even those aspects.

Schuurman recognises the gross disruption to society, environment, and individuals that technology has given rise to and, rather than simply bemoaning this and making merely general visional points, he continues with:

What needs investigating is the extent to which the lopsided and crooked condition into which things have fallen can be straightened out. It is possible that we will have to pick up—in a new way—some of the old and forgotten traditions in the history of technology. In this connection, the so-called alternative technologies (the small technologies) should be given more than passing consideration. ... (p. 363)

But this is no mere romanticism. In his exposition of this idea he demonstrates how this approach is likely to facilitate a host of aspects (identified by us in brackets):

... Modern technology should not be exclusively massive and gigantic, there should be room for a high degree of differentiation [analytic] and diversity [aesthetic] as well. Worthy of consideration in connection with this objective are technological products of greater durability [economic] that might be made with little capital investment and a minimum of energy [economic]—with natural energy and natural materials to boot [harmony-aesthetic]. Personal individual production yields satisfaction in labor [sensitive, pistic], products reflective of settled cultural patterns [social], and a minimum of pollution [biotic]. These forms of technology, despite their small scope, are decidedly difficult to realize. They require a great deal of technological fantasy and imagination [formative]. Cultural stability [social] depends on these forms of technology, however, both in developed and in developing countries. (p. 363)

A strong implication of this as well as of earlier material is that those who contribute to "progress"—and this includes all who research, develop, and use IT and IS—should live in the light of such multi-aspectual responsibility, in which all aspects are opened up rather than closed down and the relationships between them are respected and enjoyed. Taking a multi-aspectual (life-world) attitude in our research and development is no mere option, but essential to humanity's task in researching and developing IT/IS. This recognising and opening of all aspects has been a recurring theme in all Chapters IV to VII.

Schuurman then proceeds to look at the last four of Dooyeweerd's suite of aspects in more detail. **Aesthetic disclosure** should result in harmony among all persons and authorities, in selection of materials, and between nature and technology. **Juridical disclosure** involves legal institutions that see to it that technology is promoted, while at the same time people and environment are protected. Juridical disclosure also involves making compromises among the various stakeholders. We might also add that we must ensure that the technology infrastructure enhances rather than detracts from what is due to all. **Ethical disclosure**—love—"will express itself in a good choice and a proper use of available materials and objects and in a good, artisanlike finishing of technological products." We might also add that we should ensure that using and developing and dwelling in IT makes us more not less generous and self-giving, and that technology infrastructure is set up in such a way as to promote rather than hinder these. Finally, **pistic disclosure** means that "humanity is called to the task of technology, and that people are obliged to accept this mission as a responsibility before God [i.e., not just to themselves, their accountants, shareholders, or peers]. ... belief ought to anticipate the fulness of meaning and thereby be directed to the Origin of meaning" (p. 365).

If we consider the overall story of technology, Schuurman's vision may be summarised with: "The one essential condition for humanity's going on its way in freedom in technological development is that people submit in *belief* to the meaning-*dynamis* as normativity for the meaning-disclosure of creation" (Schuurman, 1980, p. 361).

8.2 Information Technology as Ecology

Schuurman presents a normative vision for technology based on Dooyeweerd's theory of progress. But to understand why IT might influence humanity in detrimental or beneficial ways we must understand the structure of the relationship we have with technology as an environment in which we live, and understand why it is not just an extension of the relationships we have with technology in the other four areas.

Parts of Dooyeweerd's theory of things are important. In the other four areas—use of IT artefacts, the nature of computers, IS development and the shaping of information technologies explored in Chapters IV to VII—it is a law-subject-object relationship with information technology which is of interest. But in this area, it is a relationship of correlative enkapsis (§3.2.6.2), in which we are denizens of an *Umwelt*, which we also form and constitute. Whether the *Umwelt* is information itself, IT or society in which these play a large part, does not need to be resolved here; what follows can be applied to either of them.

8.2.1 Correlative Enkapsis and *Umwelt*

To Dooyeweerd, an *Umwelt* is an environment that is constituted of the very things that are its denizens. A forest is *Umwelt*, to trees, insects, and so forth. Society is an *Umwelt*. An *Umwelt* has no leading aspect. In this way, society (*Umwelt*) may be distinguished from the state (a social entity led by juridical aspect).

Unlike an organisation, an *Umwelt* has no parts. Nor is it constituted of aspectual beings. Its being is that it engages in a correlative enkaptic relationship with the entities that make it up. Correlative enkapsis is the relationship that exists between a forest and its denizens (trees, other plants, animals, insects, fungi, etc.) that constitute the forest but also live in it and because of it. They could not flourish if not within their proper *Umwelt*. Likewise, society cannot exist without people, and people do not flourish as full human beings without society.

An *Umwelt* is not static, nor pre-determined in its shape (Dooyeweerd, 1984, III, p. 648):

But we can only speak of an "Umwelt" (environment) in connection with a living organism. In this enkaptic interwovenness the environment exhibits an objective biotic or objective psychic qualifying function, only opened as such by the subjective structure of the living organisms.

An *Umwelt* is of a particular radical type and radical types are defined by a particular aspect. Though Dooyeweerd discussed mainly the ecological environment, of a biotic-psychic radical type, the notion seems useful for any type of environment that is constituted of correlative enkaptic relationships. It is reasonable to suggest that IT as it is being discussed here is an *Umwelt* in which the lingual aspect is key.

8.2.2 IT as *Umwelt*

IT-as-*Umwelt* is not the same as IT resources, nor the developed IS, nor IS in use, nor our beliefs about its nature, but is something we live "in". To understand this requires an understanding of both the structural relationship and also of normative direction (Strijbos, 2006). The normative direction is set by Schuurman's liberating vision, discussed earlier. The structural relationship, which is correlative enkapsis, is a circular one:

- Humanity shapes this IT-as-*Umwelt*
- IT-as-*Umwelt* also shapes humanity, influencing the way we live and work: our habits, expectations, aspirations, and LWVs

Because an *Umwelt* is not static, we must expect IT-as-*Umwelt* to change over the years.

This is not unlike Giddens' (1993) structuration theory (ST), with its notion of the duality of structure (social structure enables and constrains human action; human action produces and reproduces social structure). But the Dooyeweerdian approach has several advantages. A full exploration of links between ST and Dooyeweerd's ideas that takes into account recent critique of ST is needed but here a few summary comments must suffice.

One is that the two arms are more fundamentally intertwined than Giddens perhaps conceives. Because, to Dooyeweerd, the human ego is not autonomous and we are part of the cosmos we fashion, both subject to the same law side, whatever we do is always-already having an impact on us and others. So the issue of *Umwelt* is not a "macro" issue, separate from the "micro", but rather is a focusing on the circular relationship which is always present. Thus Dooyeweerd provides a different basis on which to differentiate this societal relationship with technology (correlative enkapsis) from the various relationships discussed in previous chapters, especially those found in social institutions, and also relate them.

A second is that Dooyeweerd does not allow us to consider this relationship in isolation from other enkaptic relationships. In his work (1984, III, p. 652) Dooyeweerd gave the example of a farm, in which cows form various biotic enkaptic relationships with each other (copulation, mother-with-young, etc.) while at the same time relating correlatively with their natural environment—and also all these relate to human structures such as the economic functions of the farm, and beyond these to the social and other relationships that obtain. Therefore, we must always beware of treating the correlative enkaptic structure of IT-as-*Umwelt* in isolation from all other structural relationships in society. While some writers recognise this, Dooyeweerd, in identifying a number of types of enkapsis, might provide a more precise framework by which we may examine such complex interlacements as we find among state, industry, media, education, and so forth, all correlatively related to IT-as-*Umwelt*.

A third is that while there is a tendency in ST to conceive of "human action" in a very generic way, Dooyeweerd allows justice to be done to the diversity thereof, as indicated by the discussion in previous chapters. Dooyeweerd also allows us to see the enabling and constraining not as rigidly imposed by the societal structures but as something we allow to happen to us and even welcome. This suggests that both arms of the *Umwelt* relationship may be understood aspectually, which is discussed later.

As a result of this, a fourth advantage is that whereas ST focuses on human action, Dooyeweerd widens "action" to all aspectual functioning, including the pistic: aspirations and LWV.

8.2.3 Technological Determinism vs. Social Shaping of Technology

The two arms of the circular relationship have been treated by some as a dialectic involving technological determinism, which is the idea that technology defines its own path of development and also that of society, on one hand, and the approaches known as social construction of technology (SCOT) (Pinch & Bijker, 1987), actor-network theory (ANT) (Latour, 1987), and social shaping of technology (SST) (MacKenzie & Wajcman, 1999) on the other.

A useful discussion of this is found in Howcroft, Mitev, and Wilson (2004). SCOT is an interpretive approach, which argues that technology is socially constructed by "relevant social groups" within "technological frames". ANT criticises it for swapping social determinism for technological, and advocates treating human and non-human (e.g., computer system) "actants" as essentially the same insofar as organisational situations are concerned (a notion of "symmetry") and offers a set of conceptual tools for analysing such situations. Both SCOT and ANT are criticised by SST and Howcroft et al. in particular for ignoring social structures and taking an amoral and apolitical stance, especially in relation to the gendered content of IT (discussed later).

This is a topic the author is researching from a Dooyeweerdian viewpoint and hopes to discuss in more detail at a later date, but initial indications are that it can be addressed in terms of the Nature-Freedom ground-motive (§2.3.1.4)

A pattern may be detected that is similar to that seen in systems approaches by Eriksson (2006) in Chapter VI:

• Technological determinism, like hard systems thinking, adheres to the Nature pole.

• SCOT and ANT adhere to the Freedom pole, like soft systems thinking (Checkland, 1981).

• Social shaping of technology, like critical systems thinking, tries to integrate the two poles.

• (Also, we might see Schuurman's "liberating vision" as similar to multimodal or disclosive systems thinking in operating within the CFR ground-motive.)

But, as mentioned there, Dooyeweerd argued that the Nature and Freedom poles cannot be integrated under that ground-motive. So, to acknowledge insights in all these views, in a way that does not contain inner antinomy, requires shifting to a different ground-motive.

Neither pole of the NFGM offers satisfactory grounds for normativity and responsibility with regard to society's relationship with IT. The Nature pole either dismisses normativity as a category error in a determined universe or reduces normativity to control of human beings, by coercion and by reducing responsibility to the following of rules. The Freedom pole roots normativity and responsibility in the autonomous human ego and, to escape pure individualism, must resort to notions like socially constructed shared beliefs and values.

The issues encountered in this area are long-term, and so we need an approach to normativity and responsibility that will outlast currently fashionable paradigms. But we also need an approach that is usable in practice to help us take political and other decisions now for the future. While views centred on the NFGM might seem to allow us to discuss short-term normativity, none of these approaches provide any basis for this longer-term normativity which transcends society's current preferences.

Again, it is likely that we need to move to a different ground-motive. It is Dooyeweerd's espousal of the Creation-Fall-Redemption motive that has enabled him to escape these dichotomies, and adopting this ground-motive opens the way to considering the circular relationship aspectually.

8.2.4 The Nature of Both Relationships

To Dooyeweerd, human life, including social and societal life, is never absolutely determined. Therefore the impact that IT (its prevailing shape and uses) has on us is to be understood not as something that determines us absolutely, but as something we respond to.

How we respond to IT is diverse, a subject-functioning in a variety of aspects—we get angry (psychic), write letters to the newspapers (lingual), seek changes in the law (juridical), and some might be willing to sacrifice themselves (pistic). What we actually respond to is also governed by our ground-motive (usually NFGM) and by our aspectual life-and-world-view (LWV). If we adhere to the Nature pole, we might get angry at IT's failure to order life. If we adhere to the Freedom pole, we might get angry when our freedom is threatened by IT. But which particular freedom exercises us depends on which aspect(s) we value—religious freedom, emancipation from injustice, artistic freedom, economic freedom, social mobility, freedom of expression, and so forth. The influence of IT on society as such is explained by the fact that human response involves social and post-social aspects, and most of us will, much of the time, respond in the way the rest of society does. The most important aspect to be considered here is the pistic—our vision of who we are and in what we place trust—which is not only personal but very much societal.

Our LWVs are not static, especially when understood in terms of aspectual profile (§3.3.4), but can themselves be impacted, including by IT-as-*Umwelt*. Among many other influences on our LWVs, when IT benefits certain aspects of our lives, those aspects can increase in importance to us, while others decrease, changing our aspectual profiles.

But that IT itself is shaped by us, and becomes "inscribed" with our LWVs and ground-motives. Inscription with our LWVs can occur as a myriad of people, in whom certain aspectual profiles prevail, shape technology, develop artefacts, use those artefacts and believe something about the nature of computers, as discussed in Chapters IV to VII.

The influence of ground-motive is broader. Up to the 1980s, IT was shaped by the Nature pole of NFGM and its theme of control in all four areas:

- Its use was largely for controlling industrial or administrative processes.
- The machine nature of computers was emphasised in comparison with human beings.
- Its development methodologies were those that emphasised control, such as the waterfall model.
- The shape given to individual building blocks of technology assumed formal logic.

From the 1980s there began a dialectical reaction, a shift to the Freedom pole:

- Use of computers was increasingly aimed at freeing and empowering the individual, whether in business thinking, the home, or by means of computer games, with user interfaces that gave the feeling of greater freedom.
- Seeing humans as like machines became less fashionable.
- IS development became iterative and contingent.
- Types of technological building blocks developed were for the technologies of freedom, such as graphical user interfaces, virtual reality, the Internet, and the ability to cope with dynamic, uncertain information.

Thus two of the types of religious roots mentioned in §2.4.1 are shown to exercise a central, critical control over both areas of the circular relationship by which humanity shapes, and is shaped by, IT-as-*Umwelt*. A third, absolutization, is discussed later. But before that, a useful practical device will be outlined.

8.2.5 Practical Device: Aspectual Analysis of the Circular Relationship

Brandon and Lombardi (2005) used Dooyeweerd's suite of aspects as a means to defining what sustainability is (they restricted themselves to urban sustainability). They base their approach on the shalom principle (§3.4.3), which states that if we function in line with the laws of all aspects then things will go well and flourish, including socially over the longer term, because the repercussions of the later aspects are socially mediated and take a long time to reach their full impact. Therefore sustainability is not just physical (climate change), biological (pollution, depletion of rainforests), but also social (in how we relate to each other), economic, juridical, ethical, and pistic (for example in the morale and vision of society).

A similar approach is possible for understanding the "ecology" that is IT-as-*Umwelt*. If we function well in each aspect then the society is likely to prosper and flourish, and

Table 8-1. Aspectual inscription and societal impact

Aspect	Some assumptions inscribed on IT	Impact on our lifestyle (examples)
Formative	We want latest versions.	Those without latest versions cannot now do what they did before.
Lingual	GUI superior to CLI.	Marginalizes those who cannot use GUI.
Social	Virtual meetings are 'OK'.	Increased use of Internet for interaction.
Economic	We want convenience.	Have to upgrade regularly.
Aesthetic	Website appearance is more important than accessibility.	Accessing Internet becomes our lifestyle.
Juridical	Absolute right to prevent others using my ideas.	Legal structures altered to support IT.
Ethical	I want to share - the Open Source movement.	We accept minor problems and contribute to fixing them.
Pistic	Technology iwill always get better.	Technology-imposed restrictions are readily accepted.

lead to our flourishing in every way. But any negative (anti-normative) functioning in any aspect will jeopardise this. Therefore, both relationships may be understood as multi-aspectual in nature (Lombardi & Basden, 1997).

MacKenzie and Wajcman's (1999) examination of the inscription process supports this view. They uncovered "technical, economic, organizational, political, and even cultural aspects" (p. 11) and they drew attention to the impact that "obdurate physical reality" (p. 18) has on inscription. The first four are, of course, closely related to Dooyeweerd's formative, economic, social, and juridical aspects ("cultural" is ambiguous), and the "obduracy" speaks of the transcendence of the aspects (§3.1.4.1). They also mention "attitude", which might be ethical aspect, and desire to do something "monumental", which is probably the pistic aspect.

We can go beyond merely pointing out that our IT is inscribed by our assumptions and LWVs, to say that it is inscribed in each aspect. We can go beyond merely pointing out that IT-as-*Umwelt* impacts on our lives, to say it impacts in each aspect. Table 8-1 gives some examples of these, one beneficial, most detrimental. Such analysis must, of course, be deepened before it can truly inform research in this area, but it does at least indicate that aspectual analysis can remind us of aspects that might otherwise be overlooked, help us link disparate issues in new ways, and question many assumptions.

8.3 Absolutization and Idolatry in IT

Dooyeweerd traced many of the long-term problems in society to religious dysfunction. Those related to the dualistic ground-motive of Nature and Freedom (NFGM) and to life-and-world-views have been discussed earlier. A third is absolutization of that which is not absolute. The term "absolutization" itself will be used here mainly of aspects (law side), while absolutization of something on the subject side will be referred to as "idolatry", and here will apply especially to IT.

Absolutization of aspects leads us to deeply assume that only certain things are meaningful, important or Good, while other things are to be ignored, dismissed, or shunned. It is a societal rather than purely individual matter. Wherever reductionism or dialectic reaction hold sway in a community of thought and practice, we are likely to find aspects absolutized and others ignored. But other communities of thought arise which, though they might eventually absolutize some aspect, in their youth, draw attention to a wide range of aspects. Two examples will be presented of this type of critique of IT, from feminism and from a non-Western viewpoint. This is followed by a discussion of the idolatry of technology and a brief discussion of how religious dysfunction might be overcome.

8.3.1 Critique of "Masculine" Technology

Whereas Howcroft et al. (2004) draw attention to gender in IT, Adam's (1998) *Artificial Knowing: Gender and the Thinking Machine*, makes a detailed study of it. She shows how artificial intelligence (AI) is "inscribed" with masculinity, examines two "paradigm projects" in AI, Cyc and Soar, both of which seek to represent knowledge, and asks what a "feminine" artificial intelligence would look like.

A key indication of their "masculine" inscription is the propositional form knowledge takes in both of them. Both presuppose that all knowledge that is likely to be valuable or interesting can be represented in propositional form. Non-propositional knowledge, such as experience and skills-based knowledge, is either not considered for representation within this technology, or is assumed to be reducible to propositional form. But much of women's knowledge is exactly of this non-propositional type; propositional knowledge correlates with masculinity (though Adam makes clear that the correlation is not total). Not only is non-propositional knowledge deemed of lower status in a profession [e.g., "Male physicians who received medical training with no clinical component were authorized as more expert than the women midwives, despite the latters' considerable experience" (ibid, p. 114)] but the way AI technology has been designed mirrors this attitude. In this way, AI technology is inscribed with masculinity.

The reason why AI technology is like this, Adam suggests, is because it was originally given shape in research laboratories staffed by those assuming a "privileged, white, middle-class, male perspective" (p. 179). "Soar too is based on a set of experiments carried out on unrealistically bounded logico-mathematical problems carried out by a limited number of male college students in the 1960s and 1970s, with the assumption arising from this that their results can be extrapolated to apply to a wider domain of subjects and problem solving situations."

It is also inscribed with masculinity by virtue of its emphasis on the mind rather than the body. Artificial intelligence, as indeed the vast majority of information technology, processes concepts and other things of the mind, rather than the body. As many similar writers do, Adam traces this partly to Descartes:

The relation of the female to bodily things, and the male to the life of the mind, is further reinforced by Descartes' transformation of the relationship between reason and method and the radical separation of mind and body. The life of the body is seen as inimical to the life of mind and reason. The hand that rocks the cradle is unlikely to be adept at the highest cognitive exercise of the mind, namely the mathematical and natural sciences. Descartes himself seemingly had no wish to exclude women from his method. What happened rather, was the crystallizing of a number of already existing contrasts in the mind-body problem, where the soul was identified with the rational mind and the non-rational was no longer part of the soul but belonged to the body. (Adam, 1998, p. 102)

Could a body-centred AI take the form of robotics? Adam discusses this but points out that a robot may be

... physically situated but it remains to be seen whether it can be culturally situated in the appropriate sense. The embodiment that such robots possess is of a rather limited form. Their wanderings in the world, removing drinks cans, finding the centres of rooms and so on is rather aimless. To paraphrase a popular saying, we might suggest that they "get an A-life". They might find more of a purpose to artificial life if they could learn to love each other, to care for and look after one another, or indeed look after us. In other words they could take on the forms of embodiment more usually associated with women's lives, i.e., the looking after and caring for bodies, young and old. (p. 180)

She comments, "Looking at feminist visions of the future through intelligent technologies, the situation reveals some tensions. ... Feminist readings of popular cyberculture are ambivalent. It seems unlikely that the promise of Haraway's (1991b) earlier rendering of cyborg imagery can be realized through current manifestations of cyberfeminism."

"However," Adam continues, at the end of her book when we might expect her to be indicating her belief about the way forward (p. 181), "further research on women's *use* of computing technology at least offers the hope of alternative, more promising readings" (my emphasis). The inscription of technology is not only in the shape it assumes but in the types of ways it becomes widely used in everyday living. We can also detect an emphasis on use more than on shape in earlier such statements as "to love each other, to care for and look after one another, or indeed look after us."

What is particularly damning is that (p. 179) "an assumption that this does not even have to be made explicit is a way of silencing other perspectives." In other words, inscription of IT occurs because of the perspectives held, usually without realising it, by those who shape the technology and those who define our expectations of use. These include the community of research and practice which concerns itself with technology shaping (Chapter VII), that involved in ISD and use (Chapters VI, IV) and even those who lead us in understanding the nature of computers and information (Chapter V). These communities will hold various perspectives, often elevating different aspects to importance (see §3.3.4), but at a deeper level, these are all underlain by society's presuppositions.

8.3.1.1 A Dooyeweerdian View

Early feminist thought assumed it had correctly diagnosed the problem as a masculine-feminine opposition and proposed solutions based thereon, but more recent feminist thought is more nuanced about such claims. Whatever its own self-awareness, feminism may be seen from this perspective as highlighting two deeply-rooted types of religious problem, and trying in its own way to propose solutions.

One is dualistic ground-motives, which incite dialectic reaction of a religious kind. The predominant ground-motive is NFGM, and early feminists especially reacted within it against the Nature pole of control, to which early societal conception of IT assumed that the main use of computers is to control industrial and administrative processes, and the main view of computers was as machine (see earlier). But the dialectical swing from Nature to Freedom is not unique to feminism (as has been seen in the discussion in previous chapters).

Where feminism makes a more specific, and arguably more valuable, contribution is in exposing a residual presupposition of the Form-Matter ground-motive. Feminist thought dislikes the traditional emphasis on mind and rationality. Though they trace its roots to Descartes, we might see the problem as more deeply rooted, in the presumed opposition of form and matter under the FMGM (§2.3.1.1), since it more adequately accounts for the solution feminism offers: a dialectical reaction in favour of body relative to mind.

The other religious problem is absolutization of certain aspects. [Adam (1998, p. 180) even speaks of "aspects".] The "masculine" world view may be characterised as absolutizing the economic, analytic (scientific), and formative (technical) aspects in particular. "Female" thinking is more diverse, she says. Instead of theoretical or analytical thinking, intuition is emphasised, by which is meant diverse other ways of knowing which include bodily knowing, sensitive knowing, those involved in love and caring, and even aesthetical knowing. Feminism's stress on "non-Cartesian" ways of knowing may be seen as a call for reinstatement of various other ways of knowing.

This clearly echoes the Dooyeweerdian notion of multi-aspectual ways of knowing outlined in §3.3.3. To Dooyeweerd, the "Cartesian" way of knowing involves an analytic *Gegenstand* relationship (§3.3.7), whereas to truly know a thing, event, and so forth, involves engaging closely with it in a law-subject-object relationship in every aspect. In many ways, with his multiple spheres of meaning, multiple ways of knowing, multiple rationalities, and multiple normativities, Dooyeweerd prefigured what is found in feminist writing of today. A Dooyeweerdian point of view would see this not so much as "male" thinking, as that certain aspectual ways of knowing have been unduly elevated (namely, the analytical and lingual) while others, such as ethical, aesthetic, sensitive, have been suppressed.

Thus Dooyeweerd not only supports this "embodied" knowing, but provides a way of celebrating its diversity. Now that the message, that Cartesian or "masculine" modes of thinking are not the only important modes, is well understood (at least in that community of thinkers), it might now be time to appeal to a Dooyeweerdian view in order to contribute to feminism's critique by recognising and exploring the diversity of non-Cartesian ways of knowing. Dooyeweerd might also help feminism avoid becoming misled into the same error of which it accuses "masculine" thinking, of absolutizing its own preferred aspects and of adhering to a particular pole of each of the FMGM and NFGM ground-motives.

Dooyeweerd's stress on the importance of the ethical aspect of self-giving might resonate with much in feminist thought but allow that all human beings might partake of it. As mentioned in §4.4.3.1, the ethical is one aspect which has been downplayed consistently in Western life and thinking, but which is still important in non-Western life, which will be briefly considered next.

8.3.2 Critique of Western Technology

Information technology might have potential to help developing peoples "leapfrog" into the 21st century. But Pacey (1996) shows how technology, as we experience it today and especially as it was experienced in nations of the two-thirds world, is "inscribed" with Western assumptions and values. He shows how information technology is seldom a blessing to those living in the two-thirds world, because its introduction has been informed by Western assumptions and values. He identifies five component assumptions (Pacey, 1996, p. 152):

a.   "Assumptions based on academic specialisms and on boundaries between professions;

b.   The assumption that traditional communities outside the industrialized world have no technology of their own;

c.   A tendency to overlook opportunities for detailed improvements in maintenance and use and to go for technical fixes;

d.   Failure to recognize the invisible organizational aspects of technology invariably developed by users of equipment;

e.   Failure to recognize the conflicts of values and social goals which specific technological projects may entail."

The most common response to this is to call the imbalance between the rich and two-thirds worlds a digital divide, and deplore it. But that response presupposes that Western IT is normative for the two-thirds world. Dooyeweerd can help us critically explore such presuppositions. To illustrate this, a response is made to each of Pacey's components.

... (a) assumptions based on academic specialisms and on boundaries between professions;

Would Pacey recommend that we ignore the specialisms? If we are to alter the Western inscription of IT we must understand the nature of specialisms. Academic specialisms arise from scientific higher abstraction (§3.3.7), and Dooyeweerd believed that this is a proper task for humanity as long as no aspect is absolutized. The root of this problem is not the specialisms as such, but the absolutizations that are rife in Western thought. Many aspects are not only ignored but actively denied any importance at all in business and public life.

... (b) the assumption that traditional communities outside the industrialized world have no technology of their own;

The root of the problem is our arrogance that only the "latest" technology is "best" and an idolatry of technology (see later) shared by decision-makers in West and non-West alike. To Dooyeweerd, technology is a valid outcome of humanity's mandate to open up the formative aspect and thereby open up other aspects (see §3.4.6), and a major problem is when the development, adoption, and use of technology is no longer guided by the other aspects. Moreover, while one might *observe* a difference between modern technology, which depends on infrastructure, and technology in the pre-modern style, which does not, there is no *normative requirement* that makes the former superior to the latter. Thus Dooyeweerd, while supporting technology, provides a philosophical and normative basis for breaking this assumption.

... (c) a tendency to overlook opportunities for detailed improvements in maintenance and use and to go for technical fixes;

This can be attributed to two causes in combination: ignoring the multi-aspectual nature of everyday life (§3.4.5), in which many possibilities are presented to us for creative and harmonious activity, and absolutizing the formative aspect of technical functioning. Frequently, in addition, this tendency is accompanied by some personal or group agenda.

... (d) failure to recognize the invisible organizational aspects of technology invariably developed by users of equipment;

Such "invisible organizational aspects ... developed by users" are the aspects of HLC, which is multi-aspectual (Chapter IV). Pacey's concern reflects that part of Schuurman's "liberating vision" which says that IT should be led by the norms of other aspects. But Dooyeweerd can take us further because he makes the "invisible" visible and because he urges us to avoid collapsing the post-social aspects into the social. For example the pistic aspect of religion, which is so important in most of the world, has, until recently perhaps, been culpably overlooked by many Western governments, donors, and businesses.

... (e) failure to recognize the conflicts of values and social goals which specific technological projects may entail.

Participants in and stakeholders of a project all function pistically in being committed to their visions. Because these are frequently centred on different spheres of law-and-meaning (aspects), which imply irreducibly distinct spheres of values, many conflicts cannot be resolved by rational discussion alone. The Western, modern world-view elevates (even absolutizes) the economic, formative, and analytic aspects, while some non-Western world-views elevate the pistic, ethical, and social aspects. Thus conflict is most probable. As discussed in §4.4.3.1 a major way to overcome this is for all to function well in the ethical aspect of self-giving, generous-heartedness, and willingness to acknowledge error, be hurt, or be taken advantage of (example: public repentance and forgiveness, which characterised the transition in South Africa). If this insight is correct, then both competition, so universally applauded by the West, and the avoidance of personal risk, if taken too far, are anti-normative in this aspect.

Thus we see that Dooyeweerd can be used to understand more closely the assumptions that Pacey decries, especially by reference to what aspects are elevated or absolutized in the Western, modern world-view.

8.3.3 Orchestrating the Aspects

The critique of "masculine" technology by Adam and of Western technology by Pacey cannot, of course, be taken as fully representative but they are not untypical of their communities of thought. One common factor is plain in both critiques: some aspects have been elevated and even absolutized—notably the analytic, formative, and economic—and others woefully disregarded—notably the ethical. It is this aspectual imbalance that has "inscribed" our IT.

It might be no coincidence that both the type of feminism that Adam endorses, which she calls eco-feminism, and many in the development and anti-globalisation movements today align themselves with the environmental sustainability movement in which Brandon and Lombardi (2005) define sustainability in terms of shalom. These three movements all seem to point to the crucial importance, in our day, of the shalom principle in society.

It might also be no coincidence that this has also been a recurrent theme in our understanding of the other areas of research and practice in IS. Chapter VII proposed that each and every aspect deserves a technology that is shaped to its law-framework, Chapter IV proposed that in computer use we should consider every aspect, and Chapter VI likewise. Chapter V recognised the importance of every aspect in understanding the nature of computers.

This need not mean, however, a kind of vanilla-flavoured aspectual democracy in which all aspects are equal. Instead, there should be inter-aspectual harmony, like that experienced in music, which is never static and never absolute. We human beings, functioning socially, pistically, and so on in society, always and continuously have the responsibility of not just maintaining, but orchestrating the aspects of a situation. Therefore we, humankind, are called to "play a symphony" on information technology, so that in one decade or culture, one aspect may be more clearly heard, and in the next, another, but always such that every aspect may be heard as part of the long-term music. One thing that completely destroys any chance of this is idolatry.

8.3.4 Idolatry of Technology

Ellul, according to Schuurman (1980, p. 145), speaks about "worship" of technology. Shallis (1984) writes about the "silicon idol". Noble (1997) speaks of the "religion of technology". Humankind seems to treat technology as an object of religious functioning.

Noble finds rituals and myths in our attitudes to technology, which he argues are detrimental. His vision for technology is as something rational and empowering that even emancipates us from such religious attitudes, and therein he finds a paradox. From a Dooyeweerdian perspective, however, Noble's discussion seems limited. His view of what constitutes religion is rather narrow, being centred on the first of the types of religious root listed in §2.4.1. But what is perhaps of greater importance is that Noble's view can only offer a visionary direction and cannot offer a basis for understanding "the religion of technology" in a way that is useful in research and practice.

Using Dooyeweerdian thought, Goudzwaard (1984) and Walsh and Middleton (1984) have developed a notion of idolatry that is quite serviceable in understanding movements of belief in everyday life in society at large. What aspectual absolutization is to theoretical thought, so idolatry may be to everyday living. Goudzwaard (1984, p. 21) has characterised idolatry with:

Suppose we consider the worship of a wood, stone or porcelain image, a practice still common in the world today. This worship has several steps. First, people sever something from their immediate environment, refashion it and erect it on its own feet in a special place. Second, they ritually consecrate it and kneel before it, seeing it as a thing which has life in itself. Third, they bring sacrifices and look to the idol for advice and direction. In short, they worship it. Worship brings with it a decrease in their own power; now the god reveals how they should live and act. And fourth, they expect the god to repay their reverence, obedience and sacrifices with health, security, prosperity and happiness. They give the idol permission to demand and receive whatever it desires, even if it includes animal or human life, because they see the idol as their savior, as the one who can make life whole and bring blessing.

Not only a physical thing, but any type of thing, can be an idol, even an idea or cause. Goudzwaard then shows how this fits our attitude to technology. From this passage and other writings by Goudzwaard and others (e.g., Walsh & Middleton, 1984), we can summarise that an idol:

• Is set apart in privileged place, is given special esteem;
• Determines the meaning of all else;
• Determines the value of everything, and what people aspire to;
• Determines whether a thing exists or is destroyed;

- Directs people's lives, and reduces their freedom;
- Has things sacrificed to it, or for it;
- Is protected at all costs;
- Is willingly submitted to;
- Is never questioned, and questioners are deemed heretics;
- Often delivers the opposite of what it promises.

The characteristics of an idol may be applied to IT in the following manner.

- **Set apart in privileged place, is given special esteem:** Governments provide large budgets and concessions for furthering IT, which are denied to other areas of human endeavour, and we, the public, aspire to obtain technological goods.

- **Determines the meaning of all else:** Human activity is often deemed meaningless or boring if not technicised.

- **Determines the value of everything, and what people aspire to:** All activities of everyday life—of business, shopping, education, fun and even worship—are deemed superior if they employ IT.

- **Determines whether a thing exists or is destroyed:** We seem happy to accede to environmental destruction that results from our use of technology, including IT.

- **Directs people's lives, and reduces their freedom:** Our life becomes shaped by IT: social life by mobile phones, business life by e-commerce, talks and lectures, in many places, by restriction to using only computer with beamer, even when these things are inappropriate.

- **Has things sacrificed to it, or for it:** The freedom, life-choices and even lifestyle of those without (certain forms of) IT is curtailed, usually unwittingly, but when this is pointed out politicians and planners often deem it a sacrifice worth making.

- **Is protected at all costs:** Large research and other budgets are directed at IT.

- **Is willingly submitted to:** We patiently endure the time it takes to download large e-mails, and feel inferior if we complain, and we no longer mind if e-mail consumes 10 hours per week.

- **Is never questioned, and questioners are deemed heretics:** It is extremely difficult to obtain a serious critique of whether new IT is appropriate or justified; the debate, if any, is usually restricted to what kind of IT to adopt.

- **Often delivers the opposite of what it promises:** The "paperless office", "convenient" email, and so forth.

8.3.5 Overcoming Religious Dysfunction

Thus three of the types of religious roots listed in §2.4.1 are manifest in our ecological relationship with IT (all four if we include Noble's). Some might share Noble's view that a religious attitude to technology is a wholly negative thing. But Dooyeweerd would question that. Schuurman's (1980) "liberating vision for technology" is at root a religious vision and would lose its central motivation and enabling force, and would dissolve into nothing more than a dream, if robbed of its religious aspect. The problem is not religion as such, but religious dysfunction. If we allow that religion can bring immense good as well as harm, we must face the question of how to overcome religious dysfunction, especially in relation to IT.

The different types of religious roots of our attitude to IT each give rise to different types of dysfunction and demand different remedies. Elevation of aspects is no dysfunction as long as all other aspects are taken into account. Absolutization of aspects is what is dysfunctional. "Such absolutizations," remarked Dooyeweerd (1984, III, p. 161) "cannot be corrected by other absolutizations." It may however be ameliorated by abandoning absolutization altogether, that is, by deliberately recognising this and determining to take all the other aspects into account as non-reducible to the absolutized one, and to explore the inter-aspect relationships. To abandon absolutization sustainably often requires structural changes.

Idolatry of a subject-side thing like IT may be reduced by recognising that the idol cannot be absolute in either importance or value to us and should be held in proper perspective. Frequently, however, this can only occur when a few brave souls recognise the idolatry and sacrificially work to overcome it: sacrifice of time, effort, reputation, income, or anything else. The author has found that a sense of responsibility to the Creator can help enormously here.

A dualistic ground-motive leads us to locate Evil in the structure of the cosmos itself (see examples of ground-motives in Chapter II). Under such a presupposition, the remedy for Evil, and the appropriation of Good, is to be found in eschewing one half of the cosmos and adhering to the other. This is, arguably, what drives what at first seems to be a duality into a dualism because human beings have an innate sense of Good and Evil, however it is labelled. The remedy for adhering to one pole of a dualistic ground-motive is not to be found in dialectical switching to the opposing pole, as occurred in the interpretivist reaction against positivism in IS. Attempts might be made to think together the two poles, as Hegel did, but the results are seldom sustainable. If Eriksson (2006) is correct in Chapter VI, in seeing critical systems thinking as attempting that very feat, then it is unlikely to remain true to its aspirations of combining an interpretive stance with a transcending normativity. An implication from Dooyeweerd is that the only answer to the problems rooted in one ground-motive is to shift to another. Five hundred years ago such a shift was made once it had become clear that the problems resulting from the NGGM could not be resolved in its own terms of the opposing of secular and sacred, and the NFGM emerged. Five hundred years later, we have discovered many of the inner problems of NFGM itself and perhaps the time has come to shift to yet another ground-motive. Dooyeweerd suggests the CFR. Much of the discussion in this book may be seen as an exploration of the philosophical implications for IS of Creation, Fall, and Redemption, that is, transplanting our research and practice to the fertile soil of CFR.

The Creation-Fall-Redemption motive is non-polar. Creation implies that there is irreducible diversity and coherence, and that the whole of the cosmos is, at root, Good, and nothing is to be despised nor eschewed. Fall implies that Evil, the root of problems, lies not in one or other polar halves of reality (e.g., control or freedom as such) but in us. Evil is located elsewhere, namely in the supra-temporal self (§3.3.6) when it is orientated, not toward the True Absolute, but toward some pretend, false absolute, which becomes for us either our idol or our absolutization.

This is the third type of religious dysfunction, absolutization. All distorted intuition, false perspectives, harmful functioning in practice, detrimental repercussions in all aspects, distorted research questions, distorted quality criteria, and so on, follows from false orientation.

The remedy for religious dysfunction must be religious in nature. A change of thinking might change the solutions we attempt to problems, but will not change the underlying direction humanity is taking. Better economic circumstances simply results in more money to resource what we are already doing. Even a change of attitude, while important, is unlikely to be sustainable. The remedy, whatever other aspects it involves, must also be religious, a re-orientation of the self toward the True Absolute and a turning away from false ones. If Dooyeweerd is correct, then only the Divine Origin (God) is such an Absolute, and all others will eventually let us down. If this is so, then it is important to know how to shift our religious standpoint: a "conversion". Conversion is a re-orientation of the self toward a different Absolute, and is beyond rational assent; Lonergan (1992) speaks of three-fold conversion: religious, moral, and rational. The word "redemption" implies we need God's help to remedy this. Coming from a Dutch Calvinian tradition, Dooyeweerd believed that, ultimately, only God can effect such a change, but other religious (including Christian) traditions believe we might have some agency in this, by means of re-pentance. Which is valid is to be discussed in theology, not here.

Thinking philosophically rather than theologically, if the self is trans-aspectual and supra-temporal, then we can never know its dynamics nor present a theory about re-orientation of the self. We can however listen to people's stories of their experi-ence of this change, even though they are overladen with cultural manifestations and expectations. That stories are important is one more reason why a lifeworld approach is so important in both research and practice in any area of information systems.

8.4 Conclusion

8.4.1 Overview of Framework for Understanding

The starting point in this area was that a different relationship holds between humanity and information (and communication) technology: "inside" the technology rather than, as in the four previous areas, "outside" it. This invites what is sometimes seen as a broader, societal view, which considers humanity as a whole and IT as a whole, over a long timescale. The everyday experience of this must therefore take a different form. Three main questions were addressed, each of which contribute to understanding this area and its research and practice.

The first question is the validity of IT as a human endeavour: should we resist or accede to its continued development? Schuurman (1980) has already undertaken a comprehensive study of this question from three points of view: those generally optimistic, those generally pessimistic, and Dooyeweerd. He appealed to Dooyeweerd's theory of history and progress to propose a "liberating vision for technology", for which the main principle is, for us:

- IT may be seen as part of humanity's long-term opening-up of the lingual aspect, disclosing and developing its potential for blessing.

But the norm that should guide this is not that of lingual aspect of IT, nor even the formative aspect of technology generally, but those of all the other aspects:

- IT should be developed in such a way as to serve all the other aspects; it will thus bring blessing to the cosmos.

The second question concerns the nature of the relationship by which humanity is "inside" IT rather than outside it:

- The relationship between humanity and IT is that of *Umwelt* and correlative enkapsis, in which we shape IT, and it shapes the way we live and see ourselves.

This accommodates and integrates the insights from both technological determinism and the social shaping of technology, which are usually assumed to be dialectically opposed. It allows closer inspection of the dynamics and quality of both arms of this relationship:

• This relationship has two religious roots (§2.4.1): religious ground-motive and the deep commitment that is a life-and-world-view; these are both inscribed into the IT that humanity generates, and both affect how we respond to living "inside" IT, including how we change our life-and-world-views themselves.

A practical device was offered: life-and-world-views operative in both arms of the relationship may be analysed to obtain aspectual profiles (§3.3.4), and the shalom principle (§3.4.3) may be used to guide what we do.

The third question concerns what has gone wrong, or could go wrong: what could disrupt Schuurman's "liberating vision for technology" and the ecological relationship? The answer is religious dysfunction. Critiques from feminism and a non-Western viewpoint highlighted this. The two religious roots discussed earlier are involved. But a third is the deepest:

• Religious absolutization of IT or some other factor leads to idolatry, which paradoxically enslaves humanity and delivers the opposite of what it promises.

The solution to religious dysfunction must be religious in nature: the putting away of absolutization of aspects and repentance from idolatry and a re-orientation of the self toward the True Absolute.

The discussion in this chapter is perhaps more limited than in the others, in that it omits much of the extant discourse in this area. Nevertheless, the three main themes that have been interpreted from a Dooyeweerdian point of view can make a useful contribution. The benefits of this framework are that it provides better strategic direction because it provides critical understanding of the place of IT as part of modern society and life-and-world-view, whether Western or non-Western. It provides hope for IT and yet a fulfilling challenge. It overcomes the dialectic between the opposed positions of technological determinism and social shaping of technology, allowing the insights afforded by both to be integrated into a single view. It strikes at the very root of the problem that afflicts our society with regard to IT.

8.4.2 The Challenge

"The choice is ours," Leer (2000, p. 169) concludes. "Whatever course and decisions we choose to take, future generations will live with the consequences. Let's hope wisdom will prevail and that our legacy will be a good one, for never before has humankind had such a great opportunity to further democracy and build a truly global village." We might argue about the precise form for her vision, but the question that most needs posing is: On what may she base her hope?

The understanding developed in this chapter goes some way toward "wisdom", especially Schuurman's "liberating vision for technology". But if we look over the whole of our discussion of how we might use Dooyeweerd to help us construct frameworks for understanding the five areas of research and practice in IS that we have looked at, Leer's use of the word "wisdom" would seem appropriate to all of them. Dooyeweerd does in fact allow us to define "wisdom" in a way that is useful in both research and practice: in terms of shalom, or harmonious functioning in all aspects. This has been a major part of the theme in every area.

We can go a long way with that. But if we want to follow the path that Dooyeweerd has begun to beat all the way then perhaps we need to consider afresh the religious stance that he adopted but which has been rejected from theoretical thinking for 500 years and, arguably, for 1500 years. Interwoven into his philosophy is the Biblical (I do not say "Christian" though he does) notion of creation, fall and redemption. What God has created is diverse and good. It is we who have turned away from God. Yet, Dooyeweerd believed, God has acted to redeem both humankind and indeed the whole of creation by coming into the world in the person of Jesus Christ, in whom, at the end, all creation, and all our technological activity will find its completion and destiny, which is to be released to function as it was intended to in a completely renewed cosmos. But this vision, which sustained Dooyeweerd, and also sustains this author, is not to be completed this side of the eschaton.

References

Adam, A. (1998). *Artificial knowing: Gender and the thinking machine*. London: Routledge.

Brandon, P. S., & Lombardi, P. (2005). *Evaluating sustainable development in the built environment*. Oxford, UK: Blackwell Science.

Castells, M. (1996). *The information age: Economy, society and culture, volume 1: The rise of the network society*. Oxford, UK: Blackwells.

Checkland, P. (1981). *Systems thinking, systems practice*. New York: Wiley.

Dooyeweerd, H. (1984). *A new critique of theoretical thought* (Vols. 1-4). Jordan Station, Ontario, Canada: Paideia Press. (Original work published 1953-1958)

Eriksson, D. M. (2006). Normative sources of systems thinking: An inquiry into religious ground-motives of systems thinking paradigms. In S. Strijbos & A. Basden (Eds.), *In search of an integrative vision for technology: Interdisciplinary studies in information systems* (pp. 217-232). New York: Springer.

Giddens, A. (1993). *New rules of sociological method.* Cambridge, UK: Polity Press.

Goudzwaard, B. (1984). *Idols of our time.* Downers Grove, IL: Inter-Varsity Press.

Habermas, J. (1987). *The theory of communicative action volume two: The critique of functionalist reason* (T. McCarthy, Trans.). Cambridge, UK: Polity Press.

Howcroft, D., Mitev, N., & Wilson, M. (2004). What we may learn from the social shaping of technology approach. In J. Mingers & L. Willcocks (Eds.), *Social theory and philosophy for information systems.* Chichester, UK: Wiley.

Latour, B. (1987). *Science in action.* Cambridge, MA: Harvard University Press.

Leer, A. (2000). *Welcome to the wired world.* London: Pearson Education.

Lombardi, P., & Basden, A. (1997). Environmental sustainability and information systems: The similarity. *Systems Practice, 10*(4), 473-489.

Lonergan, B. (1992). *Insight: A study of human understanding.* Toronto, Ontario, Canada: University of Toronto Press.

Lyon, D. (1988). *The information society: Issues and illusions.* Cambridge, UK: Polity Press.

MacKenzie, D., & Wajcman, J. (1999). *The social shaping of technology* (2nd ed.). Milton Keynes, UK: Open University Press.

Noble, D. F. (1997). *The religion of technology: The divinity of man and the spirit of invention.* New York: Alfred A. Knopf.

Pacey, A. (1996). *The culture of technology.* Cambridge, MA: MIT Press.

Pinch, T. J., & Bijker, W. E. (1987). The social construction of facts and artifacts: Or how the sociology of science and the sociology of technology might benefit each other. In W. E. Bijker, P. T. Hughes, & T. J. Pinch, T. (Eds.), *The social construction of technological systems: New directions in the sociology and history of technology* (pp. 17-50). Cambridge, MA: MIT Press.

Schuurman, E. (1980). *Technology and the future: A philosophical challenge.* Toronto, Ontario, Canada: Wedge.

Schuurman, E. (2003). *Faith and hope in technology* (J. Vriend, Trans.). Toronto, Ontario, Canada: Clements.

Shallis, M. (1984). *The silicon idol.* Oxford, UK: Oxford University Press.

Strijbos, S. (2006). Towards a "disclosive systems thinking." In S. Strijbos & A. Basden (Eds.), *In search of an integrative vision for technology: Interdisciplinary studies in information systems* (pp. 235-256). New York: Springer.

Walsh, B. J., & Middleton, J. R. (1984). *The transforming vision: Shaping a Christian world view.* Downers Grove, IL: IVP.

Walsham, G. (2001). *Making a world of difference: IT in a global context.* Chichester, UK: Wiley.

Section III
Discussion and Conclusion

Chapter IX

Reflections

What *is* information technology (IT)? The discussions in Chapters IV to VIII have come up with not one answer to this question, but five:

- IT is an artefact we use in daily life.
- IT is a multi-aspectual meaningful whole qualified by the lingual aspect.
- IT is a representation of what we want to use it for.
- IT is an implementation of the diversity of cosmic meaning.
- IT is an *Umwelt* in which we live and for which humanity is responsible.

And there might be others that have yet to be explored. These answers are philo-sophically sound, insofar as the discussions have been sound.

The reason we have been content to emerge with five answers rather than trying to find a single over-arching one—or, contrariwise, to sink into a morass of subjectivism that disallows any answer to an "is" question from being substantive—is because we have employed the radical philosophical approach of Herman Dooyeweerd. As mentioned at the start of Chapter II, Dooyeweerd was an unusual philosopher in that, rather than merely trying to formulate philosophical theory, he tried to "clear away" what keeps us from seeing the structure of reality as it presents itself to us in the everyday attitude.

Especially he examined and tried to clear away impediments generated by presuppositions of Western thinking. One such presupposition, 2,500 years old, is that "is" (Being) is unproblematic, prior to all other questions and may be met with a single answer (see §2.4.3, §3.2.2). Another is the more recent, post-Kantian, reaction against it, which treats "is" as meaningless and to be reduced to "I think it is", so no substantive answer to "What is?" can be given. By contrast, Dooyeweerd gave priority to Meaning (cosmic) by which both Being and Thinking or Knowing are enabled to occur. Since Meaning is diverse, a variety of substantive answers may be expected to the question "What is IT?"

Dooyeweerd's approach and portions of his positive philosophy have been explained and then applied reasonably systematically to generate frameworks for understanding five areas of research and practice in information systems. This chapter reflects on this exercise and draws conclusions. It provides an overview of the frameworks developed in Chapters IV to VIII (of particular interest to researchers and practitioners in IT/IS), reviews how the issues raised in the Introduction have been addressed, discusses how Dooyeweerd's philosophy has been used and might itself be further developed (of interest to philosophers and Dooyeweerdian scholars), reflects on the process of our exploration, including its contribution and limitations, and briefly suggests directions for the future.

9.1   Overview of the Frameworks

The frameworks formulated for understanding five areas of research and practice in information systems may be summarised as:

9.1.1   Framework for Understanding Human Use of Computers

Human use of computers (HUC) is seen as multi-aspectual human functioning. The first implication of this is that none of the several mono-aspectual views on offer can do justice to HUC. Though they might offer insights related to their specific aspects, they cannot provide an understanding that is sensitive to everyday experience. HUC was seen, from an everyday attitude, to consist of several multi-aspectual functionings interwoven in an enkaptic relationship, and this intuition was shown to be valid for three of them by being able to distinguish different qualifying aspects:

- HCI, human-computer interaction, is qualified by the lingual aspect.
- ERC, engagement with represented content, is variably qualified, by the aspect which reflects the main purpose of the IS.
- HLC, human living with computers, is also variably qualified, by its meaning in use.

Each has a different structure and normativity.

The structure of HCI is a Dooyeweerdian law-subject-object relationship. In all post-physical aspects the human subject-functioning is matched by object-functioning of the user interface, while a subject-subject relationship exists in the physical aspect, which allows interaction and guarantees reliable functioning of the computer. This view reverses the traditional view in HCI, which prioritised the distal relationship of "dialogue" between human and computer and treated the proximal relationship, whereby the computer becomes almost part of the user, as an advanced, novel form. The Dooyeweerdian law-subject-object relationship allows us to see proximality as the norm, with the distal relationship as specialised, arising from a *Gegenstand* relationship. Winograd and Flores' (1986) attempt to effect a similar reversal by advocating a philosophical shift from Descartes to Heidegger was discussed.

The central norm of HCI is the lingual norm of understandability and usability, but this should never be absolutized and should be supported by the norms of all other aspects. Attention to all aspects can guide the design of user interfaces and to critique design guidelines.

The structure of ERC may be understood as the lingual aspect (which qualifies HCI) reaching out to all spheres of meaning of the domain with which the user will engage. The main norm of ERC is that what is represented should do justice to the domain meaning, whether virtual or modelled, by diversity, richness, trustworthiness, flexibility, and fitting the user. This provides a useful way of evaluating the quality of the virtual world like a game, because it provides a philosophical way of differentiating virtual from real and it was noted that virtual worlds feel more real if the law side remains intact.

The main structure of HLC may be understood as aspectual repercussion:

- Unexpected impacts are analysable by reference to cosmic law because it transcends us;
- Indirect impacts, by aspectual crossover;
- Long-term impacts by the response time of later aspects;

- Social impacts, by attention to the social aspects;
- Societal impacts by especial reference to the pistic aspect (which links to the area of technological ecology).

The main norm of HLC is shalom: the IS should enhance human living in various aspects and harm it in none. It should also maintain or restore a proper aspectual balance therein. This provides a non-reductionist, non-functionalist way of assessing success or failure of IS use that can cope with a mix of benefits and detrimental impact across all stakeholders, and can rectify distortions in the status quo. Apparent conflict between aspectual norms was discussed and the importance of the ethical aspect of self-giving was underlined. It was demonstrated how an extant FFU could be appreciated and yet critiqued and enriched.

This FFU not only provides a basis for integrating HCI, ERC, and HLC, but also provides links with other areas, from HCI to the nature of computers, from ERC to ISD and technological resources, and from HLC to technological ecology. Several practical devices were demonstrated, including several versions of aspectual analysis, aspectual checklist and the "fir tree" that gives an aspectual profile of IS use.

9.1.2 Framework for Understanding the Nature of Computers

The nature of computers is to be understood by reference to human beings. The computer functions as object, not subject, in all but the physical aspect. The being of a computer is multi-aspectual, multi-levelled; a computer is a meaningful whole constituted of a number of aspectual beings, in which the lingual aspect is key. While a part-whole relationship may be found among beings within each aspect, the aspectual beings of different aspects are bound together in the whole by foundational enkapsis, in which inter-aspect dependency plays an important part.

This enables researchers and practitioners to account for the diversity of ways computers are experienced in everyday life. It provides a sound basis for understanding the ontic status of the innards as what is "in" the computer, even though we cannot directly experience the post-psychic aspects with our senses.

A discussion of Newell's (1982) theory of levels revealed it as very like the framework developed here in many respects, and led to the conclusion that Newell was reaching for what Dooyeweerd offers. Dooyeweerd's philosophy can underpin and enrich his theory.

Dooyeweerd's notion of enkapsis was particularly helpful, especially for inter-aspect relationships. The difference between data, information, and knowledge, is also usefully understood in terms of aspects. The contrary notions of cyberspace as a reality of pure mind, and the feminist notion of embodied knowledge, may be placed in relation to each other within such an aspectual framework.

The nature of a program, as idea, script and as running may be understood in two ways, both as a law side to a virtual world that "exists" and "occurs" when this law side is activated, and also as performance art like music, which Dooyeweerd discussed at length.

Whether computers can "think", and so forth, depends on whether we see it in terms of its subject- or object-functioning. This clarifies discussion of what a computer actually "is", and also sheds new light into Searle's Chinese Room.

9.1.3 Framework for Understanding IS Development

IS development involves development of computer systems or artefacts for human use together with development of the human context of use. In ISD extant FFUs are well-developed and there has been considerable philosophical debate, especially stimulated by the objectivism-subjectivism dialectic. Analysis of a number of these indicated that a Dooyeweerdian approach might make a useful contribution. It is also ISD to which Dooyeweerd has already been most widely applied by others.

Under Dooyeweerd, ISD may be seen, not primarily as technical activity, but, like IS use, as multi-aspectual human functioning. But, unlike HUC, here aspectual normativity is directed to the future, and thus has a guiding role. Several different, but enkaptically-interwoven, multi-aspectual functionings may be distinguished, of which four were discussed.

The overall ISD project is to be guided by the shalom principle, but the aesthetic and social aspects are key, the aesthetic aspect in its focus on harmony to achieve a coherent project being its qualifying aspect, and the social because ISD is teamwork. Dooyeweerd's theory of social institutions is helpful in establishing the appropriate place for, and limits of, power-relationships. But these are by no means the only important norms in ISD, some of which come from other multi-aspectual functionings. The lingual and juridical aspects lead us to other functionings.

Anticipating use is qualified by the juridical aspect of responsibility for all outcomes in future use, and aspectual normativity as possibility is a key insight.

Creating the IS (both artefact and context) is likewise multi-aspectual functioning, and the analytic to economic aspects thereof have been long recognised. Less recognised are the four latest aspects, which can make it a delight rather than a chore if they are given their due.

Knowledge elicitation and representation are qualified by the analytic, formative, and lingual aspects and are best seen as enkaptically interwoven processes, in which each stimulates and depends on the other (which explains why separating them sequentially is detrimental). Each of these aspects reaches out to the diverse aspectual meaning of the domain of application, which should be respected. The social aspect of the relationship between the IS developer and the human expert source of knowledge must not be overlooked. Dooyeweerd's theory of knowing is helpful in this area, not just in helping to clarify what is happening in knowledge elicitation but especially in differentiating various types of tacit knowledge.

For successful ISD none of these multi-aspectual functionings may be overlooked. Through them ISD links to other areas. ISD overall links to technological ecology, anticipating use links to HUC, and knowledge elicitation and representation links to technological resources.

Practical devices for ISD include aspectual analysis as earlier discussed. But for knowledge elicitation, Winfield's (2000) MAKE (multi-aspectual knowledge elicitation) method can reveal much that is often taken for granted. It was demonstrated how Dooyeweerd can be used to critique, affirm, and enrich the extant FFU that is soft systems methodology.

9.1.4 Framework for Understanding Information Technologies

IS developers seldom start from scratch but use IT resources already available to them—tools, KR languages, libraries of code, inter-file protocols, and so forth. These have to be designed and created. The main question this area of research and practice focuses on is: what should guide the design of such resources?

A review of some influences on design led to the call for "KR to the people"—to provide IT resources that are so natural and appropriate to the diverse meaning of everyday life that representation of knowledge can, in principle, be carried out by lay people, by people who are not IT specialists. Most extant resources are not appropriate in this way, mainly because of a presupposition that all types of meaning should be able to be represented in a KR language based on a single aspect, and some of the problems that arise from this were outlined.

The notion of IT resources was reinterpreted using Dooyeweerd's notion of semi-manufactured products. These have an external rather than internal leading aspect, and in the case of IT resources, their design is led by the aspects whose meaning they are intended to represent. Some IT resources are for quantitative meaning, some for spatial, some for lingual, and so on.

This led to the proposal that a full set of IT resources (a comprehensive KR toolkit) consists of a module for each aspect, in each of which the aspect's main philosophical roles have been implemented. (Implementation is in machine code, in principle, though usually not in practice.) Each aspectual module makes available to the IS developer the basic things, properties, relatings, actions, constrains, inferences, and input-output "language" constructs which are meaningful in that aspect.

But this generated a problem: integration of all these aspectual modules. To understand inter-module links, Dooyeweerd's notion of inter-aspect relationships was useful. Foundational dependency is common, but anticipatory dependency and both kinds of inter-aspect analogy can guide how we ensure that each module is open to future expansion, and respond to unforeseen demands. Aspectual reaching-out can indicate the features of "practical" importance such as units for amounts and styles of text.

This proposal can be seen as a grand (maybe counterfactual) ideal, but is more useful as a yardstick against which to measure extant KR approaches and data models. But a different type of Dooyeweerdian critique of three extant approaches was also illustrated, in which Dooyeweerd is not assumed to be a yardstick but rather is used to uncover presuppositions that are the root of problems. Finally, Dooyeweerd was used to evaluate attempts to bring the insights of design patterns into ISD.

9.1.5 Framework for Understanding IT as Ecology

This area of research and practice differs from the rest in taking a societal or "macro" view of information and communication technology. Human life is "inside" IT, but the destiny of humanity and the cosmos ultimately transcends it. IT may be seen as part of humanity's long-term mandate to "open up" the lingual aspect, disclosing and developing its potential for blessing of the whole cosmos. Three societal issues were discussed: whether IT as a whole is a valid enterprise for humanity, the ecological relationship we have with IT, and absolutization and idolatry of IT.

Schuurman (1980) has already very adequately addressed the first question from a Dooyeweerdian point of view, making use of Dooyeweerd's view of aspectual opening and the importance of ensuring the central formative aspect of technology (or, here, the lingual aspect of IT) always refers beyond itself to serve all the others. This enabled him to define a "liberating vision for technology". This contrasts with extant optimistic and pessimistic views. His vision is worked out in enough detail to be useful in strategy planning. (Schuurman is now a member of the Dutch Senate; a useful piece of research would be to explore with him to what extent he has been able to follow his "liberating vision" in this role.)

The ecological relationship between the IT we create ("inscribe") and the way we live and the life-and-world-views we hold is circular, each influencing the other. It may be understood as correlative enkapsis, with IT being our *Umwelt*, and Dooyeweerd's belief that all things exhibit all aspects was helpful in analysing both directions of this relationship in a way that contributes to discourse in this area. A brief comparison was made with several extant frameworks. This relationship has two religious roots (§2.4.1): religious ground-motive and the deep commitment that is a life-and-world-view; these are both inscribed into the IT that humanity generates, and both affect how we respond to living "inside" IT, including how we change our life-and-world-views themselves.

The third religious root is concerned with the supra-temporal self, and the destiny of humankind and the cosmos: orientation to the true Absolute or absolutization of that which is not. Absolutization of aspects (law side) leads to reductionism, which closes down research and development, and to distorting the very nature of IT. The critiques of "masculine" technology by feminists and of Western technology were seen as versions of the this. A notion of idolatry, as the absolutization of something on the subject side, in this case IT, was helpful in understanding our everyday attitudes to IT-as-ecology. The spiritual driving force of the Nature-Freedom ground-motive was able to account for the dialectical tension in IT as both liberating and constraining-controlling. The solution to religious dysfunction goes beyond reason, dialogue, social theory, or economics, to involve relinquishing absolutizations and false polar oppositions, and might even require "conversion". This is ultimately what can deliver Schuurman's "liberating vision for technology".

9.1.6 Understanding the Whole Story

The "whole story that is IS" is not to be seen as a synonym for technological ecology. If it were then it would be impossible to do justice the detail of IS use, ISD, shaping of technology, and the nature of computers. These so-called micro-level "little things" are just as important in the "whole story" as the macro-level things of society, and just as worthy of philosophical attention.

Note the difference in the subheading title: "Understanding" rather than "Framework for Understanding". No structured framework is attempted for the "whole story that is information systems", yet we may still hope to understand it.

How may we understand it? During the exploration of area frameworks, links with each other were identified, of which we could compile a list. There are many more links than those mentioned, and it is possible that there would be no end of discovering links, so the attempt to construct a whole-story framework by identifying links would become meaningless. We could also point to the use of Dooyeweerd's aspects as a common thread in all frameworks. But simply positing Dooyeweerd's aspects as a framework for understanding the whole story would miss much. It would miss the structures of individuality important in Chapter V, the ground-motives important in several chapters, the notions of correlative enkapsis and destiny important in Chapter VIII, and the religious root, important in all.

Rather than attempt a framework for understanding the whole story, it might be better to attempt an attitude. This, the author has found over the past 30 years, is the main thing that has helped him maintain a whole-story perspective throughout all he has done in IT/IS. The attitude is first that of the lifeworld. A lifeworld attitude opens the researcher and practitioner to the wide horizons of everyday life, in all its diversity, subtleness, coherence, mystery, and glory, and to other people's views, preferences and knowledge, even while one's thinking is focused on one small point, whether this be a technical point while programming, a troublesome deadline in ISD, a philosophical point in working out the nature of information, an unexpected impact in use, or a political point of globalisation.

But the attitude is not only that of the lifeworld. It is a religious attitude, seeing all that I am involved in as part of a wider whole that transcends even humanity. It is not that I attempt to circumscribe that wider whole, just the inner awareness that I am part of it, and this provides both comfort and a sense of responsibility. It is an attitude that I might be wrong, that others also might be wrong, even the whole of humanity might be wrong, even while there is much that is right. To me (forgive me using first person singular here; it seems appropriate) the whole-story attitude involves a sense of the cosmic meaning and rightness of things (despite the presence of evil), a sense of destiny, a sense of cosmic responsibility, and a sense of cosmic joy and belonging, a sense of holiness in all I do in IT/IS, a sense of reality, and sense of relating intimately and personally to God, not because of any merits of my own but because of Who He Is and What He Does.

That is my account of the whole-story attitude that has pervaded my journey in IT/IS for over 30 years. It might not be the only possible whole-story attitude that is useful in IT/IS research and practice in all areas. Perhaps a Hindu way of thinking could yield another one. But, not being Hindu, I can only speculate about this from the outside.

9.2 Reflections on the Frameworks

To what extent has the discussion in this book addressed the issues set out in the Introduction in §1.3? They will be examined in reverse order.

9.2.1 On Multiple Frameworks and a Single Philosophy

In Chapter I we noted Lyytinen's (2003) belief that it is "hopeless" to seek any "ultimate foundation" for information systems. Have we proven Lyytinen wrong by finding an "ultimate foundation" in Dooyeweerd's philosophy? Or would those who take Lyytinen's line have to reject Dooyeweerd?

What Lyytinen was rejecting as an "ultimate foundation" appears to be something different. He was rejecting the presupposition that there is, ultimately, a single way of making theory and a single language and logic for the field of IS. What we have found using Dooyeweerd's philosophy is that no single logic is possible in any of the areas, let alone across the areas. This is because each aspect indicates a different logic, and thus language and way of making theory. No single "ultimate foundation" is possible in the sense that Lyytinen dislikes, and it is why we have not attempted to find a single over-arching conceptual framework that covers all areas. But our reasons for believing that such is not possible are different from Lyytinen's. His reasons arise from presupposing the absolute autonomy of reason. Our reasons are based on the primacy of cosmic meaning: in each area different things are meaningful, and should not be forced into a meaning-framework that is foreign to it. Clouser (2005) argued that any possible "ultimate foundation" that arises from immanence-philosophy will be reductionist in some way, and it seems to be the pretended possibility of reductionism that lies behind Lyytinen's dislike.

The five frameworks are compatible with each other conceptually: could we not just forge them into a single, mammoth framework? It might, in principle, be achievable, but doing so would be meaningless because each aspect defines a distinct sphere of meaning and in each area the aspects play their roles in very different ways. To import a load of meaning that is irrelevant to the area would confuse its practice and research.

Instead, we have heeded Lyytinen's call "to explore the content of the underlying philosophical argument in these debates [in the various areas] and what role they assume to the philosophy as a field of inquiry." But we have done so in reverse. Whereas Lyytinen assumed extant debates and calls us to explore the philosophical argument in these, we have assumed a philosophy and from that both explored (a few) extant debates and also generated new ones or new arenas for debate. In doing so, we have tried to be explicit, and even self-critical, or at least self-aware and self-reflective, about the underlying philosophical argument.

It is our desire to be able to make some sense of the "whole story", that makes a single root philosophy important. It was suggested earlier that what brings the frameworks for understanding the various areas together is not a larger framework, nor even a system of relationships between them, but an attitude in which, while working in one area we are open to all others simultaneously. It is philosophy that enables us to see attitude as something more than a disposition, an emotion or an arbitrary logical axiom. It reveals the link between attitude and presupposition, especially ground-motive. It is this kind of attitude which allows or disallows a variety of spheres of meaning, and, when allowing it, holds the diversity to be coherent. It is such an attitude that can hold all the area-frameworks together, and can also allow them to develop in ways germane to each.

Nevertheless, the fact that the concepts in the frameworks formulated for each area all derive from the same philosophical stance makes it possible that when those working in one area (say, technological ecology) want to "reach into" another area (say, the shaping of technological building blocks) then there is a chance that they can do so without finding what they reach for to be nonsense, but can understand it without undue trouble. Probably the main reason Dooyeweerd's philosophy has been able to do this is its recognition of cosmic meaning and law, which transcends all the areas—including any other areas we might wish to delineate and explore in future—and even the "whole story that is information systems".

9.2.2 Characteristics of the Frameworks

A number of characteristics were identified that FFUs should exhibit. First, the main themes or principles of the framework should be made clear in a reasonably systematic way; this has been achieved due to Dooyeweerd's ability to engage with the major issues like meaning and normativity in a systematic manner.

The second is that the frameworks generated are open to extension. For example, other types of multi-aspectual human functioning could be identified and explored in both IS use and ISD, and a basis for harmonizing these with the rest is found in Dooyeweerd's idea of enkapsis. Where the focus has been on certain aspects, other aspects could be (should be) explored in more depth. More fundamental extensions are possible; see the following.

Coherency rather than logicality is felt more than argued. But it is felt precisely because the kernel meaning of the aspects is intuitively grasped. The example given in the Introduction, of doing justice comes from our intuitive grasp of kernel meaning of the juridical aspect. In addition, every attempt has been made to avoid the three important types of incoherency outlined in §2.2.1.

Frameworks should guide. This is ensured by the intrinsic normativity of Dooyeweerd's approach, which is integrated with his notion of being rather than divorced from it. In each chapter, practical devices were described.

The "whole story that is IS" can, in principle, be tackled by Dooyeweerdian frameworks because Dooyeweerd's philosophy deals with that which transcends not just the areas of IS but even the whole story as such, seeing even this as part of the destiny of the cosmos. Because of Dooyeweerd's focus on cosmic meaning, each area's FFU is so constituted that it can be sensitive to, and respect, the issues meaningful in other areas. Points of contact with other areas have been mentioned throughout the discussion.

Finally, Dooyeweerd's grasp of everyday experience has been shown to exceed that of phenomenology and existentialism, and has been demonstrated to pervade the whole approach rather than being a specific function of it. The diverse, pre-given, shared, background character of the lifeworld is augmented with coherence and religious importance in almost every area.

Openness to extension can go further. The way Dooyeweerd has been used here need not be confined to IT/IS. For example:

- The discussion of IS use as multi-aspectual functioning could be extended to any other area of human endeavour, such as use of other technologies, management, sustainability (Brandon & Lombardi, 2005), or even farming (the Dutch farm, Eemlandhoeve, tries to organise its life according to Dooyeweerd's aspects). In any such area, identify the interwoven multi-aspectual functionings and then explore the structure and normativity of each.

- The discussion of the nature of computers as multi-aspectual meaningful wholes could be extended to understand the nature of other things—a newspaper, an organisation, and so forth. Fathulla and Basden (in press) use this to consider "What is a diagram?"

- The discussion of ISD could be extended to any activity of planning and development: seek to identify the multi-aspectual functionings therein and translate the intrinsic normativity of the aspects into guidance.

- The discussion of IT resources (languages, protocols, code) could be extended to any type of semi-manufactured products directed to the variety of aspects. For example, in the construction industry consider which aspects of the lifeworld of inhabitants are served by each type of resource and whether any are ill-served.

- The discussion of IT as ecology could be extended to any societal issue, for example to mobility, by considering (a) whether it has a valid destiny, (b) the multi-aspectual nature of the circular ecological relationship, (c) idolatry.

Such extensions are underway in a number of areas.

9.2.3 Constitution of the Frameworks

Dooyeweerd's philosophy covers all branches—ontology, epistemology, philosophical ethics, methodology, philosophical anthropology, and critical philosophy—giving an ability to incorporate all that seems necessary to formulate lifeworld-oriented frameworks for understanding. Table 9-1, based on Table 1-1, shows how Dooyeweerd can tackle each of the IS issues identified there.

Though conceptual structures have not been the major focus of this exploration, it is clear that they are either already available in each of the FFUs developed, or can be readily made available. They include, among other things, the notion of human subjects responding to cosmic aspectual law in various ways leading to repercussions (Chapter IV), aspectual levels that constitute the meaningful whole that is the com-

Table 9-1. How Dooyeweerd can tackle IS issues

Area	IS Issue	Dooyeweerd's explanation
Human Use of Computers	User-computer relationship Human activity with computer Impact in use Benefit versus detriment Variety of impacts	Law-subject-object relations L-S-O versus *Gegenstand* Aspectual repercussions Aspectual normativity Irreducibility of aspects
Nature of Computers	Computer as experienced Nature of computer, info The AI question Hardware, bits, symbols	Everyday experience Multi-aspectual being Subject-, object-functioning Seen from different aspects
Information Systems Development	Teamwork Lifecycle methods Guidelines Human creativity Requirements analysis Knowledge elicitation Conflict	Social and aesthetic aspects Multi-aspectual process Aspectual normativity Aspectual possibility Possibility, responsibility Multi-aspectual knowing Religiously held asp'l persp's
Information Technology Resources	KR languages Types and classes Inappropriateness	Lingual 'reach-out' Type laws Aspectual conflation
Information Technology as Ecology	Validity of IT IT as our environment Gender issues Modern dominance of IT	Theory of destiny, progress Multi-aspectual correlative enkapsis Shalom Idolatry

puter (Chapter V), interwoven and yet conceptually distinct multi-aspectual human functionings (Chapter VI), philosophical roles of each aspect being implemented as basic technological resources (Chapter VII), and the notions of the long-term opening up of aspects, correlative enkapsis, and idolatry (Chapter VIII).

But what is perhaps more important is that Dooyeweerd's focus on the human person who thinks, theorises, and philosophises almost guarantees that the frameworks will be able to reflect the culture, attitudes, visions, and normativity of each area. Though not always explicitly, the discussions here have been sensitive to the history and motivations of the areas as mentioned in §1.3.2. Particular attention has been given to assumptions and presuppositions.

What has not been much discussed in this work is research methodology in each area. Instead, the focus has been on ability to meet the challenge of the lifeworld of the area which is being practised within, and researched. Something of the nature of research in each area may be deduced from the frameworks developed, but this is left to another time. Most standard research methods, from surveys to action research, are likely to be applicable if certain warnings are heeded (for example the non-absoluteness of the lingual aspect means: never rely fully on surveys and interviews).

9.2.4 Compatibility of Areas with Dooyeweerd

Chapter I explained the choice of the five areas for which frameworks were formulated, and gave a reason for the order in which this has been done. Neither of these involved any reference to Dooyeweerd. But if this pre-Dooyeweerdian choice is inimical to Dooyeweerd, then a serious antinomy lurks at the root of this whole exercise. To what extent were those decisions valid in Dooyeweerdian terms? Ultimately, the answer to this is pre-theoretical, but the following indications are positive.

That the human being was placed at the centre is something with which Dooyeweerd would agree. This also supports discussing IS use first. That the areas were differentiated partly according to the relationship between humanity and IT is in accord with each being meaningful to humans in a different way. That the differentiation was made according to the lifeworld of research and practice of IS and not according to any prior theoretical framework, especially not any that is used in any one area, such as that of Burrell and Morgan (see Chapter VI), is compatible with Dooyeweerd's attempt to avoid prior commitment to any theoretical approach. His philosophical notions of spheres of meaning-and-law are not a theory but an idea that became visible when he "cleared away" that which prevents us seeing the structure of reality (as mentioned at the start of Chapter II).

If we had attempted to use a philosophy unsuited to any area, then we would expect to have experienced some discomfort, not least because we would find much that is meaningful in the area lies beyond the reach of the philosophy, or be forced to reinterpret it in ways unnatural to it. But no undue discomfort was experienced in using Dooyeweerd nor any imbalance in the amount of Dooyeweerd that seemed useful. While this cannot be taken as any absolute indication, not least because it may be that this author was blind to certain issues, it does at least suggest that the choice of areas is commensurable with Dooyeweerd.

That this author's approach before discovering Dooyeweerd was along the lines set out in this book suggests that Dooyeweerd has not overly determined the approach itself.

9.3 On Using Dooyeweerd

This part of the chapter is for those who are interested in Dooyeweerd and want to see what he has been able to do in helping us understand IT/IS. It might also be useful for those who wish either to critique the thinking in this book, or to apply Dooyeweerd's ideas in other ways or to other areas. The main part of Dooyeweerd useful in each area is summarised in Table 9-2, and how the various portions of Dooyeweerd's thought have been useful is shown in Table 9-3. The text explains them.

Table 9-2. Summary of employing Dooyeweerd's thought in each area

Area	Main Dooyeweerdian ideas
Use of Computers	Aspectual functioning, repercussions
Nature of computers	Aspectual being
IS development	Aspectual possibility, responsibility Knowing
Techn'gcl resources	Philosophical roles, ch'cs of aspects
Technological ecology	Theory of progress Correlative enkapsis Religious roots
Whole Story	Cosmic destiny Lifeworld attitude

9.3.1 On Using Dooyeweerd in Understanding Human Use of Computers

1. Dooyeweerd's notion and suite of aspects provided an avenue by which we can approach the complexity of human use of computers with critical respect.

2. Dooyeweerd's urging of sensitivity to the structure of everyday experience opened up for us several interwoven but distinct multi-aspectual functionings, of which three were examined: HCI, ERC, HLC.

3. This intuition was legitimated by Dooyeweerd's notions of qualifying aspect and enkapsis.

4. Turning to aspectual theory, Dooyeweerd's view of aspects as distinct spheres of meaning, together with our "taking" of Dooyeweerd's suite, provided several useful practical devices for analysis of complex situations of IS in use.

5. Dooyeweerd's view that aspects are spheres of law to which we are subject enabled us to understand the multi-aspectual subject-subject and subject-object relationships between human and computer in HCI in a way that affirms current thought in Winograd and Flores (1986) but overcomes its problems.

6. Dooyeweerd's notion of *Gegenstand* was useful in understanding the distal human-computer relationship.

7. That the lingual aspect "reaches out" to all others, to signify all kinds of meaning, provided a basis for understanding ERC and the difference between real and virtual worlds.

8. The notion that aspects enable functioning provided a basis for understanding what it means to live with computers (HLC).

9. The notion that aspectual functioning generates repercussions gave a basis for understanding the impact of computer use.

10. The shalom principle and the proposal that aspectual law pertains whether we know it or not helped us understand unexpected impacts.

11. It also provided an incentive for, and means of, considering stakeholders that are often overlooked.

12. The lengthening time-response of aspectual repercussions helped us understand short- and long-term impacts in a single framework.

13. The notion of aspects as normative provided an understanding of the difference between beneficial and detrimental impacts, which is neither functionalist, nor reductionist nor subjectivist, and which can address complex, mixed normativity.

9.3.2 On Using Dooyeweerd in Understanding the Nature of Computers

1. Giving primacy to Meaning rather than Being directed us away from the question "What is computer?" toward "What means computer?"

2. Aspects as spheres of meaning enabled us to answer this question in a multi-aspectual way, freeing us to give multiple answers.

3. Dooyeweerd's notion of aspectual individuality structures enabled these multiple answers to also answer the question "What is computer?"—as an enkaptically interwoven set of "aspectual beings", not unlike Newell's notion of levels.

4. Differentiating enkaptic from part-whole relationships enabled us to avoid category errors when thinking about the relationship between, for example, voltages, bits, and data.

5. It also enabled us to understand the relationship between data, information, and knowledge.

6. That Being is grounded in Meaning, but that cosmic meaning transcends us, and that everyday experience is more than psychic functioning, enabled us to pose and answer the question of why it is valid to say there are bits, data, information, and knowledge "in" the computer when we cannot see them.

7. Dooyeweerd's notion of law side enabled us to understand programs as a constructed law side.

8. Dooyeweerd's discussion of performance art (music) threw light on the relationship between program source and the running program.

9. Dooyeweerd's subject-object and subject-subject relationships (as discussed in HUC) helped us understand the ways it is valid to speak of a computer "un-

derstanding," and so forth, even though, of itself, it subject-functions only in the physical aspect. Differentiating subject- from object-functioning enabled such statements to be meaningful without metaphor or anthropomorphism.

10. This threw fresh light into Searle's Chinese Room thought experiment.

11. Dooyeweerd's notion of ground-motives helped us understand the different types of reaction to the AI question of in what way computers are like humans.

12. The determinacy of physical functioning and the non-determinancy of later aspects enabled us to understand what Newell could not: how computers can be determined while humans are free even though both might "know things".

13. Dooyeweerd's emphasis on presuppositions, especially of NFGM and CFR, and of the immanence and transcendence standpoints, helped us expose the presuppositions that Newell made that led him into complex, counter-intuitive proposals.

9.3.3 On Using Dooyeweerd in Understanding IS Development

1. Dooyeweerd's exposition of the NFGM was used by Eriksson (2006) to analyse the various systems thinking paradigms and the Burrell-Morgan model and expose the root of their problem.

2. Dooyeweerd's suite of aspects was used to place ISD generations in relation to each other within a wider picture, and also to suggest which parts of the picture were missing.

3. This made it useful to reinterpret ISD. Dooyeweerd's focus on everyday life revealed the activity that is ISD to be multi-aspectual functioning.

4. Dooyeweerd's notion that aspectual diversity coheres supported the integrality that is ISD.

5. As with computer use, Dooyeweerd's focus on everyday life allowed us to detect several intertwined multi-aspectual functionings that constitute ISD.

6. As with use, this intuition was undergirded by Dooyeweerd's notions of qualifying aspect and enkapsis.

7. In ISD, aspects are directed to future possibility rather than to present or past actuality, so their role as guiding-law was important. The shalom principle provided specific guidance.

8. The content of the suite of aspects was important in understanding ISD overall, with the aesthetic and social aspects being key.

9. Dooyeweerd's theory of social institutions enabled us to understand the proper place of authority and subordination, of mutual agreement and of friendship in ISD.

10. Dooyeweerd's notion of a juridical aspect, directed to responsibility for the future, was key to anticipating use.

11. That the juridical aspect reaches out to all kinds of meaning directed us to responsibility toward all aspects, not just legal issues.

12. Dooyeweerd's understanding of the kernel meanings of all the aspects, with the multi-aspectual conception of being fully human helped us understand how creating the IS could be both a chore and a delight.

13. Inter-aspect relatedness enabled us to understand the relationship between knowledge elicitation and representation, and the notion of enkapsis as different from the part-whole relationship revealed why these processes should not be sequentially separated.

14. That aspects transcend us directed us to the importance of what is taken for granted. It also led Winfield to devise the Multi-Aspectual Knowledge Elicitation method.

15. Dooyeweerd's theory of knowledge as including intuitive knowledge points to the possibility of eliciting taken-for-granted knowledge.

9.3.4 On Using Dooyeweerd in Understanding IT Resources

1. Dooyeweerd's emphasis on everyday life made us listen to the call for "KR to the people", who will need IT resources that can represent the whole diversity of everyday meaning naturally and appropriately.

2. Dooyeweerd's highlighting of the immanence presupposition revealed reductionist presuppositions in almost all extant KR languages.

3. Dooyeweerd's understanding of semi-manufactured products helped clarify the difference between IS for human use and IT resources, and formed the starting point for a reinterpretation of this area.

4. Dooyeweerd's distinction between law and subject sides also helped clarify the role of IT resources: to implement aspects of the law side rather than subject side.

5. Dooyeweerd's suite of aspects was useful to show the paucity of extant KRLs: most aspectual meaning remains unsupported.

6. Dooyeweerd's suite of aspects was then used to make a proposal for a KR toolkit consisting of a module for each aspect.

7. The structure of each module was governed by the philosophical roles of aspects: that each aspect enables distinct types of being, action, properties, relatings, constraints, inferences.

8. Dooyeweerd's notion of inter-aspect relationships was key in understanding how the aspectual modules can be integrated.

9. This led to a specific, practical initial proposal for a new KR toolkit.

10. Dooyeweerd's belief that aspectual meaning is grasped by intuition allays fears that such a KR toolkit would be too complex to use.

11. Dooyeweerd's notion of aspectual irreducibility revealed the nature of problems with the Relational Data Model.

12. Dooyeweerd's notion of type laws enabled us to affirm the Object-Oriented notion of classes.

13. Dooyeweerd's notion of "thing" and his non-Cartesian notion of subject and object enabled us to critically contribute to discourse over OO's notion of object, exposing the root problem in OO's notion but also in the counter-proposal of "subject-orientation".

14. Dooyeweerd's primacy of Meaning over Being, his understanding of "thing" and several other portions helped us examine the Wand-Weber ontology.

15. Dooyeweerd's notion of aspects as spheres of meaning of everyday life enabled us to explain why design patterns is attractive, and also why its use in software is problematic.

9.3.5 On Using Dooyeweerd in Understanding IT as Ecology

1. Dooyeweerd's theory of progress as humanity's mandate to open up the meaning of the aspects was used by Schuurman to define a "liberating vision for technology", in which technology is valid as a human enterprise.

2. The shalom principle places a condition on that: technological development is not to be for its own sake but should be guided by the norms of all aspects: humanity is called to "compose a symphony" by means of IT.

3. The nature of the four latest aspects as understood by Dooyeweerd provides special insight for Schuurman's vision.

4. Dooyeweerd's notion of Meaning as "referring beyond" could account for the apparent meaninglessness experienced in modern life.

5. Dooyeweerd's notion of correlative enkapsis and his understanding of the nature of *Umwelt* was central to understanding the nature of the circular relationship by which humanity shapes IT and yet is itself shaped by IT. This

Dooyeweerdian understanding seemed more penetrating and also practical than extant views.

6.   Dooyeweerd's aspectual suite may be used as a practical device to analyse both arms of this relationship.

7.   Dooyeweerd's contention that there is a religious root to all things was helpful in understanding why this circular relationship goes "wrong", leading to oppression rather than liberation and opening-up. Three types of religious dysfunction were useful: ground-motives, aspectual absolutization, and idolatry.

8.   Dooyeweerd's exposition of the NFGM threw light on the history of both arms of the circular relationship.

9.   Absolutization of aspects in the way IT is shaped, developed, and used accounted for the emergence of technology that is too "masculine" and "Western" (two characteristics that were briefly examined) which leads to considerable harm.

10.   Dooyeweerd's notion of multi-aspectual ways of knowing can be seen to prefigure much that is sought in the more recent forms of feminism. His notion of multi-aspectual normativity can help expose and understand Western assumptions inscribed into IT.

11.   The notion of idolatry, as developed by several thinkers from Dooyeweerd, may be applied to IT to give a broad yet unimpeded picture of what is wrong.

12.   Religious dysfunction demands a religious solution. Dooyeweerd's exposition of the Creation-Fall-Redemption ground-motive was found attractive.

9.3.6 On Using Dooyeweerd in Understanding the Whole Story

Several characteristics of Dooyeweerd's philosophy pervade our understanding of all areas: its wish to expose presuppositions (including its own), its lifeworld attitude, its ability to hold both meaning, being, occurrence, and normativity together within its grasp, its emphasis on meaning (cosmic) as the foundation, its notion of there being two sides to reality, law, and subject, and its espousal of the CFR ground-motive. It is these that have enabled it to usefully address every area that has been explored. Moreover, they lead at least this author to assume that Dooyeweerd's philosophy can approach the "whole story", that there is nothing in the whole story, now or in the future, that is outwith its grasp. Whether they are all necessary is a matter to be explored another time; what this exploration has shown is that they are all useful throughout.

But there is another reason for believing that Dooyeweerd's philosophy can help us understand the whole story: religious root and destiny. While religious root and destiny were important in understanding societal issues in Chapter VIII, their relevance is not exhausted therein, but extends beyond it into the micro-level "little things" that are meaningful in the other areas. What pervades all areas is the religious meaningfulness of all, whether micro or macro, and religious meaningfulness means that there is a destiny for all.

Leaf by Niggle is a tale by Tolkien (1998) of an unassuming person, Niggle by name, who tries all his life to paint a leaf perfectly, as part of a picture of a tree. He dies before he manages it. He is transported to the Real Life Beyond. There he finds his tree: no longer a mere unfinished painting, but alive, finished, and honouring even that holy realm. It is the little things we do in IT/IS, just as the big things, which are Meaningful and deserving of philosophical attention because they are Religious.

Dooyeweerd tried to recognise something of this in philosophical terms, and thus to provide humanity with the means of thinking about them in philosophical ways without the encumbrances and ideologies that bedevil theology (whether Christian, Humanist, Eastern, or other theology). This is the real joining of the micro with the macro; it is not a cycling between them, as in Latour (1987) and others, but something simultaneous: an attitude, as mentioned earlier.

9.3.7 Portions of Dooyeweerd Found Useful

It is useful for Dooyeweerdian scholars to know which parts of Dooyeweerd have proven useful, and why. Table 9-3 indicates which parts of Dooyeweerd have been useful in understanding the five areas of research and practice. The number of asterisks indicates (approximately) how important the portion has been in this exploration so far. While most portions are used positively, the row labelled "Immanence philosophy" indicates critique of extant frameworks and the degree to which an immanence standpoint seems to hinder understanding in the area in the frameworks discussed.

Such an analysis shows the importance and fruitfulness of Dooyeweerd's philosophy for IT/IS. It shows clearly that it is not just a tiny subset of Dooyeweerd that is useful. It also shows that most Dooyeweerdian ideas tend to find relevance in several areas. It would be interesting to compile such tables for other philosophers, for the purposes of comparison. This table at least lays Dooyeweerd's cards on the table!

9.3.8 Developing Dooyeweerd?

But Dooyeweerdian philosophy is itself under critique, development, and refinement. Dooyeweerdian scholars will also be interested in knowing in what way this work might contribute to Dooyeweerdian thinking as such. A few suggestions occur in the text, of which some are summarised here.

Table 9-3. How Dooyeweerdian ideas have been used

Portion of Dooy	Use	NoC	ISD	ITR	Eco
Law-Subject-Object	***	*****	*	**	*
Aspects					
Suite	*****	***	****	*****	**
Dependency	**	****	**	****	
Analogy		**		***	
Phil. Roles	*	**		*****	
Normativity	*****	*	*****	****	****
Aspects pertain	*****	*	****	**	***
Repercussions	*****		**		****
Multi-aspectual functioning	*****	**	*****	**	***
Shalom	*****		*****	***	***
Fully human	***		*****	**	**
Things					
Being from Meaning		*****		*****	*
Aspectual beings		*****		****	
Multi-aspectual whole	***	*****		***	*
Qualifying aspect	**	**	***	****	
Relationships	****	***	*	**	**
Enkapsis	***	*****	***	**	****
Knowing					
Thought + thing			*****	****	*
Intuition	***	**	*****	*****	***
Theoretical thought		*	*****	**	**
Religious Root					
Human self		**			*
Aspectual WVs	***		*****	***	****
Ground-motives		*	*****	*	*****
Absolutization	*	***	***	**	*****
Theory of Progress				***	*****
Ctq. Immanence Philosophy	***	****	***	***	***

9.3.8.1 Contributions to Dooyeweerd's Theory of Aspects

This exercise seems, at first sight, to be a massive corroboration of Dooyeweerd's notion and suite of aspects. But wider use of it, together with a more penetrating analysis, might provide grounds for a serious rethinking of many aspectual kernel meanings. To date, most proposals to modify aspects (see Chapter III) are the result of individual reflection on certain aspects rather than a trans-individual reflection on the whole suite. That is what a long-term application of Dooyeweerd to IS would offer. One suggestion, for re-evaluating the kernel meaning of the biotic aspect, has been made because of difficulties in "filling that slot" as currently understood (a difficulty Dooyeweerd himself discussed in Dooyeweerd (1984, III, p. 112ff.) but did not seem to adequately resolve. Whether or not this is accepted, our attempts to work this out in discussing the nature of computers might provide a model of how an aspect's kernel meaning could be modified.

There are issues in IT that do not seem to be easily qualified by a single aspect, such as information security and safety, without denaturing their meaning. It may be that such issues could point to new aspects or new ways of treating the aspects.

Chapter V suggested that a virtual world facility, or indeed any represented content in a program, could be seen as a man-made law side. This could be useful either for testing concrete proposals for what aspectual law is, or for exploring to what extent it might be possible for human beings to imagine a different law side, and even new spheres of meaning (if such a thing were possible, which this author doubts).

Though the notion of qualifying aspect is acknowledged as indicating the main meaning of a thing and its normative direction, this work affords it much less importance than other Dooyeweerdian thinkers do. This is because it often seems to constrain rather than stimulate insight, and because there seems to be many more ways in which aspects can be important to a thing than the limited list offered by Dooyeweerd (qualifying, founding, leading, internal leading). This is especially so in the use of IS. Instead, the shalom principle has been found of more value in addressing normative direction and the notion of multi-aspectual functioning has been found more useful in understanding the meaning and structure of a thing. Indeed, it might be argued that the notion of qualifying aspect is redundant (though that is not argued here). The case of information technology, in all its areas, might provide rich lifeworld material for a debate about the status of this notion.

Chapter III presented the philosophical characteristics and roles of aspects in a systematic way. But Dooyeweerd never did likewise and this author's interpretation might be flawed. This might assist in debate about Dooyeweerd's aspects in general. That information technologies, as developed, exhibit many of the philosophical roles of Dooyeweerd (see Chapter VII) suggests that a study of them might be used to discuss and perhaps refine our understanding of their philosophical roles.

Nowhere did Dooyeweerd provide a comprehensive list of inter-aspect analogies. Though the idea itself is simple enough, we have been given very few examples to go on, and it would be nice to have more, and to have a way of testing proposals. Gibson's (1977) idea of affordance seems not unlike these, and so its use in design of user interfaces might help provide some of this. The modelling capability of computers might also be used.

9.3.8.2 Contributions to Dooyeweerd's Theory of Things

Computer systems seem to exhibit greater complexity than any of the types of entity that Dooyeweerd himself discussed (the linden tree, the marble sculpture, utensils, books), partly because of their activity, partly because they have more aspects—and in addition, we must then take account of the applications aspects (ERC and HLC). Therefore, we may offer computer systems to Dooyeweerdian philosophers as a case study that might raise issues that Dooyeweerd himself never saw or clarify issues that he only glimpsed, and thus extend or refine Dooyeweerdian theory. It might uncover a new type of enkaptic relationship.

Dooyeweerd's discussion of semi-manufactured products is rather brief. Consideration of technological building blocks and tools in Chapter VII could significantly enrich the notion. Some of the structural relationships encountered in IS might indicate new types of enkapsis.

9.4 Reflections on the Process

The earlier discussion has reflected on the content of the exploration of using Dooyeweerd to formulate frameworks for understanding IS. It is also useful to reflect on the process.

9.4.1 Filling Slots?

Dooyeweerd seems very useful. But is this because we have just forced all into Dooyeweerd? Are we just "filling slots" when using the aspects? Yes and no.

Yes, in the sense that it is common for some thinkers in an area to make use of their favourite philosophers, such as Winograd and Flores (1986) did with Heidegger, Jackson (1991) does with critical theory, and Midgley (2000) does with process philosophy. There is certainly an element of commitment to Dooyeweerd in general, and also to his suite of aspects as a practical device. Indeed there has to be real commitment during the phase of exploring a new idea because, without some commitment to the new idea, justice cannot be done to it and, without immersion in it, it cannot be truly understood. During this phase—the phase of which this work is part—we must commit to Dooyeweerd and explore how much of reality does in fact fit his way of thinking.

If this is so, then the suite of aspects proposes spheres of meaning, and one way of testing it is to treat them as slots to fill and notice and discuss the ease (naturalness) or difficulty we experience as we do so in a myriad of situations. Our discussion has included warnings to do this sensitively and self-critically.

It is later, once his ideas have been properly understood in the context of the field, worked out, tried, tested, refined, that it is right to stand back and take a critical stance. It is then that we can see where undue force has been exerted to squeeze experience into a Dooyeweerdian way of thinking.

But something of a critical stance has already sometimes been taken here. Thus we also answer "No!".

We answer "No" in the sense that, though committed to Dooyeweerd, the author has always been aware of the possibility of limitations, and in the main taken a cautious approach, frequently appending "If Dooyeweerd is right ..." and occasionally suggesting areas where he might wish to differ from him. Continual reference to other thinking has been made.

Has there been an over-emphasis on Dooyeweerd's aspects? It might seem that his suite of aspects comes across almost as a panacea. To some extent this must be the case because, as diverse spheres of meaning-and-law, the aspects form the foundation for all other parts of Dooyeweerd's positive philosophy. They give a rich view of the cosmos with which we can undertake sophisticated analyses in which oft-overlooked aspects are brought into the light. But, though aspects have infiltrated every area of research and practice in IS, they have fulfilled different roles and have been combined with other, different, portions of Dooyeweerdian philosophy in each area.

The success of this approach lies in its fundamental understanding of naïve experience, its being based on meaning from which being, occurrence, normativity, and knowledge emerge, its ability to account for both a coherence that is diverse and a diversity that coheres, its being intuitively grasped, while at the same time its friendliness to theoretical analysis.

9.4.2 Other Aspects, Areas, Philosophies?

It may be that the reader might like to take the approach outlined here but employ a different suite of aspects, or different areas of research and practice, or even a different stream of philosophy. How should the reader proceed? The approaches worked out here might be employed as models or exemplars. But, if other aspects, areas, or philosophies are to be tried, the following guidelines might be useful.

- If we wish to change the suite of aspects, perhaps retaining the areas or the Dooyeweerdian approach, then two conditions would seem necessary. The suite should provide a wide coverage of the diversity we experience in everyday life in the area, and the aspects in the suite should be able to be treated not just as distinct categories but as spheres of law and meaning that possess a modal character and thus enable existence and functioning. The typical trio of physical-mental-social is not diverse enough, and other suites founded in a theoretical attitude are unlikely to be useful. Maslow's "hierarchy of needs" might prove capable.

- If we wish to try the approach in a different area of research and practice, then it is important to avoid taking as a starting point the current view on what are the important issues, challenges, problems, or solutions. This is because these emerge as "important" only by presupposing a framework for understanding. The problems and issues used during the development of the framework for understanding should be drawn from everyday (lifeworld) experience in the area, both practice and research. Once a framework has been constructed, then it can be applied to currently-important problems, partly by way of testing it to see whether it can address them and to what extent it can throw new light on them, and how it might engage with extant frameworks can be explored. The results of this might be fed back to refine the framework.

- It is likely that we could not achieve what we have if we presupposed Meaning is derived from Being, Process, or the autonomous ego. This is because each framework constitutes a horizon of meaning, in which certain things are meaningful while others are not. It is thus useful if the process is sensitive to the issue of meaning so that such boundary decisions are clearly visible.

- The human being has been central in relation to IT. If, instead, either technology, society, language, logic, or anything else were to be made central, it is likely that the frameworks we create for some areas would be incommensurable with those we create for others, and that for some areas it would prove difficult to create a framework of sufficient richness.

- If we wish to attempt this exercise with a different philosophy, then it is likely that we must find some way in which that philosophy acknowledges all of normativity, ontology, epistemology, methodology, anthropology, and critical philosophy because they have all been important. In Dooyeweerd, none of these are reduced to the others, but many philosophies have difficulty with one or more of these. Such philosophies would have, perhaps, to be modified to derive the missing elements from those that it espouses, but doing this is unlikely to prove entirely satisfactory because it is likely that certain areas would be invisible to us and for those that are visible we would create rather thin frameworks which impose a theoretical position on our understanding and rob us of the lifeworld approach.

The degree of success with which any of these may be carried out might provide a useful test for the validity, utility, and power of Dooyeweerd.

9.4.3 The Effect of Dooyeweerd's "Christian" Philosophy

Dooyeweerd's Christian stance has been clearly mentioned but, following Dooyeweerd, theological issues have been kept at bay so that people of all religious persuasions, including humanism, should be able to accept most of this approach to IS.

Dooyeweerd held that his was an attempt not only to reform philosophy according to criteria that philosophy would itself recognise, but also uncover the necessary conditions for a "Christian" philosophy which, he believed, has never been adequately discussed in 2,000 years. Does this mean that a Dooyeweerdian approach has at its heart a repressive dogma that should be shunned? Does it mean that one must be a Christian believer to benefit from Dooyeweerd? Does it mean that all others must, or are entitled to, reject his philosophy along with any approach to IS based on his philosophy?

Since it is, at least logically, possible that a similar philosophical approach could emerge from a different non-dualistic ground-motive, this means there is no logical reason why those without a Christian belief should shun Dooyeweerd's philosophy. In practice, the author has found that it is non-Christians rather than Christians who have found Dooyeweerd of interest, mainly perhaps because of the help he gives us in coping with diversity and interdisciplinarity. For example, in Basden and Wood-Harper (2006) it is made clear that one of the authors has a Christian faith while the other does not.

But this question cannot be answered adequately without understanding what Dooyeweerd meant by a "Christian philosophy", and dispelling misunderstandings.

- Scholastic or Thomistic philosophy, which has for long been thought to be "Christian" and against which Humanistic philosophy was pitted for many years, Dooyeweerd argues, is not Christian philosophy because it is based on the Nature-Grace ground-motive.

- Dooyeweerd was very careful to differentiate philosophy from theology. Within theology, seen by Dooyeweerd as a science of the pistic aspect, it is valid to adhere to a religious belief and to defend it. But philosophy, including any possible "Christian" philosophy, must be critical and self-critical. Therefore, the tendency of Christian believers to engage in apologetics is not a valid exercise in philosophy, whether Christian or any other. (Christians are not alone in engaging in apologetics: believers in positivism, interpretivism, feminism, for example, do also.) Dooyeweerd was always careful to avoid apologetics, while still being clear about Christian content where it is different from other content.

- What Dooyeweerd meant by a Christian philosophy is one that begins with the Creation-Fall-Redemption ground-motive, as it is understood to include the cosmic Christ, and works out the philosophical, rather than theological, implications of the presuppositions that attend it. For example, if the cosmos is created then a number of things follow, including: it has an existence and an occurrence that is separate from the Divine and not part of the Divine, it cannot be self-dependent, it has the character of Meaning, and that Meaning refers beyond the cosmos to its Creator. It also means that diversity and coherence can be brought together, since there is no philosophical pressure to reduce either for the sake of the other. A summary comparison Dooyeweerd made between the Humanist Nature-Freedom ground-motive and the Creation-Fall-Redemption ground-motive may be found in Dooyeweerd (1984, I, pp. 501-508).

9.4.4 Contributions of the Exercise

A new way of looking at, and understanding, information systems and IT has been explored. The exploration has taken the form of generating philosophical frameworks for understanding five areas of research and practice in IT/IS, but it has been undergirded by an overall approach geared to the everyday lifeworld of each area and employing a philosophy that is uniquely capable of addressing everyday issues.

Most books that cross area boundaries (such as Mingers & Willcocks, 2004; Walsham, 2001; Winder, Probert, & Beeson, 1997) still neglect certain areas of IS. This book addresses a wider selection of areas, both technical and non-technical: human use of computers, IS development, the nature of computers (including artificial intelligence), the shaping of basic technologies, and our ecological relationship with IT.

The exploration has also tried to show how to put areas of interest together to orchestrate a whole story of a discipline. The framework generated for each area relates to all the others, so that research in each area need no longer be divorced from that in others. This provides a sound basis for interdisciplinary research and practice. It might yield new strategic directions for research in all areas of information systems and IT. It can also help those who work in a given area to be more aware of the issues and concerns of all other areas, without denigrating the meaningfulness of those issues.

Because each area has been reinterpreted in the light of everyday experience of both researchers and practitioners and under the influence of a different ground-motive, fresh insights emerged in each area. Some new things have been said, but also familiar things have been said with a different tone or from a different foundation. These pose new questions, indicate new directions for research and suggest new practical devices, as well as helping to ameliorate problems in extant frameworks by transplanting them into the different ground-motive where they are enriched.

In the course of the exploration, the following extant frameworks have been augmented, undergirded, or otherwise discussed, because the aim has not been to denigrate and replace existing approaches so much as to critique, support, and enrich them:

- Computer use

    o   Winograd and Flores
    o   Walsham

- Nature of computers:

  o Newell's computer systems levels
  o Chinese Room debate

- ISD

  o Burrell-Morgan model and systems-thinking paradigms
  o Soft systems methodology

- Technological resources

  o Object orientation
  o Wand and Weber
  o Use of design patterns

- Technological ecology

  o Structuration theory, technological determinism, social construction of technology, actor-network theory, social shaping of technology (briefly)
  o Feminism
  o Critique of Western assumptions

It also been demonstrated how to seek an understanding in each area, which is both grounded in philosophy and sensitive to everyday experience. Whereas in some areas philosophical attempts at understanding are already well-established, in other areas they are not. Even where they are, most discussion of how to understand an area of research and practice adopts a theoretical stance. This book addresses the life-world of each area of research and practice in IS, and uses philosophy to formulate a framework for understanding each. The frameworks generated here are able to embrace the diversity, coherence, meaning, and normativity of everyday life. They are also more systematic than most.

As a result the FFUs formulated should be able to support thinking that goes outside the box in most areas, for example:

- Not only business, professional, organisational use, but also computer games, art, music, and home use;

- Not only accepting the battle between the two camps in AI as either inevitable or irrelevant, but having a sound way to accept insight from both camps;

- Not only professional IS development, but also amateur;

- Not only not only technologies based on objects or text, but also those suited to spatial, aesthetic, and other domains;

- Not only the battles between technological determinism and various social constructionist theories, between Western and non-Western views, or between feminist and "Cartesian" thinking, but a "liberating vision for technology";

- Throughout all, a recognition that IT/IS can no longer be treated as a religion-free zone.

A number of practical devices have been introduced, to assist exploration of ill-structured, interdisciplinary domains of interest, notably various forms of multi-aspectual analysis.

Most reference to philosophy in IS finds it must use different types of philosophy in different areas (e.g., ontology for technical areas, epistemology for IS development, philosophical ethics for use). This book shows how it might be possible to use a single philosophy for all areas.

This book has introduced the philosophy of a little-known but interesting thinker, Herman Dooyeweerd. It has explained, in Chapter II, the ways in which Dooyeweerd's philosophy is different from and similar to many others, and has provided, in Chapter III, a systematic brief exposition of most portions of his positive philosophy. This book may be used as an introduction to Dooyeweerd and as a reference work about Dooyeweerd when trying to understand his ideas, at least to a depth sufficient for use in IT/IS.

Finally, through the work, indications have been given of future research and philosophical exploration that might be fruitful.

9.4.5 Limitations of the Exercise

The main weakness of this work is that it might sound pompous, as though the author is looking down from some Olympian height and calling out, "Here is the one way in which IT/IS can become perfect." That is certainly not intended. The intention is to recommend Dooyeweerd as one little-known alternative that is worth exploring, as a way to understand and look at IT/IS rather than as a ready-made answer. Dooyeweerd might help us address the questions set out on the first page of the Introduction, but will not answer them. The way to recommend Dooyeweerd is to show exciting or interesting possibilities. The way this author has chosen to do so is to set out the ways he has found Dooyeweerd helpful in understanding IT/IS thus far, and to discuss why Dooyeweerd has been so helpful. Some of his own excitement might have come through in the process. If this excitement sounds Olympian, the author can only crave the reader's indulgence.

There are also a number of more specific weaknesses in the work. One is that research methodology has not been discussed, but only the content and strategic direction of research. The contribution Dooyeweerd might make to IS research methods is being explored.

This study of how Dooyeweerd might help us formulate frameworks for understanding the areas of research and practice in information systems has necessarily been brief. For example, there are many current issues in IT that have been overlooked or merely mentioned in passing, such as learning to use computer systems, information security, the whole fields of e-learning, e-commerce, e-government, and so forth, legacy systems, information technologies like case-based reasoning, induction, robotics, multimedia, whole technologies like mobile technology, ubiquitous computing. However, this work does at least indicate a process by which such issues could be examined.

The breadth and depth have been rather inconsistent. For example, object orientation was treated in a rather cavalier fashion in Chapter VII whereas the suggestion for aspectual modules was presented in detail. The reason for this was that whereas other information about OO is widely available, no other source of information is yet available about aspectual modules. The amount of detail in which some things are explained is much greater than for others. This is because, for the purposes of this work, it is important to give a few exemplars of how Dooyeweerd could be worked out, which provide guidance of sufficient detail so that others could follow the approach in working out other issues. The purpose is not to conclusively argue the superiority of Dooyeweerd but merely to indicate he offers an alternative that is worth the effort of exploring.

For a similar reason, the reference to philosophy other than Dooyeweerd has been patchy. Thought that deserves closer scrutiny from a Dooyeweerdian point of view includes, for example, those of Churchman, systems theory, Midgley, application of Habermas' theory of communicative action, feminism. Two major omissions from general philosophy are postmodernism and the systematic philosophy of Bernard Lonergan; relating these to Dooyeweerd in the context of IS must be future work. Moreover, the author's interpretation of those that have been mentioned might be partial or open to question. However, it was not deemed appropriate to engage in full discussion of all such strands of philosophic thinking in this work; pointers have been given for those who wish to explore these.

Even the author's own understanding of Dooyeweerd is still imperfect. In some places it may be that it is his own (mis-)interpretation, built up over a decade or more of trying to apply to IS, rather than Dooyeweerd's, that has been presented. For example, the notion of aspectual beings, introduced in Chapter V, though one that validly emerges from Dooyeweerd's thought, owes a lot to Newell's levels, and is not a notion Dooyeweerd himself used. But this was admitted when the notion was introduced, and it is assumed that Dooyeweerd's thought may be extended in such manners, of which this might be one. Such deficiencies must await the critique that comes from exposure of this work to public scrutiny.

The impression might be obtained that Dooyeweerd's suite of aspects is almost a panacea. The reasons for, and validity of, this was presented earlier. But there is perhaps a more fundamental weakness in this work: it has been shown in many places that various portions of Dooyeweerd's philosophy can account for issues in IS, but it has not been shown that Dooyeweerd is necessary, providing the only, or even the best, account. It was not the intention to do this because it is too early for such attempts. To overcome this weakness requires a mature, widespread understanding of Dooyeweerd in all areas of IS, which itself requires the IS community to learn, truly understand, adopt, test, and refine the application of his philosophy. Until that time, it is appropriate to present positive accounts such as are found in this work, as a stimulus to interest in Dooyeweerd.

9.5 The Future

What do we do now? Despite what I have just said, let me finish with a bold claim and challenge.

Our re-examination of each of the five areas has identified new problems and questions, which cry out to be addressed in research and practice:

- How can we differentiate beneficial from detrimental impact of IS in use, especially when many stakeholders are overlooked, and when many of the impacts are hidden, indirect, unexpected, or long-term?
- How can we integrate the many ways in which the computer can exist in our experience, and link our understanding of its nature to the normativity of its use? Can we break the stalemate in the AI debate?
- How can we transform ISD from a single-stranded chore into a fully satisfying human activity in which the several interwoven strands are all given their due respect for their diversity?
- How can we deliver "KR to the people", that is KR languages and other IT resources which are so natural and appropriate to the wide diversity of meaning encountered in everyday experience that the "lay person" can undertake their own IS development?
- How can we achieve a "liberating vision for technology" in which IT is a sustainably beneficial "ecology" for human life and planet, shaping us in ways that are beneficial?

(Linked to these questions, are many subsidiary ones, which emerged during our exploration.)

As has become clear, each of these questions is rather challenging for existing frameworks for understanding, even in the area in which it arises. When collected together as representing something of "the whole story that is IS" they present an almost insurmountable challenge.

Even if we might not wish to adopt the specific positive Dooyeweerdian frameworks worked out for each area, it is clear that the Dooyeweerdian approach, based on an everyday attitude and a different ground-motive, has at least enabled us to identify issues across the vast spectrum of the "whole story" which it is all too easy to miss. We tend to miss them, not only because we fail to take a "whole story" approach—we miss much of them even when we focus on one area, because we have usually looked at our areas through the lens of immanence thinking. It seems Dooyeweerd has kept his promise to "clear away" that which stops us seeing the structure of reality.

We might stop there. Having had the issues made visible and given form by Dooyeweerd, we could abandon Dooyeweerd and attempt to find other new philosophical frameworks which address those questions. We have briefly discussed how this might be achieved.

But thus far this author has not encountered any other framework that identifies all these as questions. He has not even encountered any other philosopher who could help us formulate frameworks that can do so.

But Dooyeweerd has done more. He has opened up paths to explore for answers. Using his approach we have been able to formulate frameworks not only for seeing what is a central question that demands an answer, but also for understanding the structure, normativity, and meaning of each area, as it presents itself to us in the everyday attitude of thought. And in all areas this has led to at least one practical device.

Come on, people: listen to Dooyeweerd. Understand him fully. Try him out. Apply him. Feel his strengths. Discover and uncover his weaknesses. Debate and research which of these can be overcome and which cannot. Refine his ideas. Then let us, together, either reject or adopt him, to help us understand and guide every area of research and practice in "the whole story that is information systems".

References

Basden, A., & Wood-Harper, A. T. (2006). A philosophical discussion of the root definition in soft systems thinking: An enrichment of CATWOE. *Systems Research and Behavioral Science, 23*, 61-87.

Brandon, P. S., & Lombardi, P. (2005). *Evaluating sustainable development in the built environment*. Oxford, UK: Blackwell Science.

Clouser, R. (2005). *The myth of religious neutrality: An essay on the hidden role of religious belief in theories* (2nd ed.). Notre Dame, IN: University of Notre Dame Press.

Dooyeweerd, H. (1984). *A new critique of theoretical thought* (Vols. 1-4). Jordan Station, Ontario, Canada: Paideia Press. (Original work published 1953-1958)

Eriksson, D. M. (2006). Normative sources of systems thinking: An inquiry into religious ground-motives of systems thinking paradigms. In S. Strijbos & A. Basden (Eds.), *In search of an integrative vision for technology: Interdisciplinary studies in information systems* (pp. 217-232). New York: Springer.

Fathulla, K., & Basden, A. (in press, July 2007). What is a diagram? In *Proceedings of the 11th International Conference on Information Visualisation, IV07 Zurich*. Los Alamitos, CA: IEEE Computer Society Press.

Gibson, J. J. (1977). The theory of affordances. In R. Shaw & J. Bransford (Eds.), *Perceiving, acting and knowing* (pp. 67-82). Hillsdale, NJ: Erlbaum.

Jackson, M. C. (1991). *System methodology for the management sciences*. New York: Plenum Press.

Latour, B. (1987). *Science in action*. Cambridge, MA: Harvard University Press.

Lyytinen, K. J. (2003). Information systems and philosophy: the hopeless search for ultimate foundations. In J. I. DeGross (Ed.), *Proceedings of the Americas Conference on Information Systems: AMCIS 2003*. Atlanta, GA: Association for Information Systems.

Midgley, G. (2000). *Systemic intervention: Philosophy, methodology and practice*. New York: Kluwer/Plenum.

Mingers, J., & Willcocks, L. P. (Eds.). (2004). *Social theory and philosophy for information systems*. Chichester, UK: Wiley.

Newell, A. (1982). The knowledge level. *Artificial Intelligence, 18*, 87-127.

Schuurman, E. (1980). *Technology and the future: A philosophical challenge*. Toronto, Ontario, Canada: Wedge.

Tolkien, J. R. R. (1998). *Tales from the perilous realm*. London: Harper Collins Publishers.

Walsham, G. (2001). *Making a world of difference: IT in a global context*. Chichester, UK: Wiley.

Winder, R. L., Probert, S. K., & Beeson, I. A. (Eds.). (1997). *Philosophical aspects of information systems*. London: Taylor and Francis.

Winfield, M. (2000). *Multi-aspectual knowledge elicitation*. Unpublished doctoral thesis, University of Salford, UK.

Winograd, T., & Flores, F. (1986). *Understanding computers and cognition: A new foundation for design*. Reading, MA: Addison-Wesley.

Glossary and Abbreviations

Key:

[D] In Dooyeweerdian thought
[E] In everyday use
[IS] In information systems and related disciplines
[P] In philosophy
[number] Optional chapter number, or part of chapter

Amiga: An innovative multi-tasking, multi-media computer platform of the 1990s, which was particularly suited to home and games use; the platform used by this author. [IS]

ANT: Actor Network Theory [IS]

Anticipation: An aspect refers forward to a later one. [D3]

Antinomy: A deep inconsistency or incoherency that cannot be overcome by logic. [D2]

Archimedean point: A hypothetical vantage point from which an observer can objectively perceive the subject of inquiry, with a view of totality. (From Wikipedia article of the same name). [D]

Aspect: 1. [D] Synonymous with "sphere of meaning-and-law"; this is a central concept in Dooyeweerd's thought [§3.1] 2. [E] A general kind of property that is to be distinguished from others.

Aspectual law: Cosmic law (q.v.) that enables and guides the functioning of the whole cosmos. [D3]

Assumption: A belief that makes a proposition true or false (as opposed to meaningful; see Presupposition). [P]

CATWOE: A simple checklist to help with problem solving as part of Soft Systems Methodology; see §6.8. [IS6]

CCA: Client Centred Approach to ISD. [IS6]

CFR: Creation, Fall, and Redemption ground-motive [D2]

Conflate: merge together; combine two into one in a way that is probably inappropriate. [E]

Cosmic law: The transcending law-framework that enables the cosmos to be and occur, but which can never be fully known. [D2]

Cosmic meaning: The transcending framework of spheres of meaning which stamps the entire cosmos as Meaning. [D2]

Cosmic time: Dooyeweerd's notion of Time. [D]

Cosmonomic philosophy: A name for Dooyeweerd's positive philosophy. [D2]

CR: Critical Realism. [P]

CST: Critical systems thinking. [IS6]

Dialectic: The opposing of one central idea to another in a way that is deeper than logic and involves deep commitments; thought tends to swing from one pole to the other. [P,D2]

DST: Disclosive Systems Thinking [IS6]

EISD: Emancipatory information systems design/development. [IS6]

Enkapsis: A relationship in which two "wholes" are joined in a structural relationship in which both are necessary. [§3.2.6.2]

Entity side: See Subject side [D]

ERC: Engagement with represented content [IS4]

Everyday attitude: An attitude of thinking or reflecting that is open to the richness of everyday experience. (= "lifeworld attitude", "naïve attitude"). Contrasted with theoretical attitude. [P]

Everyday experience: Human experience of life in all its richness. Contrasted with theoretical observation and analysis of a narrow focus. [P]

Extant: Existing, used of current thinking or ideas or discourse; still relevant and applicable. [E]

FFU: Framework for understanding.

FMGM: Form-Matter ground-motive [D2]

Foundational dependency: Each aspect (q.v.) depends on earlier ones for its positivisation (q.v.); example: social functioning cannot occur without lingual functioning. [D3]

FPB: Fixed point binary number [IS7]

Functioning (in an aspect): Response to aspectual law (q.v.). Example: in reading this you are functioning in the lingual aspect of understanding symbols, and also the psychic aspect of vision. Can be either object- or subject-functioning (q.v.). [D3]

Gegenstand: 1. [P] In immanence-philosophy, that which is valid and objective in our experience. 2. [D3] Relationship in which we "stand over against" an object of our thought or action, rather than engage intimately with it; in contrast to (1), this reduces validity.

GIS: Geographic information systems [IS]

Ground-motive: Deep presupposition about the nature of reality; it is a "moving power or spirit at the very roots of man, who so captured works it out with fear and trembling, and curiosity". [D2]

HCI: Human-computer interaction [IS4]

HLC: Human living with computers [IS4]

HST: Hard systems thinking [IS6]

HUC: Human use of computers, an area of research and practice. [IS4]

ICT: Information and communication technology.

Immanence-standpoint: A deep presupposition that the fundamental Principle of all temporal reality that is may be found within temporal reality itself. It usually leads to reductionism. Example; materialism. Clouser (2005) explains it well. [D2]

Immanence-philosophy: Philosophy based on the immanence-standpoint (q.v.) [D2]

Immanent critique: Criticism of a position or stream of thought in terms of what it itself finds meaningful and important and seeks to achieve, usually to expose presuppositions it makes to show they are contradictory. Used especially by Habermas and Dooyeweerd. [P]

Internal structural principle: The internal structural principle is a cosmic law which governs how a thing of a given type responds to the various aspects; also called "type law". [D3]

IS: Information systems.

ISD: Information systems development: an area of research and practice. [IS6]

IT: Information technology.

ITE: Information technology as ecology: an area of research and practice. [IS8]

ITR: Information technology resources: an area of research and practice. [IS7]

KM: Knowledge management. [IS]

Knowledge elicitation: Coming to distinguish and know what is relevant in a domain of application for encapsulation in or as a computer system. [IS6]

Knowledge representation: Representing or expressing what is relevant in a domain in a computer language. [IS6,7]

KR: knowledge representation.

KR formalism: A general kind of KR language; it embodies a theory or idea of how knowledge can be represented. [IS7]

KR language: A formal language that is used to program computers; it can be graphical as well as textual (for example box-and-arrows diagrammers). [IS7,6]

Law: 1. [D] Aspectual law 2. [E] Legal requirement.

Law-promise: Law, emphasising repercussions. [D]

Law-side: Adjective derived from "law side"). [D]

Lifeworld (= Life-world): The "world" of everyday life as distinct from the scientific "worlds" of e.g. physics, psychology, social science. Synonym for "everyday", "naïve", and "pre-theoretical". [P1,3]

LOFFU: Lifeworld-oriented framework for understanding. [1]

LWV: Life-and-world-view

Meaning: To Dooyeweerd, this is almost a synonym for reality: all created reality *is* meaning; Meaning has the character of referring beyond. Four uses of "meaning". [D2]

Meaning-and-law: The law side: the cosmic framework that enables the Cosmos to be and occur. [D2]

Meaningful-functioning: Functioning in an aspect as either subject or object. [D]

MMORPG: Massively-multiplayer online role-playing games: computer game played over the Internet. [IS]

MST: Multimodal Systems Thinking [IS6]

MUDs: Multi-User Dungeon computer game played over Internet. [IS]

Naive: Simple (with negative connotation); contrast with 'naïve).

Naïve: An adjective used for experience or attitude; A synonym for everyday, pre-theoretical, lifeworld; It does not have a negative connotation (contrast "naive"). [P]

NCI: National Cancer Institute.

NFGM: Nature-Freedom ground-motive. [D2]

NGGM: Nature-Grace ground-motive. [D2]

NoC: Nature of computers: an area of research and practice. [IS5]

Normativity: The branch of philosophy concerned with good and evil; also the right versus wrong of a situation or thing. [P]

Noumenon: Something whose existence can be reasoned but not perceived. [P]

Object: (multiple meanings) 1. Thing 2. [D] Something affected or created by something else's subject-functioning 3. [IS6,7] Representation of domain meaning.

Object-oriented: An approach to programming [IS7]

Object-functioning: When something functions in an aspect as part of something else's subject-functioning. [D2,3]

Ontic: Of, relating to, or having real being or existence. [P]

Ontology: 1. [P] A branch of philosophy concerned with studying ontic issues 2. [IS7] (usually "KR ontology") a statement in a KR language about such ontic issues in a domain of meaning.

OO: Object-oriented.

Phenomenon: Anything perceived by the senses or how we are conscious of something as distinct from the nature of the thing itself. [P]

Pistic: (Greek, "pistis": faith) Usual name of fifteenth aspect in Dooyeweerd's suite. [D3]

Positivisation: Law is positivised when some subject responds to it in time; the Cosmos is positivisation of cosmic law. [D]

Presupposition: An assumption that makes a statement or belief meaningful, rather than true; contrast Assumption (q.v.). [P]

Qualifying aspect: The aspect which expresses the main meaning of a thing as that type of thing; Example: the qualifying aspect of a pen is the lingual. [D3]

RDM: Relational Data Model [IS7]

Reduction(ism): The attitude or action of trying to explain one type of meaning fully in terms of another, usually with connotations that this is undue. [P]

Religion: "The innate impulse of human selfhood to direct itself toward the true or toward a pretended absolute Origin of all temporal diversity of meaning, which it finds focused concentrically in itself." (Dooyeweerd, 1984, I, p. 57). NOT to be confused with particular creeds or religious practices. [D2]

Religious root: See §2.4.1. [D2]

Repercussion: Result of aspectual functioning. [D3]

Represented content: Meaning of a domain that is represented in a computer [IS4]

Retrocipation: Referring back to earlier aspects [D3]

S-O: Subject-object relationship

S-S: Subject-subject relationship

SDM: System development method [IS6]

Shalom principle: If we function well in every aspect then things will go well, but if we function poorly in any aspect, then our success will be jeopardised. See §3.4.3. [D3]

SMP: Semi-manufactured product. [D]

Sphere of law, or meaning: Aspect (q.v.) [D]

SSM: Soft Systems Methodology [IS6]

SST: Soft systems thinking [IS6], and social shaping of technology [IS8]

ST: Structuration Theory (Giddens)

Subject: 1. [D] Agent that responds to and is subject to aspectual law 2. [E] Topic.

Subject side: That which is subject to the law side (q.v.); synonym for the cosmos as it actually exists and occurs, including the conceptual and social worlds. The subject side, also called entity side or fact side, comprises all that exists or occurs in the cosmos, as concrete reality, including concrete meanings that are ascriptions we make, and includes all our experience, past, present, future and potential. [D2]

Subject-functioning: When a thing (or person) actively responds to aspectual law, as agent rather than as object. [D]

Subject-object relationship: See Chapter II "Escaping Descartes and Kant". Dooyeweerd proposed a notion of subject and object which is radically different from that inherited from Descartes. [P,D]

Transcendent critique (Not to be confused with transcendental critique, q.v.): Criticising a position from the perspective and value-system of a completely different position (e.g. criticising Marxism from Capitalist perspective or vice versa.) Opposite of Immanent critique (q.v.). [P]

Transcendental critique (Not to be confused with transcendent critique, q.v.): Critical analysis of a position immanently to expose the conditions that are necessary for it to be possible. [P]

Two-thirds world: A better name for developing world or 'third world'. [IS8]

Type law: See "Internal structural principle" (q.v.) [D3]

UI: User interface [IS]

VR: Virtual reality [IS]

VSM: Viable Systems Model (Beer) [IS]

Weltanschauung: World-view [P]

World-view: (= World view) Set of assumptions, aspirations, etc. that shape the way we see, understand, and react to things. [P]

About the Author

Dr. Andrew Basden is professor of human factors and philosophy in information systems at the University of Salford, UK. After obtaining a first class honours degree in electronics and a Doctor of Philosophy in computer-aided design at the University of Southampton, he spent 14 years in computer programming in applications and expert systems in the pharmaceutical, health, chemical and construction sectors, before returning to academic life in 1987. Since then he has taught and published in a wide range of fields, including knowledge-based systems, knowledge acquisition, development methodologies, databases, multimedia, business information systems, human-computer interaction, human factors, critical studies, and philosophy. He is a founding member of the International Centre for Philosophy, Technology, and Social Systems. He is also a follower of Jesus Christ and a green activist, and has always tried to integrate all areas of his life into a coherent life-and-world-view.

Index

A

aboutness 201
absolute truth 66, 79
absolutization 70, 115, 313, 321
actants 319
actor-network theory 160, 318, 367
aesthetic
  aspect 235, 237, 246, 249
  disclosure 315
affordance 271
Africa 165, 284
agriculture
  British 243
ALGOL, computer language 267
algorithms 183, 205, 265, 272, 287
analog computers 186
anthropology 17, 26, 104, 116
anticipatory dependency 204, 285, 286, 287,
    304, 345
APL, computer language 267
appropriateness 269, 273
area boundaries
  crossed by books 366
Aristotle 17, 39, 50, 82, 85, 110
artificial intelligence (AI)
    4, 12, 19, 174, 265, 323
  debate 210, 218, 370
aspect 62, 67, 74
  aesthetic 148, 236, 326
  analytic 71, 98, 187, 245, 325
  biotic 205

economic 242, 325
ethical 152, 243, 328
formative 236, 242, 249, 290, 325
juridical 153, 236, 242
kinematic 64, 71, 80, 155
lingual 163, 193, 236, 245, 252
physical 205
pistic 71, 155, 163, 328
psychic 187, 205
quantitative 290
social 237, 242
spatial 290
tree 156–157
aspectual
  analysis 127, 153, 155, 162, 163, 253,
    299, 300, 302, 321
  being 186, 188, 221, 342, 369
  crossover 145
  diversity 355
  functioning 76, 85, 90, 98–99, 101–102
  harmony 69, 102, 152, 188
  law 54, 56, 99, 101, 106, 108, 350, 354
    , 361
  law-framework 312
  meaning 54, 186
    intuitive grasp of 74, 97, 109, 256, 349
  modules 277, 284, 304
  normativity 73, 234, 259
  norms
    conflicts between 148
  profile 95, 106, 320, 335
  rationality 277